SUSTAINABILITY AND RESOURCES

Theoretical Issues in Dynamic Economics

SUSTAINABILITY AND RESOURCES

Theoretical Issues in Dynamic Economics

Mukul Majumdar

Cornell University, USA

World Scientific

NEW JERSEY · LONDON · SINGAPORE · BEIJING · SHANGHAI · HONG KONG · TAIPEI · CHENNAI · TOKYO

Published by

World Scientific Publishing Co. Pte. Ltd.

5 Toh Tuck Link, Singapore 596224

USA office: 27 Warren Street, Suite 401-402, Hackensack, NJ 07601

UK office: 57 Shelton Street, Covent Garden, London WC2H 9HE

Library of Congress Cataloging-in-Publication Data

Names: Majumdar, Mukul, 1944–   author.

Title: Sustainability and resources : theoretical issues in dynamic economics /
    Mukul Majumdar (Cornell University, USA).

Description: USA : Worldscientific, 2020. | Includes bibliographical references and index.

Identifiers: LCCN 2019049841 | ISBN 9789811210204 (hardcover)

Subjects: LCSH: Renewable natural resources. | Nonrenewable natural resources. |
    Sustainable development.

Classification: LCC HC85 .M34 2020 | DDC 333.7--dc23

LC record available at https://lccn.loc.gov/2019049841

British Library Cataloguing-in-Publication Data

A catalogue record for this book is available from the British Library.

For any available supplementary material, please visit
http://www.worldscientific.com/worldscibooks/10.1142/11548#t=suppl

Desk Editors: Vishnu Mohan/Karimah Samsudin

Typeset by Stallion Press
Email: enquiries@stallionpress.com

To
Aveek and Jennifer

"Limitless is the ocean of knowledge,
Since life is short and impediments galore,
One has to capture the essentials..."
— The Panchatantra

Preface

The literature on *sustainability* is huge. It is the culmination of efforts from researchers in many disciplines over a few generations. It is profoundly influenced by the attempts of visionaries to identify the direction of our journey and the challenges that lie ahead. There is no universally acknowledged formal definition of "sustainable development". The book begins with an informal introduction to (a) the widely accepted "definition" of sustainable development and (b) the attempts on the world stage to identify some of the more fundamental issues that, if left unattended, will threaten "planet, people, prosperity, peace and partnership" (the 2015 New York Summit). Concerns over the devastating impact of degraded environment due to human activities are nothing new. Rachel Carson's *Silent Spring* (1962) stands out as a classic: a warning against the indiscriminate use of chemicals that can stifle man and nature slowly but surely. Further research (somewhat controversial in matters of detail and prediction) has led to a better understanding of the long-run effects of global warming that may affect both rich and poor countries. Similarly, threats to achieving prosperity are painfully recognized in periods of sharply increasing prices due to depletion of natural resources.

Subsequent chapters explore a select class of dynamic models dealing with renewable and exhaustible resources. A few of these are descriptive: well-known examples from mathematical biology that explored the possibility (or inevitability) of extinction. Others have deep roots in inter-temporal welfare economics and dynamic optimization. Each model addresses themes of continuing interest.

The approach is "formal". The reader is led to some basic results with proofs by using simple *discrete time* dynamical systems. No attempt is made to prove the most general results, which may be readily available or achieved with modest efforts, in the frameworks chosen. A formal presentation demands a precision in thinking and encourages a search for the most direct route from a set of assumptions to a conclusion. Despite its stark simplicity, a model may dramatically confirm or reject an "intuitive" perception. For example, with the identification of tipping points, we see that even small changes may set the stage for *inevitable* rather than *possible* extinction. We are forced to recognize that simple dynamic models may display highly complex, essentially unpredictable evolutions. We often gain valuable insights into the perennial debates involving "prices versus direct intervention".

The book is intended for advanced undergraduate and graduate courses on applied microeconomics, capital theory, dynamic welfare economics and, as the title suggests, resource economics and sustainable development. Several chapters are based on lecture notes prepared for two courses primarily attended by advanced undergraduate students at Cornell. The exposition in the first 10 chapters assumes a command over advanced calculus, while the last two chapters presuppose an understanding of the basic concepts of probability theory. There are plenty of examples to elucidate and exercises (with "hints") which invite the readers to complete some proofs. Other exercises, a bit more difficult, are separately listed. In the technically demanding Chapter 12, two fully worked-out examples (in Section 12.2) capture the essence of the main Theorem 12.2.3. Similarly, in Section 12.3, a self-contained exposition of a special case, as well as the two fully worked-out Examples 12.3.1 and 12.3.2 convey the range of questions that are attacked in more abstract mode. In Complements and Details section, the reader is referred to both primary and secondary sources to proceed farther. Chapter 13 collects most of the frequently used techniques/results from analysis and probability for ready reference. For using the book at the graduate level, the instructor can collect supplementary material from comprehensive reviews and pioneering articles: the three volumes of the *Handbook* (edited by Kneese and Sweeney (1993)), a collection on exhaustible resources (edited by Heal (1993)), and the two volumes of the UNESCO *Encyclopedia of Life Support Systems* (edited by Majumdar, Willis, Sgro and Gowdy (2010)). For drawing attention to the frontiers, in addition to *Journal of Environmental Economics and Management*, the students should be encouraged to turn to *Mathematical Biosciences, Journal of Mathematical Biology* and *Journal of Applied Probability*.

I have drawn directly from my research with Venkatesh Bala, Talia Bar, Partha Dasgupta, Rabi Bhattacharya, Nigar Hashimzade, Geoffrey Heal, Tapan Mitra, Manfred Nermuth, Yaw Nyarko, and Roy Radner. I have learned so much from them. My lectures on sustainability at Ashoka University (2016), Indian Statistical Institute (2017) and The University of Arizona (2017) were based on parts of the book. I am grateful to Malabika Sarkar, Vijay Kelkar, Sanghamitra Bandyopadhyay and Edward Waymire for organizing the presentations and many of the participants for their comments.

In planning the scope of the book, and guidance at various stages, I had the benefit of advice from Santanu Roy. His cheerful participation defies adequate acknowledgment. On many occasions, I have received valuable inputs from Shubhashis Gangopadhyay, Ali Khan, James Foster, Nancy Chau, Arnab Basu, Jayanta Roy, Jennifer Wissink and Seung Han Yoo. Ram Dubey, Debi Mahapatra, Duk Gyoo Kim, Sanket Roy and Abhishek Ananth, at Cornell helped completing the manuscript. Financial support from the H.T. and R.I. Warshow endowment and the Department of Economics at Cornell University is gratefully acknowledged. In this context, the initiative of Larry Blume was uplifting. Also, I am especially obligated to Yongmiao Hong and Nancy Chau who arranged for research assistance at crucial stages.

Quotes attributed to Maya Angelou and Werner von Siemens are readily available on the internet (for example, at Brainyquote.com and at www2.pvlighthouse.com.au [The first solar cells]). Similarly, one can easily access numerous discussions on S.T. Coleridge (*The Rime of the Ancient Mariner*), Leo Tolstoy (*How Much Land Does a Man Need*) or Rachel Carson (*Silent Spring*). Other sources of quotes are in the list of references.

When the manuscript was in its final stage, Tapan Mitra passed away on February 3, 2019. The imprint of his definitive research on themes that some of the chapters deal with is clear. With sadness, I realized that he would not enjoy the book in print.

Finally, I would like to thank Rochelle Kroznek and Karimah Samsudin for supporting the publication of this book.

About the Author

 **Mukul Majumdar** is H.T. and R.I. Warshow Professor of Economics Emeritus at Cornell University. He has been a Ford Rotating Research Professor at the University of California, Berkeley, Guggenheim Fellow, Visiting Oskar Morgenstern Professor at New York University, Erskine Fellow at the University of Canterbury, Overseas Fellow at Churchill College, Cambridge University. He is a Fellow of the Econometric Society, an Economic Theory Fellow, and the recipient of a College de France Medal. He is a major contributor to wide ranging research on deterministic and stochastic models of dynamic economics. His book (with Rabi Bhattacharya) on *Random Dynamical Systems*: *Theory and Applications* is a definitive contribution to the study of probabilistic methods and their applications to economics.

Contents

Chapter 1

Introduction

My mission in life is not merely to survive, but to thrive: and
to do so with some passion, some compassion, some humor and
some style.

— Maya Angelou

Sustainable Development is ...

considering that the concept of sustainable development is fea-
tured on 399,000,000 webpages (searched on July 4, 2018, mm),
and enmeshed in the aspirations of countless programs, places,
and institutions, it should be easy to complete the sentence. But
the most widely accepted definition is creatively ambiguous.

— R.W. Kates, T.E. Parris and A.A. Leiserowitz

1.1 In Search of a Definition

A convenient point of departure is the widely discussed "definition"
of *sustainable development* introduced by the *World Commission
on Environment and Development* (Brundtland Commission, to be
abbreviated as BC, 1987, p. 43):

> *"Humanity has the ability to make development sustainable —
> to ensure that it meets the needs of the present without com-
> promising the ability of future generations to meet their own
> needs."*

1

Although *environment* does not appear explicitly in this definition (BC, 1987) argued:

> "*The environment does not exist as a sphere separate from human actions, ambitions, and needs, ... the "environment" is where we live; and "development" is what we all do in attempting to improve our lot within that abode. The two are inseparable.*"

Moreover:

> "*The concept of sustainable development does imply limits — not absolute limits but limitations imposed by the present state of technology and social organization on environmental resources and by the biosphere to absorb the effects of human activities.*"

The perception of *needs* has changed from one generation to the next with the evolution of cultural norms, values, institutions, and more importantly, with the contributions of science and technology. Appearance of new commodities and services have had stunning impact on the quality of life all around the world. Some of the *necessities* and *needs* of the *present* were not in the dreams even of the *recent past*. Hence, it is not easy to anticipate or quantify the needs of a distant future generation. Nevertheless, it is fair to say that the definition attempts to capture a commitment on the part of the present generation to avoid actions/policies that will severely limit or perhaps irrevocably eliminate consumption possibilities of some future generations. Such a commitment, reflecting a concern for *intergenerational equity* immediately suggests conservation and efficient use of natural resources, as well as prevention of irreversible environmental degradation caused by human activities. Quite rightly, these have been recurrent themes in the literature.

There are other definitions that are particularly appealing in specific contexts. In the second half of the last century, many economies committed to planned "development" ("the process by which a poor and largely illiterate agrarian society, which is plagued by high morbidity and mortality, is transformed into an affluent and literate urban society, whose members can expect long and healthy lives"

(Bell, 2010)). These economies would like to sustain the momentum of transformation for a significantly better future. The recognition of limits to growth suggests a definition, which requires that, although the reserves of natural resources will diminish, through appropriate changes in methods of production ensuring productivity improvements, an index of social welfare should not decline in the foreseeable future. Yet another definition involves the need for some characteristics (ecosystems, emission levels, etc.) to be maintained within specific limits.

Whatever be the convenient definition, alternative views on *"what should be sustained and how"* have been gained from various *summits* that have called for global cooperation on sustainability. These public engagements have identified a broad range of complex and interlinked issues that, if unaddressed, will frustrate the aspirations for a better quality of life. Some of these, involving management of resources, are central to the mathematical models that we explore in the following chapters.

The following two sections provide some comments on the major proclamations that emerged out of these meetings. The reader who is not interested in a historical narrative may proceed directly to Section 1.4.

Tolstoy (1886) thought about the question in his short story: *How Much Land Does A Man Need?* Pahom, the central character, collapsed when he overextended himself in his quest to grab bigger and bigger plots of land.

> *"His servant picked up the spade and dug a grave long enough for Pahom to lie in, and buried him in it.*
> *Six feet from his head to his heels was all he needed."*

1.2 Stockholm, Rio and Johannesburg

Periodic meetings on the world stage kept sustainability at the center of (often acrimonious) debates by stressing an approach that integrated development needs, environment concerns and resource management. Given the global political climate, the contributions of these summits in attempting to seek common grounds, in finding an acceptable language of discourse and in urging international collaboration cannot be underestimated. Sustainability was contemplated in terms of specific goals and priorities in policy making. Some challenges to attaining these goals were best handled at a local level, others called for a national consensus, and some clearly needed global efforts.

Of particular importance were the Stockholm Conference (SC) in 1972 on the Human Environment, United Nations Conference on Environment and Development (the "Earth Summit") in Rio in 1992, the United Nations Millennium Development Goals (MDG, 2000), the "World Summit" in Johannesburg in 2002, Rio +20, and the United Nations "Sustainable Development Summit" in New York (NYS) in 2015. These, in turn, led to numerous forums with a narrower focus (for example, initiatives on climate change). In contrast with the earlier sweeping declarations on objectives and priorities, there has been a move towards defining sustainable development in terms of quantifiable goals and specific targets, and numerical indicators to monitor progress.

The Stockholm Conference stated "the common conviction" on 26 "principles", many of which were confirmed or refined subsequently. There was a contentious exchange between the developed and developing countries about the priorities involving *development* and *environment*. A widely discussed statement by Indira Gandhi, the then Indian Prime Minister, noted that affluent countries had gone through a process of industrialization and contributed to environmental degradation. She asserted:

> "*On the one hand the rich look askance at our continuing poverty — on the other, they warn us against their own methods. We do not wish to impoverish the environment any further and yet cannot for a moment forget the grim poverty of large numbers of people. Are not poverty and need the greatest polluters?*"

In the end, the declaration was remarkable for its stress on the concept of *global environmental cooperation* ("international matters concerning the protection and improvement of environments should be handled in a cooperative spirit by all countries, big and small, on an equal footing" — SC Principle 24) and on the needs for states to design *"an integrated and coordinated approach to development planning in order to achieve a more rational management of resources and thus to improve the environment"* (Principles 13, 24).

With admirable persistence, the summits, following the lead of SC Principles 1, 8, stressed the need for avoidance of conflicts and attainment of a just social order as preconditions for improving the quality of life or protecting the environment, if not for the survival of mankind. In NYS 2015, we have a drastic warning:

> *"There can be no sustainable development without peace, and no peace without sustainable development."*

The recognition of resource management came next in the SC list of principles.

- **(SC Principle 2):** *The natural resources of the earth, including the air, water, land, flora and fauna, especially representative samples of natural ecosystems, must be safe guarded for the benefit of present and future generations through careful planning or management, as appropriate.*
- **(SC Principle 4):** *Nature conservation, including wildlife, must receive importance in planning for economic development.*
- **(SC Principle 5):** *The non-renewable resources of the earth must be employed in such a way as to guard against the danger of their future exhaustion.*

The conference in Rio, well attended by many heads of states, was a true "Earth Summit". It generated five documents: the "Rio Declaration", a statement of 27 broad principles to guide national conduct on environmental protection and development; treatises on climate change and biodiversity; a statement on forest principles; and "Agenda 21", a massive effort in itself that provided work-plans covering goals, responsibilities, estimates for funding, etc.

First, the Stockholm declaration was resoundingly reaffirmed: global cooperation and the need to avoid conflicts appeared in several contexts (Rio Principles 2, 7, 9, 12, 13, 18, 24–27).

Next, it took a strong position on "fulfilling the right to development" (Rio Principle 3) and on the *universal* and essential task of *"eradicating poverty . . . in order to decrease the disparities in the standards of living and better meet the needs of the majority of the people of the world"* (Rio Principle 5).

The following statement provides the motivation of some of the basic themes in the following chapters:

> (Rio Principle 8): *"States should eliminate unsustainable patterns of consumption and production and promote appropriate demographic policies."*

It was stressed that an effective implementation of a wide range of policies would be easier in a climate of *inclusion*: embracing the entire society (women, the young, indigenous people, etc.) (Rio Principles 20, 22, 23, respectively).

The 2002 "World Summit" at Johannesburg emphasized the need to go beyond a narrow view of economic development. It stressed the need to face up to:

> *"a collective responsibility to advance and strengthen the interdependent and mutually reinforcing pillars of sustainable development — economic development, social development, and environmental protection — at local, national and global levels."*

The somewhat poetic description of sustainability found an enthusiastic response in the subsequent literature and public engagements. But there were no ready agreements on quantitative or qualitative details of the structure or dimensions of these "pillars".

The rhetoric on an ever-expanding notion of sustainability also evoked sharp responses. *Sustainability, The Economist* (2002) retorted, *risks being about everything and, therefore, in the end, about nothing.*

1.3   Goals, Targets, Indicators: A Quantitative Approach

An alternative approach, starting with the UN Millennium Development Goals (MDGs) announced in 2000, involved a more precise specification of *goals* and *targets* that sustainable development ought to achieve. The MDGs were as follows:

- (MD Goal 1) *Eradication of poverty and hunger*;
- (MD Goal 2) *To achieve universal primary education*;
- (MD Goal 3) *To promote gender equality and empower women*;
- (MD Goal 4) *To reduce child mortality*;
- (MD Goal 5) *To improve maternal health*;
- (MD Goal 6) *To combat HIV/AIDS, malaria and other diseases*;
- (MD Goal 7) *To ensure environmental sustainability*;
- (MD Goal 8) *To develop global partnership for development*.

Of course, MD Goals 1 and 3 are just restatements of Rio Principles 5 and 20, and Goals 7 and 8 remind us of the Stockholm summit. What distinguishes this (and the subsequent more elaborate document generated by the New York Summit in 2015) is the detailed list of *targets* and *indicators* for monitoring progress. The numerical measures and time frames spelled out were meant to provide a more concrete context and for systematic, periodic evaluations of progress in different directions.

The first six goals listed above focused on the *quality of life and equality*. For example, to achieve Goal 1, two targets and five indicators were proposed. These two targets were to halve between 1990 and 2015: (1) the proportion of people whose income is less than one dollar (PPP) a day and (2) who suffer from hunger. The year 2015 was also relevant for some of the targets to achieve MD Goals 3–6.

For achieving MD Goal 7 on environment, three targets and eight indicators were suggested, while associated with MD Goal 8 on global partnership were 9 targets and 16 indicators.

The list of indicators included widely used statistics/indices in the literature on health and human development. For targets on the quality of life, some of these were as follows: the proportion of population below \$1 (PPP) per day, the share of the poorest quantile in national consumption, the proportion of population below minimum level of dietary energy consumption, literacy rates (male and female), child mortality rates, etc. Indicators for targets on environmental sustainability include the following: the proportion of land covered by forest, the ratio of land protected to maintain biological diversity to surface area, energy use (kg oil equivalent) per \$1 GDP (PPP) carbon dioxide emissions per capita, etc. In short, a fairly comprehensive and detailed road map relying on numbers was provided.

The agenda of the New York Summit announced 17 goals and 169 targets: a significant amplification of MDGs. It urged actions over the next 15 years of critical importance involving "People, Planet, Prosperity, Peace and Partnership".

The summits (as well as the BC) recognized the intimate interlinking of the issues that arose out of the quest for prosperity and the resultant strain on the planet/environment as a *source* to support and a *sink* to absorb the waste generated by this quest. Given the wide-ranging objectives and targets dealing with the economic, environmental and social aspects of development (three pillars introduced in Johannesburg), perhaps the most remarkable is a statement on priorities which appeared in the NYS as well as an earlier Rio +20 document in a virtually identical manner:

> "*Eradicating poverty is the greatest global challenge facing the world today and an indispensable requirement for sustainable development. In this regard we are committed to freeing humanity from poverty and hunger as a matter of urgency.*"

But this enunciation reminds us of the exchange on priorities at the Stockholm conference and Rio Principle 5!!!

1.4   The Future We Want: A Summing Up

In many ways, the summits over the years struggled to portray a picture of the *future we want* (the title of a Rio +20 document prepared in 2012). Participants interpreted the notion of *development* broadly so as to make the concept meaningful to the vast majority of the citizens of the world. Drawing from the literature, we can (somewhat arbitrarily!) think of the major challenges to sustainability as related to the following:

- *population*: growth, urbanization, quality of life (poverty and inequality);
- *environment*: degradation (air pollution, acid rain, contamination of ground water, disposal of waste/toxic waste, etc.); global warming and possible changes in weather pattern (melting glaciers; rise of the sea levels; loss of vital ecosystems);
- *resources*: depletion of cheap energy resources; extinction of species, deforestation, loss of biodiversity; receding ground water levels and shortage of drinking water; development/adoption of new technology to minimize the use of exhaustible resources as inputs.

1.5 Mathematical Models

The focus of Part A of the book (Chapters 2–6) is on the management of *renewable* resources. Part B (Chapters 7–10) explores the role of *exhaustible* resources. While we study deterministic models in the first two parts, simple *stochastic* models are introduced in Part C (Chapters 11 and 12).

A simple (crude) way to capture a notion of intergenerational equity is to study consumption (resource harvesting) plans that assign the same quantity $c > 0$ to every period (can be interpreted as per capita consumption in some models). Thus, when plans are made over a finite number (T) of time periods, one defines an equitable program as a T-vector $c = (c_1, \ldots, c_T)$ where $c_t = c$ for some $c > 0$ for all $t = 1, 2, \ldots, T$. In an infinite horizon economy, an equitable program is a stationary sequence $\mathbf{c} = (c_t)$, where $c_t = c > 0$, for all $t \geq 1$. The existence of an equitable program is one of the fundamental questions appearing in a number of contexts in all three parts.

In Chapter 2, we begin with some well-known population models from mathematical biology. Think of a renewable resource, say, a plant or some animal population that has the capacity of natural reproduction and growth, *but only when a positive stock is available for such reproduction.* The evolution of a population is governed by an exogenously specified rule, a regeneration (reproduction) function (i.e., a rule α which specifies for each population size x [non-negative] in any period, the corresponding size $\alpha(x)$ [non-negative] in the next period, with the property that $\alpha(0) = 0$). Formally, we study the first-order difference equation:

$$x_{t+1} = \alpha(x_t), \tag{1.5.1}$$

where x_t is the size of the population in period t.

The aim is to understand (qualitatively) whether from an initial positive stock, the population is heading toward a phase of sustained natural growth, perhaps reaching an equilibrium, or to total extinction, or is displaying a more complex behavior. We can also ask whether/how the long-run behavior is sensitive to the initial condition. The regeneration behavior captured by α may be affected by forces beyond the control of the population ("exogenous" elements). To study this, it is convenient to explore a family of difference equations:

$$x_{t+1} = \alpha(x_t; \theta), \tag{1.5.2}$$

where θ is a parameter (say, some positive numbers); again, we assume that $\alpha(0, \theta) = 0$ for all admissible values of θ (see, e.g., (1.5.4) and (1.5.5)). Questions of comparative dynamics arise naturally: how does the qualitative behavior of the process respond to a change in the parameter? Is the possibility or inevitability of extinction affected by such changes?

After developing the basic mathematical framework, we turn to some examples that are worked out in detail. These are invaluable in understanding the alternative scenarios of evolutions. The logistic model (see Example 2.3.5) is one of the earliest to introduce a parameter ($K > 0$) to be interpreted as *the maximum sustainable stock* or the *carrying capacity* of the population. Another parameter ($r > 0$) is the intrinsic or natural growth rate when the population size is small. Treating time as a continuous variable, denote the population size at time t by $x(t)$ and its rate of change over time by $\dot{x}(t)$. The evolution of the population is described by

$$\dot{x} = r \cdot x(t)[1- x(t)/K]. \tag{1.5.3}$$

The model predicts that $x(t)$ approaches K in the long run monotonically: decreases to K if the initial size $x(0) > K$, or increases to K if the initial size $x(0) < K$.

On the other hand, the discrete-time Verhulst models reveal alternative possibilities. As an example, consider the following process:

$$x_{t+1} = \theta_1 x_t^2/[x_t^2 + \theta_2], \tag{1.5.4}$$

where the parameters θ_1 and θ_2 are positive numbers. The qualitative behavior of the process depends on the values of θ_1 relative to θ_2. If $\theta_1 < 2\sqrt{\theta_2}$, the population decreases and *extinction is inevitable*. If $\theta_1 > 2\sqrt{\theta_2}$, there is a *critical stock* or a *tipping point*, say $s^* > 0$, such that for all initial $x_0 < s^*$, the population decreases to extinction, whereas, for all $x_0 > s^*$, the population increases to the maximum sustainable stock.

The processes generated by the "quadratic family" of regeneration functions,

$$x_{t+1} = \theta \cdot x_t(1 - x_t), \quad 0 \le x_0 \le 1, \ 1 \le \theta \le 4, \tag{1.5.5}$$

display a wide spectrum of behavior which depends so sensitively on the initial conditions that the prediction of long-run behavior may be practically impossible.

Next, we turn to the possibility of extinction due to "overconsumption" or "excessive harvesting". One approach (keeping in mind Rio Principle 8, Rio +20, paragraph 58, NYS, paragraph 9) is to characterize consumption programs that are equitable (assign the *same* quantity of consumption $c > 0$ to each period) and can be indefinitely sustained given the regeneration function and the initial stock. This direction leads to the examination of a process such as:

$$x_{t+1} = f(x_t) - c, \qquad (1.5.6)$$

where f (the reproduction function) belongs to a class of functions (increasing, strictly concave with derivatives meeting suitable end-point conditions). A complete analysis is given in Chapter 2.4 and is referred to in the subsequent chapters.

While we have examined a single population, there is a voluminous literature on the interaction of populations. We sketch the famous Lotka–Volterra model that captures many pervasive predator–prey relationships (frogs–mosquitoes, bats/snakes–rats, etc.). Such relationships are very much relevant in understanding natural balance and in controlling insect-borne diseases through biological methods rather than chemicals (to which the insects develop resistance over time).

In dealing with natural resources, the policy makers stress "conservation" and "prudent long-run management". Turning to the dictionaries, we obtain the following definitions:

- "to conserve" is "to keep in a safe or sound state; especially, to avoid wasteful or destructive use of (natural resources)" [Webster's New Collegiate];
- "to keep from harm, decay, or loss, for future use" [Oxford American].

So, it is natural for Clark (2010, p. 2) to observe: "Roughly speaking, conservation is saving for the future. The theory of resource conservation can therefore be addressed as a branch of the theory of capital and investment. A resource stock, such as a forest or a population of fish, is a capital asset — natural capital if you like." Following this lead, in Chapters 3–6, we rely upon the literature on the theory of efficient and optimal intertemporal allocation (beginning with Ramsey, with notable contributions from Malinvaud, Samuelson, Radner, Uzawa, Srinivasan, Cass, Gale, McFadden and others) to explore a number of themes. The idea that we are planning for *a long time* can be captured in several ways: a finite T-period model (with a presumably large T: Chapter 3); a rolling plan that is generated by a succession of T-period models (Chapter 4) or an infinite horizon model (Chapters 5 and 6).

Chapter 3 deals with the problem of optimal harvesting with a finite (presumably long) horizon. The analytical framework is "classical": the regeneration function satisfies the usual strict monotonicity, strict concavity, smoothness and Uzawa conditions. The economy begins with an initial resource stock $x_0 > 0$ and the planner sets a target stock $x_T > 0$. The return from harvesting is also represented by a strictly increasing, smooth, strictly concave felicity (utility, welfare, etc.) function u satisfying a boundary condition near 0. The optimization criterion is the (possibly discounted) sum of returns over all periods subject to technological feasibility. Some of the questions studied are as follows: (a) existence and uniqueness of an optimal program — the Ramsey–Euler condition characterizing optimal intertemporal trade-off; (b) the link between the optimal programs and price-guided competitive programs; (c) "equitable" (or stationary) programs that generate the same yield in every period; (d) the "golden rule" or maximal equitable program; (e) implications of raising the target stock (comparative dynamics); (f) qualitative properties of optimal programs.

In Chapter 4, we review a model of optimal resource consumption in which the stocks evolve through a succession of finite horizon plans. The particular framework is of interest for two reasons. First, it has been shown to be an example of a decentralized, "evolutionary" mechanism in the sense of Hurwicz–Weinberger. Secondly, it introduces a constraint that stresses conservation as plans are revised. Informally, given an initial stock x_0, the economy contemplates a T-period plan in which the objective is the maximization of a sum of returns subject to the constraints imposed by the production function and $x_0 = x_T$ (a "conservation of inheritance" clause). The optimal first period consumption c_1^* takes place and the economy moves to the stock level $x_1^* = f(x_0) - c_1^*$ (here, f is the regeneration function). Next, a new T-period plan (with the same return and production function) is made, *but with initial and terminal stocks set equal* *to* x_1^*. Again, the optimal consumption according to the revised plan takes place, and the story is repeated. The sequence of stocks generated this way is called a *rolling plan*. This rolling plan is shown to be intertemporally (Malinvaud) efficient (i.e., there is no alternative program from the same initial stock that ensures at least as much consumption in every period and strictly more in some period). The concept of Malinvaud efficiency captures avoidance of waste, but has no built-in considerations for intergenerational equity. Interestingly enough, it is also proved that the rolling plan maximizes the long-run average return.

In Chapters 5 and 6, we clarify the link between discounting the future and the possibility or inevitability of extinction in a large class of infinite horizon models of optimal harvesting. In both of these chapters, we allow for the production/regeneration function to have an initial phase of increasing returns (Knightian). This takes us beyond convex optimization and forces us to design new methods for deriving qualitative properties. In his paper, a landmark in capital theory, Ramsey (1928) observed that "we do not discount later enjoyments in comparison with earlier ones, a practice which is ethically indefensible and arises merely from the weakness of the imagination." We first provide a self-contained exposition of the "undiscounted" case. Recall that u denotes the (one period) return function. A program of resource management that generates a consumption (or yield) sequence $\mathbf{c}^* = (c_t^*)$ is optimal if

$$\limsup_{T \to \infty} \sum_{t=1}^{T} [u(c_t) - u(c_t^*)] \leq 0 \qquad (1.5.7)$$

for all consumption sequences (c_t) generated by programs from the *same* initial stock. Note that (1.5.6) implies (see Apostol (1957, p. 384)) that for any $\varepsilon > 0$, there is some $\hat{T}$ such that

$$\sum_{t=1}^{T} [u(c_t)] \leq \sum_{t=1}^{T} [u(c_t^*)] + \varepsilon \quad \text{for all } T \geq \hat{T}. \qquad (1.5.8)$$

It is shown that an optimal program exists and converges to the unique maximal equitable (optimal stationary) program.

Next, we turn to discounting. First, the objective function in the *linear* case $(u(c) = c)$ is $\sum_{t=1}^{\infty} \delta^{t-1} c_t$, where δ is the discount factor. A qualitative analysis of the optimal policy reveals a conflict between conservation and impatience. More precisely, there are numbers $\delta_1 < \delta_2$ (determined by the production function f), such that if $0 < \delta < \delta_1$, the immediate extinction program is optimal. If $\delta_2 < \delta < 1$, an optimal program *attains* an equitable program. In the intermediate range, corresponding to each δ, there is a tipping point $k(\delta)$ such that if the initial x_0 is below $k(\delta)$, the immediate extinction program is optimal, whereas if x_0 is above $k(\delta)$, the optimal program attains an equitable program. The details follow directly from the analysis in Chapter 6, when we interpret the problem of possible or inevitable extinction of a competitive fishery. Next, a similar classification when the return function is *strictly concave* is sketched.

In Part B, the reader is introduced to some topics in the literature on exhaustible resources. Chapter 7 reviews the partial equilibrium analysis of optimal extraction of an exhaustible resource, a discrete-time version of Hotelling's model (1931).

A resource is *exhaustible* if its supply cannot be augmented given the known technology. Let $S > 0$ be the initial stock of such an exhaustible resource. A sustainable consumption (or extraction) program is a non-negative sequence $\mathbf{c} = (c_t)$, satisfying $\sum_{t=0}^{\infty} c_t \leq S$. It is shown that $c = (c_t)$ is (Malinvaud) efficient if $\sum_{t=0}^{\infty} c_t = S$. The return function $u(c)$ is assumed to have the usual properties indicated earlier.

First, let T be finite, and consider the following optimization problem:

$$\text{Maximize} \sum_{t=0}^{T} u(c_t) \text{ subject to } \sum_{t=0}^{T} c_t = S. \qquad (1.5.9)$$

It is not difficult to see that the equitable program $c_t^* = 1/(T+1)$ is the unique solution to the problem. However, if T is infinite, and no discounting is introduced, analytical issues involving existence arise.

Hence, we introduce discounting and are led to

$$\text{Maximize} \sum_{t=0}^{\infty} \delta^t \cdot u(c_t) \text{ subject to } \sum_{t=0}^{\infty} c_t = S, \text{ where } 0 < \delta < 1.$$

$$(1.5.10)$$

We characterize some properties of the unique optimal program and the supporting competitive prices. Two cases are distinguished: (i) the utility function satisfies $\lim_{c \to 0} u'(c) = \infty$; (ii) $\lim_{c \to 0} u'(c) \leq M$ for some positive finite M. Comparative dynamic results on the sensitivity of optimal programs with respect to changes in the initial S and the discount factor δ are spelled out.

In Chapter 8, there is a producible good which can be used either in consumption or as an input (capital) in production.

The technology is described in terms of a production function

$$V = G(k, r, z), \quad (k, r, z) \geq 0, \tag{1.5.11}$$

where the triplet (k, r, z) denotes, respectively, the inputs of capital, an exhaustible resource and labor (a primary factor of production, the supply of which is given exogenously in each period). We assume that capital does not depreciate, so the total output is given by

$$F(k, r, z) = G(k, r, z) + k, \quad (k, r, z) \geq 0. \tag{1.5.12}$$

The supply of labor is the same constant $z_- > 0$ in every period, taken as $z_- = 1$. The primary objective is to characterize conditions under which there is a consumption program that is *both* equitable and (Malinvaud) efficient. It is shown that if the technology can generate some equitable program with positive consumption, then there is an efficient equitable program (even when the exhaustible resource is an "important" input). Whether the technology can generate some positive equitable program depends on the possibility of substitution among the inputs, and this question is examined in more detail in Chapter 2.

In Chapter 9, the net output function $G(k, r, z)$ is assumed to have the Cobb–Douglas form, and by direct calculations, we settle some issues when the supply of labor is *not* treated as a constant.

We identify the conditions under which the economy can generate an equitable positive per capita consumption program. In the special case where labor is treated as a constant, this condition can be expressed as an inequality involving the exponents of capital and resource in the Cobb–Douglas form. When the utility function is of the form $u(c) = -(1/c^\sigma)$ (where $\sigma > 0$ and c is per capita consumption), necessary and sufficient conditions for the existence of an (undiscounted) optimal program are established.

In Chapter 10, we present a discrete-time variation of Vernon Smith's model (1974) of an "optimistic" transition to a "green" technology. There is a substantial literature based on continuous-time models that attempt to capture the adoption of the "backstop" technology. Perhaps not unexpected, "sharp" results on the switching points that support an intuitive scenario are very much dependent on the structure of the model. This is also the case with the simple framework we explore.

Chapters 11 and 12 deal with stochastic models of evolution and harvesting. The first-order difference equation,

$$x_{t+1} = \alpha(x_t), \qquad (1.5.13)$$

provided the foundation of our mathematical analysis in Chapter 2. Here, given an initial x_0, the law of motion α determines the entire trajectory

$$(x_0, \alpha(x_0), \alpha^{(2)}(x_0), \ldots, \alpha^{(j)}(x_0), \ldots).$$

As in (1.5.6), we are led to the following process:

$$x_{t+1} = f(x_t) - c, \qquad (1.5.14)$$

where f is the regeneration function and $c > 0$ is the target level of consumption. We can write

$$x_{t+1} = \alpha(x_t), \quad \text{where } \alpha(x) = f(x) - c.$$

One simple way to introduce uncertainty is to treat the law of motion α in (1.5.13) as random. To this effect, we consider a *random dynamical system* described by a triplet (S, Γ, Q) where S is the state space (some subset of R^l_+, for example), Γ an appropriate family of maps from S into itself (interpreted as the set of *all admissible laws of motion*), and Q is a probability distribution on (some sigma-field of) Γ. The evolution of the system is described informally as follows: initially, the system is in some state x in S; an element α_1 of Γ is chosen by Tyche according to the distribution Q, and the system moves to the state $X_1 = \alpha_1(x)$ in period one. Again, independent of α_1, Tyche chooses α_2 from Γ according to the same Q, and the state of the system in period two is obtained as $X_2 = \alpha_2(X_1)$, and the story is repeated. The sequence X_{t+1} so generated is a Markov process:

$$X_{t+1} = \alpha_{t+1}(X_t) \equiv \alpha_{t+1}\alpha_t \ldots \alpha_1(x_0) \quad (t \geq 0), \tag{1.5.15}$$

with a stationary transition probability $p(x, dy)$ given as follows: for $x \in S$, and any (measurable) subset C of S,

$$p(x, C) = Q(\{\gamma \in \Gamma : \gamma(x) \in C\}). \tag{1.5.16}$$

As an example, we can go back to (1.5.14) and start with an exogenously specified set

$$\Gamma = \{f_1, \ f_2, \ldots, f_N\}$$

of all possible regeneration functions. Given the initial state x_0, Tyche ("nature") chooses some element $\tilde{f}$ from Γ according to a probability distribution Q (i.e., $\tilde{f} = f_j$ with probability $Q_j > 0$, $\sum_{j=1}^N Q_j = 1$), and we have the transition

$$X_1 = \alpha_1(x_0) \quad \text{where } \alpha_1(x_0) = \tilde{f}(x_0) - c.$$

Clearly, X_1 is a random variable (with N possible values $f_j(x) - c$, $j = 1, 2, \ldots, N$), the distribution of which is determined by Q.

Our focus is a characterization of the long-run behavior of the process: examples of alternative possibilities are discussed, and some general results on convergence to an invariant distribution are stated.

We present two representative examples of models on identifying sustainable consumption patterns under uncertainty. The first deals with the problem of *living off a wealth fund*. The second analyzes a *water management problem.* Now, as Governor Oeystein Olsen of the Central Bank of Norway observed in 2016: "We have always known that the long oil–driven expansion would come to an end. We have now reached that juncture. The Norwegian economy has enjoyed an exceptionally long summer. Winter is coming." To prepare for such winter, many oil–natural gas exporting economies have developed a wealth fund (oil fund) that can be tapped to meet a target consumption level for the future generations. Informally, an economy starts with a positive initial stock y (initial reserve of a sovereign wealth fund). From this, a positive quantity, a target consumption c is subtracted (a withdrawal to meet the needs of citizens). If the remainder $x_0 = y - c$ is negative, the economy is ruined. If the remainder is non-negative, it is interpreted as an *input* into some productive activity (i.e., as an investment). The *output* of this activity is then the reserve at the beginning of the next period and is given by $y_1 = g(x_0) = g(y - c)$, where $g : R \to R_+$ [$g(x) = 0$ for $x \leq 0$; $g(x) > 0$, for $x > 0$]. This g is the return function. We introduce uncertainty by letting g to be chosen by Tyche in every period (independently, according to a common distribution Q) from a class Γ of functions. Then, Y_1, the reserve in period 1 is the random variable $Y_1 = g(x - c)$. Again, in period 1, if $Y_1 - c < 0$, the economy is ruined in period 1; otherwise, the investment $X_1 = Y_1 - c$ generates the output $Y_2 = g(Y_1 - c)$, and the story is repeated. Now, the period T is the (random) first time such that $Y_t < 0$; otherwise, $T = \infty$, i.e., the equitable program $\mathbf{c} = (c, c, \ldots)$ is sustainable. We say that the economy sustains c with probability $\rho(y, c)$ if $\Pr[T = \infty] = \rho(y, c)$. Intuitively, such probability depends on the initial stock y, the target consumption c and the distribution of the random function g. Characterization, approximation or computation of $\rho(y, c)$ are the main issues. Despite the simplicity of the model, the tasks are not easy, and open questions remain.

The second model (the Lindley–Spitzer process) studies the problem of sustaining a target of water consumption from a reservoir when the augmentation in each period (net increment due to rainfall) is random. Let $X_t \geq 0$ be the stock (level) at the end of period t, $\{\mathfrak{R}_{t+1}\}$ be a sequence of independent and identically distributed (i.i.d.) non-negative real-valued random variables (new supply, "recharge" or augmentation of the stock during the period $t+1$) and $c > 0$ be a target level of consumption. At the end of period $t + 1$, if $X_t + \mathfrak{R}_{t+1} > c$, the planner withdraws (consumes c), leaving $X_{t+1} = X_t + (\mathfrak{R}_{t+1} - c)$ as the stock at the end of period $t + 1$. If, on the other hand, $X_t + \mathfrak{R}_{t+1} \leq c$, the entire available stock is withdrawn, leaving $X_{t+1} = 0$ (the economy fails to meet the target c and faces a *crisis*). Writing $Z_{t+1} = \mathfrak{R}_{t+1} - c$, we obtain

$$X_{t+1} = (X_t + Z_{t+1})^+ \quad \text{where } z^+ = \max(z, 0). \qquad (1.5.17)$$

Our focus is on the case where $EZ_t < 0$, i.e., the reservoir is "stressed". A relatively non-technical exposition, which does not require any integration theory at all, of the dynamic behavior of the process is first presented in Section 12.3.1 of Chapter 12. The variables X_t and c are assumed to take only non-negative integer values, and the net increments Z_t are $+1$ and -1 with probabilities $p > 0$ and $(1 - p) > 0$, respectively. The more general case is then taken up in Section 12.3.2 of Chapter 12. The process converges to a stochastic equilibrium, and a fundamental problem is to determine the expected number of periods for a crisis to return. Two examples in the general case in which such issues can be resolved by direct computation are discussed in detail.

The book does not offer a unified analytical framework that can throw light on all the major links between resources and sustainability. As we strive for rigor in a formal presentation, the limited scope and inadequacies of a model appear in robust relief. I am reminded of the following quote:

> *"But a Grand Unified Theory will remain out of reach of economics, which will keep appealing to a large collection of individual theories. Each one of these deals with a certain range of phenomena that it attempts to understand and to explain."*
>
> — Gerard Debreu

1.6 Complements and Details

1.6.1 Overpopulation

"Overpopulation" relative to limited resources and food supply has been a theme of continuing interest. For some countries, the size, rate of growth and density are challenging the efforts to improve the quality of life. It is interesting to recall Schumpeter (1954, pp. 254–255, and the references cited therein):

> *"divested of non-essentials, the 'Malthusian' Principle of Population sprang fully developed from the brain of Botero in 1589: populations tend to increase, beyond any assignable limit, to the full extent possible made by human fecundity; the means of subsistence, on the contrary, and the possibilities of increasing them are definitely limited and therefore impose a limit on that increase, the only one there is; this limit asserts it self through want, which will induce people to refrain from marrying (Malthus' negative check, prudential check, 'moral restraint') unless numbers are periodically reduced by wars, pestilence, and so on (Malthus' positive check). This path-breaking performance — the only performance in the whole theory of population to deserve any credit at all — came much before the time in which its message could have spread . . . But about two hundred years after Botero, Malthus did no more than to repeat it, except that he adopted particular mathematical laws . . . population was to increase 'in geometric ratio or progression' — that is, as a divergent geometric series — food in 'arithmetic ratio or progression'. But the 'law of geometric progression' though not in Botero's work, was suggested by Petty (1686), by Süssmilch (1740), by R. Wallace (1753), and by Oates (1774), so that within this range of ideas, there was nothing left for Malthus to say that had not been said before . . . Botero, then was the first to sound that note of pessimism which was to become so famous a bone of contention in the days of Malthus."*

Early warnings on the limits to growth and environmental calamities came from, among others, Carson (1962), Boulding (1966), Meadows, Meadows, Randers, and Behrens (1972).

The article by Kates, Parris, and Leiserowitz (2005) is a very useful overview of the literature on the search for a definition. The authors draw attention, in addition to the Brundtland report, to *Our Common Journey: A Transition Toward Sustainability* (1999), an important study by National Research Council, the Board on Sustainable Development of the US National Academy of Sciences. The authors also provide a list of webpages that are insightful.

1.6.2   Supplementary readings

There are numerous sources on the summits that are readily available on the internet. Some examples are as follows:

(a) The proclamations of the Stockholm Conference on the Human Environment are available in UN publication (1972).
(b) The Rio proclamations are in UN Document A/Conf.151/26 (1992). A useful summary is presented in Parson, Haas and Levy (1992).
(c) The Johannesburg Conference is reported in UN publication (1992).
(d) The details on MDGs are to be found in: www.un.org/ millenniumgoals, also in https://sustainabledevelopment.un. org/post2015/transformingourworld.
(e) The details on 2015 NY Summit are available from "Transforming our world/the 2030 Agenda for Sustainable Development", in *Sustainable Development: Knowledge Platform* on the web: https://sustainabledevelopment.un.org/post2015/trans formingourworld.

From the same platform, one can access the documents, *The Future We Want* and *Agenda 21 Earth Summit* (1992).

A comprehensive list of health indicators is provided in WHO (2018).

The book by Dasgupta and Heal (1979), covering a wide range of issues, is still a useful supplementary reference. For appraisals of important themes touched upon (or casually treated or not mentioned at all!) in this book, see Solow (1974a,b), Roy (2010), Krautkraemer (2010), Cogoy and Steininger (2010) (international perspective), Cleveland (2010) (biophysical constraints), Hussen (2000) (environment). Hanely, Shrogen and White (2001) (environment).

Chapter 2

Evolution, Extinction
and Sustainability

The nature of dynamic processes can best be appreciated from
a study of concrete examples. Moreover, if one agrees that the
common core of dynamic process analysis consists of its formal
method, and recognizes the intrinsic technical difficulties of that
method, then the advantages of a case treatment of the subject
are reinforced.

— Paul Samuelson

Nature is complex. Despite this acknowledged fact, there is a
rationale for constructing oversimplified mathematical carica-
tures of reality: one hopes to capture the essence of observed
patterns and processes without becoming enmeshed in the
details.

— Robert M. May and George F. Oster

2.1   Introduction

In this chapter, we study the evolution of renewable resources by
drawing on simple "population" models from mathematical biology.
Recall that a renewable resource has the capacity of natural regeneration, *but only when* a positive stock is available for reproduction.
Various species of insects, plants, birds, animals, fish, etc. are typical examples. The basic concept of the models is a *reproduction*
or *regeneration function* that specifies the "output" (the stock in

any period) generated by the "input" (the sock involved in the production) in the previous period. Formally, the dynamic evolution of the process is described by a difference equation of the following type:

$$x_{t+1} = \alpha(x_t). \tag{2.1.1}$$

The variable x_t is the size of the resource stock ("population" size of a species, of a generation, etc.) in period t involved in reproduction, leading to the stock x_{t+1} according to the mapping (rule) α (from non-negative real numbers into itself), satisfying $\alpha(0) = 0$. Thinking of a resource as a "natural capital", one can write

$$\alpha(x) = \mu \cdot x + \hat{\alpha}(x), \tag{2.1.2}$$

where $\mu \cdot x$ is the size of the population in any period that survives through the following period (μ is the depreciation factor for capital, $0 \leq \mu < 1$) and $\hat{\alpha}(x)$ is the *net* reproduction from x, satisfying $\hat{\alpha}(0) = 0$. If the generations do not overlap (there are many such models, see details in May (1974, 1976)), then $\mu = 0$, and $\alpha(x) = \hat{\alpha}(x)$.

To emphasize the role of exogenous forces that can affect reproduction, it is often convenient to introduce one or more parameters explicitly. Thus, we study a family:

$$x_{t+1} = \alpha(x_t, \theta), \tag{2.1.3}$$

where θ is a positive number (or, a collection of such numbers). Again, it is assumed that $\alpha(0, \theta) = 0$ for all θ.

The first step is to identify an equilibrium of the process (2.1.1) and see whether that equilibrium has some stability properties. Naturally, in the more general framework (2.1.3), the equilibrium and stability properties typically depend on the parameter value(s) and this immediately leads to questions of comparative statics and dynamics: the impact of a change in the parameter on an equilibrium and its stability. It turns out that the *nonlinearity* of the regeneration function poses many subtle and challenging issues: multiplicity of equilibria allowing for possibilities of extinction and indefinite sustainability of population, emergence of "thresholds" or "tipping points" that mark a change from growth to a stunning inevitability of extinction, and

> *"a rich spectrum of dynamical behavior, from stable points (equilibria, m.m) through cascades of stable cycles, to a regime in which the behavior (although fully deterministic) is in many respects "chaotic", or indistinguishable from the sample function of a random process."* (May (1976, p. 459))

The sensitive dependence of trajectories on initial conditions may make long-run predictions virtually impossible (given the constraints on the accuracy of measurement and collection of data). It is clear that a successful application (in terms of description and prediction of an event) of a model depends crucially on understanding the properties of the regeneration function.

In Section 2.2, we introduce a few basic definitions and some results from the theory of dynamical systems. This is followed by a fairly detailed exploration of five classes of descriptive models in Section 2.3. Perhaps, no account can be complete from a historical perspective without a look at Verhulst models (Examples 2.3.1 and 2.3.2). Emergence of complexity and chaos in the presence of nonlinearity is captured in Example 2.3.3. Some important implications of *monotone* laws of motion (without specification of functional forms) are spelled out in Example 2.3.4. Finally, in Example 2.3.5, the famous logistic equation is introduced.

In Section 2.4, we examine the implications of *constant yield harvesting* (a plan to harvest a constant $c > 0$ in every period, the population size so permitting). The possibility or inevitability of extinction depends on the size of harvest relative to the maximum sustainable consumption. When the stock ("population") size is small, an outright ban on harvesting is called for. If the harvest is too high, an immediate intervention to lower it is needed to stave off extinction. If the harvest level keeps increasing, careful monitoring is essential.

There is quite naturally a substantial and growing literature on capturing the interaction among different populations. Even a cursory review is beyond the scope and size of this book. I conclude with just one example of such an interaction: the celebrated "predator–prey" model of Lotka–Volterra.

2.2   Dynamical Systems: Formalities

This section develops the analytical framework. We bring together the basic definitions and a few mathematical results from the literature on dynamical systems.

We consider a dynamic process described formally by a difference equation:

$$x_{t+1} = \alpha(x_t). \tag{2.2.1}$$

Here, x_t belongs to S, the set of all possible *states* of the process, and $\alpha : S \to S$ is the *law of motion* of the process. The pair (S, α) is referred to as a dynamical system. We typically choose S to be a non-empty subset of real numbers ($S = R$, R_+, a closed interval $[a, b]$). In general, S is a metric space and α is a *continuous* map from S into S. Given an initial state x_0, we write $\alpha^0(x) \equiv x, \alpha^1(x) = \alpha(x)$, and, for every positive integer $j \geq 1$,

$$\alpha^{j+1}(x) = \alpha(\alpha^j(x)).$$

We refer to $\alpha^j(x)$ as the jth iterate of α. For any initial $x \in S$, the *trajectory* from x is the sequence $\tau(x) \equiv \{\alpha^j(x)_{j=0}^\infty\}$. It is the sequence of states, starting from x, that the law of motion α generates: an evolution of the dynamic process governed by the law of motion α. A point $x^* \in S$ is a *fixed point* of α if $x^* = \alpha(x^*)$. A point $y^* \in S$ is a *periodic point* of period k if $\alpha^k(y^*) = y^*$ and $\alpha^j(y^*) \neq y^*$ for $1 \leq j < k$. Thus, to prove that y^* is, say, a periodic point of period 3, one needs to show that y^* is a fixed point of α^3 and y^* is *not* a fixed point of α and α^2.

Clearly, if x^* is a fixed point of α, the trajectory from x^* is given by $\tau(x^*) = (x^*, x^*, \ldots, x^*, \ldots)$. It is an *equilibrium* or a *steady state* of the dynamic process. Here is an important result on the existence of fixed points.

Theorem 2.2.1. *Let $S = R$, and α be a continuous map from $S \to S$. If there is a (non-degenerate) closed interval $I = [a, b]$ such that (i) $\alpha(I) \subset I$ or (ii) $I \subset \alpha(I)$, then there is a fixed point of α in I.*

Proof. (i) If $\alpha(I) \subset I$, then $\alpha(a) \geq a$ and $\alpha(b) \leq b$. If $\alpha(a) = a$ or $\alpha(b) = b$, then the conclusion is immediate. Otherwise, $\alpha(a) > a$ and $\alpha(b) < b$. This means that the function $\beta(x) \equiv \alpha(x) - x$ is positive at a and negative at b. Using the Intermediate Value Theorem, $\beta(x^*) = 0$, for some x^* in (a, b). Hence, $\alpha(x^*) = x^*$.

(ii) By the Weierstrass theorem, there are points x_m and x_M in I such that $m \equiv \alpha(x_m) \leq \alpha(x) \leq \alpha(x_M) \equiv M$ for all x in I. Then, by the Intermediate Value Theorem, $\alpha(I) = [m, M]$. Since $I \subset \alpha(I)$,

$$m \leq a \leq b \leq M.$$

In other words,

$$\alpha(x_m) = m \leq a \leq x_m,$$

and

$$\alpha(x_M) = M \geq b \geq x_M.$$

The proof is now completed by an argument similar to that in case (i). $\square$

It should be stressed that *no claim of uniqueness is made in the above proposition.*

[▲ *Exercise*

(i) Let $S = [0, 1]$. Consider the maps (a) $\alpha(x) = x$, (b) $\alpha(x) = (1/2)x$ and (c) $\alpha(x) = 2x$ for $0 \leq x \leq 1/2$, $\alpha(x) = 2(1 - x)$ for $1/2 \leq x \leq 1$. Calculate the fixed points of α in each case.

(ii) Let $S = (0, 1)$. Consider the map $\alpha(x) = 1/x$. Show that α is continuous but does not have a fixed point.

(iii) Let $S = R$, and $\alpha(x) = x^2 - 1$. Compute α^2 and α^3.

(iv) Let $S = R$, and $\alpha(x) = 1 - x^2$. Show that the fixed points are $x = (1/2)[-1 \pm \sqrt{5})$ and the periodic points of period 2 are $\{0, 1\}$. ▼]

A fixed point x^* of α is (locally) *attracting* or (locally) *stable* (a "*sink*") if there is an open interval U containing x^* such that for all $x \in U$, the trajectory $\tau(x)$ from x converges to x^*. A fixed point x^*

is *repelling* (a *"source"*) if there is an open interval U containing x^* such that for any $x \in U$, $x \neq x^*$, there is some $k \geq 1$, such that $\alpha^k(x) \notin U$.

Consider a dynamic process (2.2.1) where S is a non-degenerate closed interval $[a, b]$ and α is continuous on $[a, b]$. Assume that α is continuously differentiable on (a, b). A fixed point $x^* \in (a, b)$ is *hyperbolic* if $|\alpha'(x)| \neq 1$.

Proposition 2.2.1. *Let $S = [a, b]$ and α be continuous on $[a, b]$ and continuously differentiable on (a, b). Let $x^* \in (a, b)$ be a hyperbolic fixed point of α.*

(i) *If $|\alpha'(x^*)| < 1$, then x^* is locally stable.*
(ii) *If $|\alpha'(x^*)| > 1$, then x^* is repelling.*

Proof. (i) There is some $u > 0$, and some $\mathbf{m} < 1$ such that $|\alpha'(x)| < \mathbf{m} < 1$ for all x in $I = [x^* - u, x^* + u]$. By the Mean Value Theorem, if $x \in I$,

$$|\alpha(x) - x^*| = |\alpha(x) - \alpha(x^*)| \leq \mathbf{m}|x - x^*| < \mathbf{m}u < u.$$

Hence, α maps I into I and, again, by the Mean Value Theorem, is a uniformly strict contraction on I. The result follows from the Contraction Mapping Theorem; (ii) is left as an exercise. □

2.3   Population Models

A qualitative description of all the trajectories of a dynamical system is not easy even when the law of motion is assigned a particular functional form. This will be clear as we work through the examples of this chapter. However, with the insights gained from an example (and simulation and computations), it is often possible to identify a pattern of behavior shared by a much larger class of models (see Example 2.3.4).

In many contexts, the growth of "population" of a species is described by a dynamic process in discrete or continuous time. The models are primarily descriptive and can be used for prediction.

In fact, a prominent issue has been the validity or applicability of such models for *long-run* prediction. Although we focus mostly on discrete-time models, we shall touch upon a few continuous-time models that have played a distinctive role in the development of analytical theory and its applications.

Example 2.3.1: A Discrete-Time Verhulst Model:
Consider a model of a population process described by

$$x_{t+1} = \alpha(x_t), \tag{2.3.1}$$

where the reproduction function $\alpha : R_+ \to R_+$ is given by

$$\alpha(x) = \frac{\theta_1 x}{x + \theta_2}.$$

Here, both the parameters θ_1 and θ_2 are positive real numbers. The following properties of α are important:

P1. $\alpha'(x) = \theta_1 \theta_2/(x + \theta_2)^2$; if $x > 0$, $\alpha'(x) > 0$ (α is strictly increasing: if $y > z$, $\alpha(y) > \alpha(z)$).
P2. $\alpha''(x) = -2\theta_1\theta_2/(x + \theta_2)^3 < 0$; if $x > 0$, $\alpha''(x) < 0$ (α is strictly concave).

Next, verify the following property:

P3. If $\theta_1 \leq \theta_2$, α has a unique fixed point $x^* = 0$. If $\theta_1 > \theta_2$, α has two fixed points $x_1^* = 0$, $x_2^* = \theta_1 - \theta_2$.

We shall now analyze the behavior of the trajectories from $x_0 \geq 0$.

Case (i): $\theta_1 \leq \theta_2$. Here, the trajectory from 0 is clearly $\tau(0) = (0, 0, \ldots, 0, \ldots)$. Now, for $x \neq 0$, $\alpha(x)/x = \theta_1/(x + \theta_2) < \theta_1/\theta_2 \leq 1$. So,

$$\text{when } x > 0, \ x_{t+1} = \alpha(x_t) \leq x_t \quad \text{for all} \quad t \geq 0.$$

Since $x_t \geq 0$ for all $t \geq 0$, the sequence (x_t) converges to some $\bar{x} \geq 0$. Now, (x_{t+1}) also converges to the same limit $\bar{x}$. By using the continuity of the function α, from

$$x_{t+1} = \alpha(x_t),$$

we get, in the limit as $t \to \infty$,

$$\bar{x} = \alpha(\bar{x}).$$

This means that $\bar{x}$ is a fixed point of α. But in this case, α has a unique fixed point x^*. Hence, $\bar{x} = x^*$.

To summarize,

> *for any $x > 0$, the trajectory $\tau(x) = (x, \ldots, x_t, \ldots)$ is decreasing and converges to the unique steady state $x^* = 0$.*

Case (ii): $\theta_1 > \theta_2$. The two fixed points of α, denoted by $x_1^* = 0$, $x_2^* = \theta_1 - \theta_2$, satisfy $\alpha'(x_1^*) > 1$ (hence, the steady state x_1^* is a "source", it is locally "repelling"), $\alpha'(x_2^*) < 1$ (hence, the state x_2^* is a "sink", it is locally "attracting"). As before, $\tau(x_1^*) = (0, \ldots, 0, \ldots)$ and $\tau(\theta_1 - \theta_2) = (\theta_1 - \theta_2, \ldots, \theta_1 - \theta_2, \ldots)$. Consider any x_0 satisfying $0 < x_0 < \theta_1 - \theta_2$.

Since $\theta_1 > \theta_2$, it is certainly true that for $0 < x < \theta_1 - \theta_2$, $[\alpha(x)/x] = \theta_1/[x + \theta_2] > 1$. Hence, $x_1 = \alpha(x_0) > x_0$. Iterating,

$$x_{t+1} = \alpha(x_t) > x_t \quad \text{for all } t \geq 0.$$

Since α is strictly increasing, $x_t < x_2^* = \theta_1 - \theta_2$ for all $t \geq 0$. Hence, the sequence (x_t) converges to some $\bar{x} \geq 0$. Now, (x_{t+1}) also converges to the same limit $\bar{x}$. By using the continuity of the function α, from

$$x_{t+1} = \alpha(x_t),$$

we get, in the limit as $t \to \infty$,

$$\bar{x} = \alpha(\bar{x}).$$

This means that $\bar{x}$ is a fixed point of α. But, $x_0 > 0$. Hence, $x_t > 0$ for all $t \geq 0$. So, $\bar{x} = x_2^*$.

To summarize,

> *when $0 < x < \theta_1 - \theta_2$, the trajectory $\tau(x) = (x, \ldots, x_t, \ldots)$ is increasing and converges to x_2^*.*

Consider $x_0 > \theta_1 - \theta_2$. Note that for $x > \theta_1 - \theta_2$, $[\alpha(x)/x] = \theta_1/[x + \theta_2] < 1$. Modifying the arguments made in the other case, one shows that [▲ *Exercise* ▼]

when $x > \theta_1 - \theta_2$, the trajectory $\tau(x) = (x, \ldots, x_t, \ldots)$ is decreasing and converges to x_2^*.

Thus, the qualitative properties of the trajectories depend critically on whether $\theta_1 \leq \theta_2$ or $\theta_1 > \theta_2$.

Example 2.3.2: S-shaped Reproduction Functions: Tipping Points:

Consider a population process:

$$x_{t+1} = \alpha(x_t),$$

where the reproduction function $\alpha : R_+ \to R_+$ is given by

$$\alpha(x) = \theta_1 x^2 / [x^2 + \theta_2]. \tag{2.3.2}$$

Again, both θ_1 and θ_2 are positive numbers.

To obtain the steady states x^*, one needs to solve the following equation:

$$\alpha(x^*) = x^* \quad \text{or} \quad x^* = \theta_1 x^{*2} / [x^{*2} + \theta_2]. \tag{2.3.3}$$

The solutions are given by

$$x_1^* = 0, \quad x_2^* = \left[\theta_1 - \sqrt{(\theta_1^2 - 4\theta_2)}\right]/2, \tag{2.3.4}$$

$$x_3^* = \left[\theta_1 + \sqrt{(\theta_1^2 - 4\theta_2)}\right]/2,$$

$$x_2^* + x_3^* = \theta_1, \quad x_2^* \cdot x_3^* = \theta_2. \tag{2.3.5}$$

We verify that $\alpha'(x) = 2\theta_1 x \theta_2 / [(x^2 + \theta_2)^2]$. Note that $\alpha'(0) = 0 < 1$ so 0 is a *sink* (*locally attracting, locally stable*)

Thus,

Case (i). For $\theta_1 > 2\sqrt{\theta_2}$, there are three steady states $x_1^* < x_2^* < x_3^*$;

Case (ii). For $\theta_1 = 2\sqrt{\theta_2}$ there are two $x_1^* = 0$, $x_2^* = \theta_1/2$;

Case (iii). For $\theta_1 < 2\sqrt{\theta_2}$, there is a unique steady state $x_1^* = 0$.

We begin with the following case.

Case (iii). By continuity of $\alpha'(x)$, one can find an interval $I = (0, \vartheta)$ such that $\alpha'(y) < 1$ for all $y \in I$. Then, for any $x \in I$,

$$\alpha(x) = \alpha(0) + x\alpha'(y), \text{ where } 0 < y < x, \text{ or } \alpha(x) < x \text{ for } x \in I.$$

But, then, it must be true that for all $x > 0$, $\alpha(x) < x$; otherwise, if $\alpha(x') = x'$ for some $x' > 0$, this x' is a fixed point of α, contradicting the fact that in this case 0 is the unique fixed point of α. Similarly, if $\alpha(x') > x'$ for some $x' > 0$, by the Intermediate Value Theorem, there will be some $z > 0$ such that $\alpha(z) = z$, i.e., there will be a steady state $z > 0$, again contradicting the fact that in this case 0 is the unique fixed point of α. We conclude the following:

> *For any $x > 0$, the trajectory $\tau(x) = (x, \ldots, x_t, \ldots)$ is strictly decreasing $(x_{t+1} = \alpha(x_t) < x_t$ for all $t)$ and converges to 0.* [▲ *Exercise*: Complete the proof of the assertion on convergence. ▼]

Next, we turn to the following case.

Case (i). First, note the following:

> *For any $0 < x < x_2^*$, the trajectory $\tau(x) = (x, \ldots, x_t, \ldots)$ is strictly decreasing $(x_{t+1} = \alpha(x_t) < x_t$ for all $t)$ and converges to 0.* [▲ *Exercise* ▼]

We now consider trajectories from x where $x_2^* < x < x_3^*$. We shall prove we following:

> *For any $x \in (x_2^*, x_3^*)$, the trajectory $\tau(x) = (x, \ldots, x_t, \ldots)$ is strictly increasing and converges to x_3^*.*

First, observe that since α is strictly increasing $(\alpha'(x) > 0$ when $x > 0)$, $x_2^* < x_t < x_3^*$ for all t.

$x_2^* < x < x_3^*$ implies $\alpha(x_2^*) < \alpha(x) < \alpha(x_3^*)$ leading to $x_2^* < x_1 < x_3^*$. We can proceed by induction on t.

Now, $\alpha(x_2^*) = x_2^* < x_3^*$. Also, from (2.3.3), $x_2^* = \theta_1(x_2^*)^2 / [(x_2^*)^2 + \theta_2]$, or $[(x_2^*)^2 + \theta_2] = \theta_1 x_2^*$.

$$\alpha'(x_2^*) = \frac{2\theta_1 x_2^* \theta_2}{[((x_2^*)^2 + \theta_2)^2]} = \frac{2\theta_1 x_2^* \theta_2}{(\theta_1 x_2^*)^2} = \frac{2\theta_2}{(\theta_1 x_2^*)}$$

$$= \frac{2(x_2^* \cdot x_3^*)}{(x_2^* + x_3^*)x_2^*} = \frac{2x_3^*}{(x_2^* + x_3^*)} > 1.$$

By using the continuity of $\alpha'(x)$ at x_2^*, one can find an interval $I = (x_2^*, x_2^* + \vartheta)$ such that $\alpha'(y) > 1$ for $y \in I$. Now, for any $x \in I$, $\alpha(x) = \alpha(x_2^*) + (x - x_2^*)\alpha'(y)$, where $0 < y < x$. This leads to for any $x \in I$.

$$\alpha(x) = \alpha(x_2^*) + (x - x_2^*)\alpha'(y),$$

$$\text{or} \quad [\alpha(x) - \alpha(x_2^*)] > (x - x_2^*),$$

$$\text{or} \quad [\alpha(x) - x_2^*] > (x - x_2^*),$$

$$\text{or} \quad \alpha(x) > x.$$

This means that $\alpha(x) > x$ for all $x \in (x_2^*, x_3^*)$. Otherwise, there is a fixed point of α in (x_2^*, x_3^*), a contradiction to the fact that the only positive fixed points of α are x_2^* and x_3^*. So, $x_1 = \alpha(x) > x$ and $x_{t+1} = \alpha(x_t) > x_t$ for all $t \geq 1$.

Hence, the trajectory $\tau(x) = (x, \ldots, x_t, \ldots)$ is strictly increasing, and, as $x_t < x_3^*$, the sequence x_t converges to some $\tilde{x} > 0$. But, this means that x_{t+1} also converges to $\tilde{x}$, as $t \to \infty$.

Now,

$$x_{t+1} = \alpha(x_t) \text{ implies, by the continuity of } \alpha,$$

$$\tilde{x} = \alpha(\tilde{x}).$$

This means that $\tilde{x} > x > x_2^*$ is a fixed point of α. Hence, $\tilde{x} = x_3^*$, completing the proof.

Next, consider $x > x_3^*$. Show that $\alpha'(x_3^*) < 1$, and with appropriate modifications of the above proof, verify that *the trajectory* $\tau(x) = (x, \ldots, x_t, \ldots)$ *is strictly decreasing and converges to* x_3^*. [▲ *Exercise* ▼]

Thus, x_2^* is a *tipping point*: a remarkable change in the qualitative properties of trajectories emerges: for all initial stocks below x_2^*, extinction is inevitable, and the trajectories approach 0 in a monotonically decreasing manner. For stocks above x_2^*, trajectories approach the maximum sustainable stock $x_3^* > 0$.

Finally, we consider

Case (ii). Here, first, show that $\alpha'(x_2^*) = 1$.

For $x < x_2^*$, the trajectory $\tau(x) = (x, \ldots, x_t, \ldots)$ *is strictly decreasing and converges to* 0. [▲ *Exercise* ▼]

For $x > x_2^*$, *the trajectory* $\tau(x)$ *is strictly decreasing and converges to* x_2^*. [▲ *Exercise*: hint: Use the fact that $\alpha''(x_2^*) < 0$.

Hence, for all $x > x_2^*$, $\alpha'(x) < 1$. So, $[\alpha(x) - \alpha(x_2^*)] < (x - x_2^*)$, or, $[\alpha(x) - x_2^*] < (x - x_2^*)$. ▼]

Here again, x_2^* is a tipping point.

Example 2.3.3: Complexity: The Quadratic Family:

In his lucid review, May (1976) noted the following:

> "*in many contexts, and for biological populations in particular, there is a tendency for the variable x to increase from one generation to the next when it is small, and for it to decrease when it is large. That is, the nonlinear function α has the following properties:* $\alpha(0) = 0$; $\alpha(x)$ *increases monotonically as* x *increases through the range* $0 < x < A$ (*with* $\alpha(x)$ *attaining its maximum value at* $x = A$); *and* $\alpha(x)$ *decreases monotonically as* x *increases beyond* $x = A$. *Moreover,* $\alpha(x)$ *will usually contain one or more parameters which "tune" the severity of this nonlinear behavior . . . these parameters will typically have some biological, economic and sociological significance.*"

The canonical example that was studied is the quadratic family of maps. Here, $S = [0, 1]$, and the family is described by

$$\alpha_\theta(x) = \theta \cdot x(1 - x), \qquad (2.3.6)$$

where the parameter $\theta \in [1, 4]$. One can verify that the map (2.3.6) sends S into itself for the range of values of θ that we have specified (▲ *Exercise*: hint: $\alpha_\theta(x)$ attains its maximum $\theta/4$ at $x = 1/2$. ▼) May and Oster (1976) and May (1983) contain lists of many functional forms that have similar qualitative properties.

The family (2.3.6) inspired intensive research on going beyond functional forms and identifying a broad class of "unimodal" laws of motion with similar qualitative patterns. It was noted that for values of θ close to 4, the qualitative behavior of a trajectory is "virtually impossible" to predict due to its extreme sensitivity to the initial condition. Indeed, the innocent looking process,

$$x_{t+1} = 4x_t(1 - x_t), \tag{2.3.7}$$

qualifies as "complex" according to most (if not all!) of the commonly used formal definitions of complexity. John von Neumann suggested using this map as a random number generator.

Taking informality to an extreme, let us note that the process (2.3.6) is "closely related" ("topologically semi-conjugate") to the process generated by the "tent map" $T : [0,1] \to [0,1]$ defined as

$$\mathbf{T}(x) = \begin{cases} 2x & \text{for } 0 \le x \le 1/2, \\ 2(1-x) & \text{for } 1/2 \le x \le 1. \end{cases} \tag{2.3.8}$$

This relation is exploited to show that if one process is chaotic according to a particular definition, so is the other. Some computations are particularly simple for the tent map (see Section 2.6). The precise definition of semi-conjugacy and the connection between (2.3.6) and (2.3.8) are spelled out in Hirsch, Smale and Devaney (2013, pp. 340–344).

To introduce the first notion of complexity, I shall recall the celebrated Li–Yorke theorem (1975) and then summarize a few calculations involving the quadratic family that illustrate the "period doubling route" to chaos. There are numerous sources with varying degrees of technical sophistication which the interested reader can explore. Calculations become tedious, if not overwhelming, soon enough as one has to confront the iterates α^k but there is no dearth of computer programs to assist us (see, e.g., Lynch (2003)). Observe that the Li–Yorke theorem does not depend on a particular functional form assigned to α.

Theorem 2.3.1. *Let I be an interval and $\alpha : I \to I$ be continuous. Denote by $\wp(I)$ the set of all periodic points of I and by $\aleph(I)$ the set of non-periodic points.*

Assume that there is some point a in I for which there are points $b = \alpha(a)$, $c = \alpha(b)$ and $d = \alpha(c)$, satisfying

$$d \le a < b < c \quad (\text{or } d \ge a > b > c). \tag{2.3.9}$$

Then,

(1) *For every positive integer $k = 1, 2, \ldots$, there is a periodic point of period k, $x^{(k)}$, in I.*

(2) *There is an uncountable set $\aleph' \subset \aleph(I)$ such that*

  (i) *for all x, y in $\aleph'$, $x \neq y$,*

$$\limsup_{n \to \infty} |\alpha^n(x) - \alpha^n(y)| > 0, \qquad (2.3.10)$$

$$\liminf_{n \to \infty} |\alpha^n(x) - \alpha^n(y)| = 0, \qquad (2.3.11)$$

  (ii) *if $x \in \aleph'$, and $y \in \wp(I)$*

$$\limsup_{n \to \infty} |\alpha^n(x) - \alpha^n(y)| > 0.$$

We say that a dynamical system satisfying the properties (1) and (2) displays "Li–Yorke chaos". The theorem makes it clear that, if there is a periodic point of period 3, prediction of the qualitative behavior requires the exact knowledge of the initial x_0 and the level of precision demanded may not simply be attainable in practice. It may not be possible after calculating a finite (but arbitrarily large) number of distinct terms to say whether a trajectory aperiodic or periodic! The periodic points of period 3 of the tent map are $\{2/7, 4/7, 6/7\}$, $\{2/9, 4/9, 8/9\}$ (see Section 2.6 for calculation and it is easy to confirm by direct substitution); hence, by semi-conjugacy, we can conclude that the map (2.3.5) also displays Li–Yorke chaos. More is revealed when we study the implications of changes in the parameter θ: as θ crosses threshold values, *the number as well as stability* of fixed points change significantly. This is at the heart of the "bifurcation theory".

Stepping back, recall that a periodic point z of period k satisfies the condition $\alpha^k(z) = z$, i.e., z is a fixed point of the kth iterate [and $\alpha^j(z) \neq z$ for $j < k$]. Also, recall that the fixed points of α are necessarily fixed points of α^j for $j \geq 2$. Note that $x = 0$ is a fixed point of α_θ for all $\theta \in [1, 4]$. Show that if $\theta = 1$, the trajectory (x_t) from any $x_0 \in [0, 1]$ converges to 0, the unique fixed point [▲ *Exercise*: hint: in this case, $x_{t+1} < x_t$ ▼]: it is "globally stable".

But as θ increases beyond 1 (even so slightly), the fixed point 0 loses its stability. Formally, we have the following proposition.

Proposition 2.3.1. *For $\theta = 1$, the unique fixed point $x^* = 0$ is (locally) stable. For $\theta > 1$, there are two fixed points $x_1^* = 0$ and $x_2^* = [1 - (1/\theta)]$. The fixed point x_2^* is stable for $1 < \theta < 3$, but unstable for $\theta > 3$.*

Proof. Solving $x = \theta x (1 - x)$, we get $x_1^* = 0$ and $x_2^* = [1 - (1/\theta)]$ (the two coincide when $\theta = 1$. Taking the derivative, we get $\alpha'_\theta(x) = \theta(1 - 2x)$. Use Proposition 2.2.1 to get the result. □

The fixed point $x_2^* = [1 - (1/\theta)]$ loses stability as θ increases beyond 3 and a periodic point of period 2 (briefly, a 2-*cycle*) emerges, giving us a "period doubling".

Proposition 2.3.2. *For $\theta > 3$, the logistic map has a 2-cycle that is (locally) stable if $\theta < 1 + \sqrt{6}$.*

Proof. For a law of motion α, a 2-*cycle* is a couple of distinct points, say, p and q in $[0, 1]$ such that $\alpha(p) = q$, $\alpha(q) = p$; both are fixed points of α^2. Solving

$$\alpha_\theta^2(x) = \theta^2 x(1 - x)[1 - \theta x(1 - x)] = x$$

yields four fixed points of $\alpha_\theta^2(x)$:

$$x_1^* = 0, \quad x_2^* = [1 - (1/\theta)];$$
$$x_3^* = [\theta + 1 + \sqrt{(\theta - 3)(\theta + 1)}]/2\theta,$$
$$x_4^* = [\theta + 1 - \sqrt{(\theta - 3)(\theta + 1)}]/2\theta.$$

The last two solutions are new and form a 2-*cycle* if $\theta > 3$. So, we can take $p = x_3^*$ and $q = x_4^*$. To get the local stability, we need to compute

$$\lambda \equiv d(\alpha_\theta^2(x))/dx|_{x=p}. \tag{2.3.12}$$

By the chain rule,

$$\lambda = \alpha'(q) \cdot \alpha'(p)$$

and it is clear that

$$\lambda \equiv d(\alpha_\theta^2(x))/dx|_{x=q}.$$

Substituting the values $p\,(= x_3)$ and $q\,(= x_4)$,

$$\lambda = \theta^2(1 - 2q)(1 - 2p) = 4 + 2\theta - \theta^2.$$

Hence, in the interval $(3, 1 + \sqrt{6}), |\lambda| < 1$, and local stability follows.   □

As we see, as θ crosses the threshold 3, the fixed point x_2^* loses its stability and a stable 2-*cycle* is born, which in turn loses its stability as $1 + \sqrt{6}$ is crossed and a stable 4-*cycle* is created. These values where 2^n-*cycles* are created have been identified with computer experiments, and analytically in some cases: they involve studying roots of higher order polynomials. Identifying values of θ such that α_θ has a stable 3-*cycle* has also been the subject of lively discussion (see Saha and Strogatz (1995), Bechhoefer (1996) and Gordon (1996)). We note: *for* $1 + 2\sqrt{2}$ (~ 3.828427) $\leq \theta \leq 3.8415499\ldots$ *there is a stable* 3-*cycle*. For a more elaborate guide to the period doubling route, see Drazin (1992, Chapter 3), or Lauwerier (1986).

In fact for the range $(1, 3)$, all trajectories starting from $x \in (0, 1)$ converge to the fixed point $x_2^* = [1 - (1/\theta)]$. But the proof is long and examines several cases and is not pursued here.

The examples reviewed above, and many others, specify the reproduction functions in terms of particular functional forms. These

> "*are attempts to give quantitative expressions to rough qualitative ideas about the biological laws governing the population. For this reason, we should be skeptical of the biological significance of any deduction from any specific model that holds only for* that *model. Our goal should be to formulate principles that are* robust, *that is valid for a large class of models, (ideally for all models that embody some set of qualitative hypotheses).*"

— Brauer and Castillo-Chavez, 2001, p. 54

Example 2.3.4: Monotonicity:

Now, we explore a class of dynamical systems without specification of functional forms. *Let* $S = R_+$ *and* $\alpha : S \to S$ *be a continuous, non-decreasing function that satisfies the following condition*:

There is a unique $x^* > 0$ such that

$$\alpha(x) > x \quad \text{for all } 0 < x < x^*,$$
$$\alpha(x) < x \quad \text{for all } x > x^*. \tag{2.3.13}$$

In this case, if the initial $x_0 \in (0, x^*)$, then

$$x_1 = \alpha(x_0) > x_0.$$

But note that

$$x_0 < x^*$$

implies that $x_1 = \alpha(x_0) \leq \alpha(x^*) = x^*$. Repeating the argument, we get

$$x^* \geq x_{t+1} \geq x_t \geq x_1 > x_0 > 0. \tag{2.3.14}$$

Thus, the sequence $\{x_t\}$ is non-decreasing, bounded above by x^*. Hence, $\lim_{t \to \infty} x_t = \hat{x}$ exists, and, using the continuity of α, we conclude that $\hat{x} = \alpha(\hat{x}) > 0$.

Now, the uniqueness of x^* implies that $\hat{x} = x^*$.

If $x_0 = x^*$, then $x_t = x^*$ for all $t \geq 0$.

If $x_0 > x^*$, then $x_1 = \alpha(x_0) < x_0$. Now,

$$x_1 \geq x^*,$$

and, repeating the argument,

$$x^* \leq x_{t+1} \leq x_t \leq \cdots \leq x_1 \leq x_0.$$

Thus, the sequence $\{x_t\}$ is non-increasing and bounded below by x^*. Hence, $\lim_{t \to \infty} x_t = \hat{x}$ exists. Again, by continuity of α, $\hat{x} = \alpha(\hat{x})$, so that the uniqueness of x^* implies that $x^* = \hat{x}$.

To summarize,

Proposition 2.3.3. *Let $S = R_+$ and $\alpha : S \to S$ be a continuous, non-decreasing function that satisfies (2.3.13). Then, for any $x > 0$, the trajectory $\tau(x)$ from x converges to x^*. If $x < x^*$, $\tau(x)$ is a non-decreasing sequence. If $x > x^*$, $\tau(x)$ is a non-increasing sequence.*

Example 2.3.5: The Logistic Growth Model:
Let $x(t)$ be the size of the population at time t, and its rate of change is denoted by x' (or $\dot{x}$, or dx/dt). We look at early efforts to link the growth rate to the size of the populations. Consider the *logistic* differential equation as a formal description of the dynamic process of population change:

$$\frac{dx}{dt} \equiv \dot{x} = rx(t)\left[1 - \frac{x(t)}{K}\right]. \qquad (2.3.15)$$

Here, the parameters r and K are positive and will be interpreted shortly. Informally, when x is "rather small" relative to K, one has $\dot{x} \approx rx$, whereas when x is "close to" K, one has $\dot{x} \approx 0$. In other words, when population size x is small, it experiences exponential growth, whereas when x is near K, its change is negligible. Of course, from the point of view of interpretation, we need $x(t) \geq 0$. Let us denote the initial population size (i.e., size at time $t = 0$) $x(0)$ by $x_0 > 0$. Observe that if $x_0 > 0$, any solution of (2.3.15) must satisfy $x(t) > 0$ for all $t > 0$.

Claim 2.3.1. *The solution to* (2.3.15) *given* $x_0 > 0$ *is given by*

$$x(t) = \frac{Kx_0 e^{rt}}{[K - x_0 + x_0 e^{rt}]} = \frac{Kx_0}{[x_0 + (K - x_0)e^{-rt}]}. \qquad (2.3.16)$$

Verification. [▲ *Exercise*: hints: Since there are general theorems on the existence and uniqueness of a solution given the initial value $x(0) = x$, in order to verify that the given function (2.3.16) is the unique solution, it is enough for us to compute $\dot{x}$ from the functional form (2.3.16) and then check that (2.3.15) is satisfied upon substitution of $\dot{x}$ from (2.3.16) on the left and $x(t)$ from (2.3.16) on the right. ▼] Equation (2.3.15) can be solved by the method of separation of variables (see, e.g., Hirsch, Smale and Devaney (2013, p. 5), or Brauer and Castillo-Chavez (2001, pp. 8–10).   □

Of course, if $x_0 = 0$, $x(t) = 0$ for all t. Similarly, if $x_0 = K$, $x(t) = K$ for all t. Thus, 0 and $K > 0$ are two steady states. Given equation (2.3.16), we see that the population size $x(t)$ approaches the limit K as $t \to \infty$ if the initial $x_0 > 0$. The value K is often called the *carrying capacity* of the population: it represents the population size that the environment (available resources) can *carry* or *sustain*. The value r is called the *intrinsic growth rate* because it represents the per capita growth rate achieved when the population size is small enough to make resource constraints non-binding.

In the absence of harvesting, then, there is no possibility of extinction from any initial $x_0 > 0$. We shall return to this example when constant harvesting is contemplated.

2.4 Constant Yield Harvesting and Extinction

We shall now turn to the implications of constant yield harvesting in a class of models without any specification of functional forms. Consider an economy which starts with an initial stock y of a renewable resource. In each period t, the economy is required to *consume* or *harvest* a positive amount c of the beginning of the period stock y_t. The remaining stock $x_t = y_t - c$ is the input that generates the (gross) output $y_{t+1} = f(x_t)$ where f is the regeneration function. Assume the following:

(F.1) $f : R \to R_+$ is continuous on R_+, $f(x) = 0$ for $x \leq 0$.

(F.2) $f(x)$ is twice continuously differentiable at $x > 0$, $f'(x) > 0$, $f''(x) < 0$.

(F.3) $\lim_{x \to 0} f'(x) = 1 + \mu_1$ where $\mu_1 > 0$, and $\lim_{x \to \infty} f'(x) = 1 - \mu_2$ where $\mu_2 > 0$.

Example 2.3.1 (Continued). $f(x) = \theta_1 x / [x + \theta_2]$ where $\theta_1 > \theta_2 > 0$.

Remark 2.4.1.

(1) The assumptions imply that there is some $\bar{k} > 0$ such that $f(x) > x$ for all x satisfying $0 < x < \bar{k}$, and $f(x) < x$ for $x > \bar{k}$. By continuity of f, it follows that $f(\bar{k}) = \bar{k}$.

(2) The assumptions imply that $f'(x)$ is decreasing on R_{++}.

(3) Also, note that, by the Mean Value Theorem, $f(\bar{k}) - f(0) = (\bar{k} - 0)f'(z)$ where $0 < z < \bar{k}$. Hence, $\bar{k} = \bar{k}f'(z) > \bar{k}f'(\bar{k})$, or $1 > f'(\bar{k})$.

(4) There is a unique k^*, satisfying $0 < k^* < \bar{k}$, such that $f'(k^*) = 1$. For all $0 < x < k^*$, $f'(x) > 1$, and for all $x > k^*$, $f'(x) < 1$.

Proof of Remark 2.4.1. We want to show the following.

 (i) $A(x) = [f(x)/x]$ is decreasing in $x > 0$.
 (ii) For some $\bar{x} > 0$, $A(\bar{x}) > 1$.
(iii) For some $x' > \bar{x} > 0$, $A(x') < 1$.

To show (i), take $x_1 > x_2 > 0$. Then, for some $t \in (0,1)$, $x_2 = t \cdot x_1 + (1-t) \cdot 0$. Since f is strictly concave (from (F.2), see Chapter 13), we have

$$f(x_2) = f(t \cdot x_1 + (1-t) \cdot 0)$$
$$> t \cdot f(x_1) + (1-t) \cdot f(0)$$
$$= t \cdot f(x_1).$$

Hence,

$$\frac{f(x_2)}{x_2} > \frac{[t \cdot f(x_1)]}{x_2} = \frac{[t \cdot f(x_1)]}{t \cdot x_1} = \frac{f(x_1)}{x_1}.$$

Thus, we have established (i).

To establish (ii), observe that there is some $\bar{x} > 0$ such that $f'(x) > 1$ for all $x \in (0, \bar{x}]$. By the Mean Value Theorem,

$$f(\bar{x}) = f(0) + \bar{x}f'(z), \quad 0 < z < \bar{x},$$
$$= \bar{x}f'(z) > \bar{x}f'(\bar{x}) > 1,$$
$$\text{or} \quad \frac{f(\bar{x})}{\bar{x}} = A(\bar{x}) > 1.$$

Thus, we have established (ii).

To establish (iii) note that if $f(x)$ is bounded, i.e., if there is some $N > 0$ such that $f(x) \leq N$ for all $x > 0$, then $[f(x)/x] \leq [N/x]$ for all $x > 0$. Hence, there is some $x' > 0$ ("sufficiently large") such that $[f(x')/x'] \leq [N/x'] < 1$. Otherwise, $f(x) \to \infty$ as $x \to \infty$, and

$\lim_{x \to \infty} [f(x)/x] = \lim_{x \to \infty} f'(x) = 1 - \mu_2 < 1$. Hence, there is some $x' > 0$ such that $[f(x')/x'] = A(x') < 1$.

By the Intermediate Value Theorem, there is some $\bar{k}$ satisfying $\bar{x} > \bar{k} > x' > 0$ such that $A(\bar{k}) = 1$, i.e., $f(\bar{k}) = \bar{k}$. From (i), we get (a) $A(x) > 1$ for $0 < x < \bar{k}$, or, $f(x) > x$ for $0 < x < \bar{k}$, and (b) $A(x) < 1$ for $x > \bar{k}$, or, $f(x) < x$ for $x > \bar{k}$. $\qquad\square$

The evolution of the system when $c > 0$ is harvested or consumed is given by

$$y_0 = y, \quad x_t = y_t - c \quad \text{for } t \geq 0,$$

$$y_{t+1} = f(x_t) \quad \text{for } t \geq 0.$$

Let T be the first period (if any) such that $x_t < 0$; if there is no such t, then $T = \infty$. If T is finite, we say that the target c is sustained up to (but not including!) period T. We say that the target c is sustainable (forever) if $T = \infty$.

Define the net return function $h(x) = f(x) - x$. It follows that h satisfies

$$h(x) \begin{Bmatrix} > 0 \\ = 0 \\ < 0 \end{Bmatrix},$$

according as

$$\begin{Bmatrix} 0 < x < \bar{k} \\ x = 0, \bar{k} \\ x > \bar{k} \end{Bmatrix}.$$

The maximum sustainable consumption (or harvest) is

$$H = \max_{[0, \bar{k}]} h(x).$$

Observe that $h(x) > 0$ on $(0, \bar{k})$. Hence, $H > 0$, and the maximum is attained at some point(s) in $(0, \bar{k})$, as $h(0) = 0 = h(\bar{k})$. Let ξ^* be some point in $(0, \bar{k})$, satisfying $h(\xi^*) = H$. Then

$$h'(\xi^*) = 0 \quad \text{or} \quad f'(\xi^*) = 1.$$

This implies that $\xi^* = k^*$, and $h(k^*) > h(x)$ for all $x \in [0, \bar{k}]$ with $x \neq k^*$. Also, $h'(x) > 0$, for $0 < x < k^*$, and $h'(x) < 0$, for $k^* < x < \bar{k}$.

We write

$$x_{t+1} = x_t + h(x_t) - c.$$

If $c > H$, $x_{t+1} - x_t = h(x_t) - c < H - c < 0$. Hence, x_t will fall below 0 after a finite number of periods.

If $0 < c < H$, we show that there will be two roots ξ' and ξ'' of the equation

$$c = h(x),$$

which have the properties

$$0 < \xi' < k^* < \xi'' < \bar{k}.$$

First, observe that $h(0) = 0 < c < H = h(k^*)$. By the intermediate value theorem, there is some ξ', $0 < \xi' < k^*$, satisfying $c = h(\xi')$. Since $h'(x) > 0$, for $0 < x < k^*$, there cannot be two distinct points ξ', ξ_1' in the interval $(0, k^*)$ satisfying $c = h(\xi') = h(\xi_1')$. If there are, and, say, $\xi' > \xi_1'$, we have an immediate contradiction, by using the Mean Value Theorem,

$$0 = h(\xi') - h(\xi_1') = (\xi' - \xi_1')h'(z) > 0, \quad \text{where } \xi' > z > \xi_1'.$$

By a similar argument, there is one and only one point ξ'' satisfying $k^* < \xi'' < \bar{k}$ and $c = h(\xi'')$.

We write $y' = c + \xi'$. If $y_0 = y'$, set $x_0 = \xi'$. Since $c = h(\xi')$, we get $x_1 = x_0 + h(x_0) - c = \xi' + h(\xi') - c = x_0 = \xi'$. Repeating the argument, we see that $x_t = \xi'$, $y_t = y' = c + \xi'$. Hence, the target c is sustainable forever. It follows that whenever $y_0 > c + \xi'$, the target c is sustainable forever. Indeed, note that $c = h(\xi')$ is equivalent to $c + \xi' = f(\xi')$. For any $y_0 > c + \xi'$, set $x_0 = y_0 - c > \xi'$. Hence, $y_1 = f(x_0) > f(\xi') = c + \xi' > c$. Again, $x_1 = f(x_0) - c > \xi'$.

This leads to $y_2 = f(x_1) > f(\xi') = c + \xi' > c$. Repeating the steps we see that $y_t > c$ for all $t \geq 1$, confirming that the target c is sustainable forever. We can, in fact, show that x_t converges monotonically to ξ''.

On the other hand, consider $y_0 < c + \xi'$. Then, $x_0 = y_0 - c < \xi'$, If $x_0 > 0$, writing $c - h(x_0) = \beta > 0$

$$x_1 = x_0 + h(x_0) - c = x_0 - \beta.$$

Either $x_1 \leq 0$, or, $x_1 > 0$. In the latter case, $h(x_1) < h(x_0)$, and we get

$$x_2 = x_1 + h(x_1) - c < x_1 + h(x_0) - c = x_1 - \beta = x_0 - 2\beta.$$

Repeating the argument, we see that $x_t \leq 0$ after a finite number of periods.

The implications of the above discussion for survival and extinction are summarized as follows.

Theorem 2.4.1. *Let $c > 0$ be the planned consumption for every period and H the maximum sustainable consumption.*

(1) *If $c > H$, there is no initial y from which c can be sustained forever.*
(2) *If $0 < c < H$, then there is $\xi' > 0$, with $h(\xi') = c$, such that c can be sustained forever if and only if the initial stock $y \geq \xi' + c$.*
(3) *If $c = H$, then there is $\xi' > 0$ with $h(\xi') = H$, and ξ' tends to 0 as c tends to 0.*

Despite the simplicity of the example, two implications stand out for the policy maker/social planner concerned about extinction: first, if the population size is already small (namely below $\xi' + c$), an immediate intervention, a ban on harvesting, is called for. Similarly, if $c > H$, an immediate intervention is needed to lower c. Secondly, while a modest harvesting is consistent with growth (provided the initial size is above the threshold stock $\xi' + c$), monitoring is essential if the harvest level keeps increasing.

Example 2.4.1. Here, we look at numerical examples to illustrate some of the issues in a more "concrete" way. In both cases, the regeneration functions are taken from the Verhulst model of Examples 2.3.1 and 2.3.2.

(i) Let $S = R_+$ and consider $f(x) = 3 \cdot x/(x+2)$.

The steady states are given by $x_1^* = 0$, $x_2^* = \bar{k} = 1$.

$f'(x) = 6/(x+2)^2 > 0$ for all $x \geq 0$.

$f''(x) = -12/(x+2)^3 < 0$ for all $x \geq 0$.

$h(x) = f(x) - x = [x(1-x)]/(x+2)$.

Note that $h(x_1^*) = h(x_2^*) = 0$.

$h'(x) = (2 - x^2 - 4x)/(x+2)^2$ and clearly, $h''(x) = f''(x) < 0$ for all $x \geq 0$.

$h'(0) > 0$ and $h'(1) < 0$.

Since $h'(x) = 0$ at a unique $x \simeq 0.45$, $H = \max_{[0,1]} h(x) \simeq 0.1$

Hence, $c > 0.1$ is not sustainable.

(ii) Again, let $S = R_+$ and consider $f(x) = r.x^2/[x^2 + A]$.

Here, $f'(x) = 2xrA/(x^2 + A)^2$ and $f''(x) = [2rA(x^2 + A) - 8x^2rA]/(x^2 + A)^3$.

Take $r = 4$, $A = 3$. Clearly, $r = 4 > 2\sqrt{3} = 3.46$. Then, $x_1^* = 0$, $x_2^* = 1$, $x_3^* = 3$.

$f'(0) = 0$; $f'(1) = 1.5$, $f'(3) = 0.5$.

We see that if $x_0 \leq 1$, no $c > 0$ is sustainable.

Verify that when $x > 1$, $f''(x) < 0$. On $(1, 3)$, $f(x) = 4x^2/[x^2 + 3]$ is strictly concave.

Now, consider $h(x) = f(x) - x$. Observe that $h(1) = h(3) = 0$. Moreover, $h(x) > 0$ on $(1, 3)$. From the first- and second-order conditions, $h'(x) = 0$, and solving, we get that $h(x)$ attains a maximum $H = 0.286044$ at a unique point $\hat{x} = 1.96779$. Hence, any $c > H$ is not sustainable. If $c = 0.1$, the critical stock ξ_1 is approximately 1.2055.

One can draw the graph of $h(x)$ and complete the analysis by following the arguments spelled out in the proof.

Example 2.3.5 (Continued). We return to the logistic growth model and study the effect of *constant yield harvesting*, described formally by the differential equation:

$$\dot{x} = rx(t)[1 - (x(t)/K)] - \Theta, \qquad (2.4.1)$$

where Θ is a positive constant. The steady states or equilibrium values of (2.4.1) are given by the condition $\dot{x} = 0$, and these may be found by solving the quadratic equation

$$rx[1 - (x/K)] - \Theta = 0,$$

$$or, x^2 - Kx + (K\Theta)/r = 0.$$

There are two equilibria:

$$x_L = [K - (K^2 - (4\Theta K/r))^{1/2}]/2 \quad \text{and}$$

$$x_U = [K + (K^2 - (4\Theta K/r))^{1/2}]/2, \qquad (2.4.2)$$

provided $K^2 - (4\Theta K/r) \geq 0$ or $\Theta \leq rK/4$. If $\Theta > rK/4$, $\dot{x} < 0$ for all initial $x_0 \geq 0$, and *every* solution hits 0 in finite time. If a solution "hits" zero at a (finite) time T, we say that the population is *extinct at time* T. If $0 \leq \Theta \leq rK/4$, the upper equilibrium x_U decreases from K to $K/2$ as Θ increases from 0 to $rK/4$, and the lower equilibrium x_L increases from 0 to $K/2$ as Θ increases from 0 to $rK/4$. Writing $\hat{\Theta} = rK/4$, a critical value, we can summarize the qualitative properties of the trajectories and the possibilities of extinction as follows. For any harvest rate $\Theta < \hat{\Theta}$, the population size tends to the equilibrium size x_U, provided $x_0 \geq x_L$. For all $x_0 < x_L$, the population size hits zero in finite time. For $\Theta > \hat{\Theta}$, the population size hits zero in finite time for *all* initial population sizes (i.e., extinction is inevitable due to overharvesting). When Θ is close to $\hat{\Theta}$, a small change in the rate may have a disastrous impact on the population.

2.5 Natural Balance: Lotka–Volterra Model

Here is an informal description of the celebrated Lotka–Volterra model, a convenient framework for studying predator–prey relationships (interdependence of species and "natural balance"). For a more complete account, see Brauer and Castillo-Chavez (2001) and Hirsch, Smale and Devaney (2013).

Let $x(t) > 0$ be the fish population and $y(t) > 0$ be the shark population at time t. Assume that plankton, which is the food supply for the fish, is unlimited and thus that the per capita growth rate of the fish population in the absence of sharks would be constant. Thus, if there were no sharks, the dynamic behavior of the fish population would satisfy a differential equation of the form $\dot{x}/x(t) = a > 0$. The sharks, on the other hand, depend on fish as their food supply, and we assume that in the absence of fish, the sharks would have a constant per capita death rate; thus, in the absence of fish, the dynamic behavior of the shark population would satisfy a differential equation of the form $\dot{y}/y(t) = -c$ (where $c > 0$). We next assume that the presence of fish increases the per capita growth rate of sharks, changing the rate from $-c$ to $-c+dx$ ($d > 0$). The presence of sharks reduces the per capita growth rate of fish from a to $a - by$. Thus, we are led to the celebrated Lotka–Volterra equations (to simplify typing, we drop (t) repeatedly):

$$\dot{x} = (a - by)x, \qquad (2.5.1)$$

$$\dot{y} = (dx - c)y. \qquad (2.5.2)$$

Here, all parameters a, b, c, d are positive parameters: a and c are the two "natural growth rate parameters", whereas c measures the impact of predation and d measures the impact of the prey population on predator growth. Since we are dealing with population size, we restrict $x, y \geq 0$.

There are two steady states $x = 0, y = 0$ and $x = c/d$ and $y = a/b$ (coexistence).

An equilibrium is *stable* if solutions from nearby initial points stay nearby for all future times. Formally, suppose that $X^* \in R^n$ is an equilibrium point for the differential equation

$$\dot{X} = F(X(t)). \qquad (2.5.3)$$

Then X^* is a *stable equilibrium* if for every neighborhood $\mathcal{O}$ of X^* in R^n, there is a neighborhood $\mathcal{O}_1$ of X^* in $\mathcal{O}$ such that every

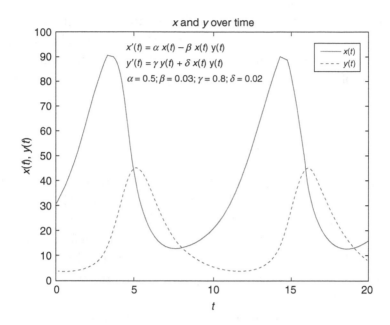

Fig. 2.1.

solution $X(t)$ with $X(0) = X_0$ in $\mathcal{O}_1$ is defined and remains in $\mathcal{O}$ for all $t > 0$.

To emphasize the dependence of a solution on its initial value x_0, we write the corresponding solution by $\phi_t(x_0)$. A *periodic solution* (*"closed orbit"*) occurs for (3) if there is a non-equilibrium point X and a time $\tau > 0$ such that $\phi_\tau(X) = X$. It follows that $\phi_{t+\tau}(X) = X$ for all t. The least such τ is called the period of the solution (Fig. 2.1).

A fundamental result on the Lotka–Volterra model is as follows.

Theorem 2.5.1. (a) *The steady state* $Z = (\frac{c}{d}, \frac{a}{b})$ *is stable.*

(b) *Every solution is periodic (except the equilibrium* Z *and the coordinate axes).*

Proof. (a) (Sketch) We construct a Lyapunov function.

$$L(x, y) = d \cdot x - c \cdot \log x + b \cdot y - a \cdot \log y.$$

(i) We can verify directly that $\dot{L} = 0$ on solutions of the system when x, $y > 0$. (ii) We can also verify directly that the equilibrium point $Z = (c/d, a/b)$ is a global minimum for L. It follows that $L - L(Z)$ is a Lyapunov function for the system. The conclusion follows from a fundamental result on stability (see Section 2.6).

A proof of (b) involves several steps: one has to rule out a limit cycle (see Hirsch, Smale and Devaney (2013, p. 240)). □

2.6 Complements and Details

For extremely lucid introductions to simple mathematical models in mathematical biology, see May (1974, 1976, 1983), May and Oster (1976). More technical treatments of mathematical models dealing with single as well as interacting populations can be found in Brauer and Castillo-Chavez (2001) and Hirsch, Smale and Devaney (2013). There are works on nonlinear dynamics that provide extensive coverage of the quadratic family: see Devaney (1986), Drazen (1992) and Day (1994). Majumdar, Mitra and Nishimura (2000) present a collection of articles, exploring the possibility of chaotic behavior arising in a variety of contexts involving dynamic optimization in economics. The exposition of Section 2.4 is a simplified version of the more general model in Majumdar and Radner (1991).

Extinction, tipping points, response of regeneration patterns to changes in parameters and interaction of populations are major issues that have challenged researchers. Habitat destruction and climate change figure prominently as causes for a rapid and alarming decline of specific populations. A lucid study of a community of elephants is reported in Moss (1988). In *The Sixth Extinction*, Kolbert (2014) provides examples of dramatic declines of specific populations (frogs, bats) [the article by Kolbert in the May 25, 2009 issue of the New Yorker has a succinct account]. Needless to say, there is no lack of academic debates on measurements and predictions. It is pointed out that nature has been "the alloy of organic chemists" and has helped Man's efforts to develop "wonder drugs". Hence, the implications of large-scale deforestation and loss of animal/insect populations "are staggering. Humans have always engaged in the search for drugs from nature, and the quest has a long history of paying off" (see Eisner

(1991)). The price mechanism cannot be relied upon to ensure survival, a point that is taken up in more detail in Chapter 6.

2.6.1 Complexity (Example 2.3.4)

For the tent map $\mathbf{T}$, to search for periodic points of period 2, we can start by calculating the functional form (and it is useful to draw the graph):

$$\mathbf{T}^2(x) = \begin{cases} 4x, & 0 \leq x \leq 1/4, \\ 2 - 4x, & 1/4 < x \leq 1/2, \\ 4x - 2, & 1/2 < x \leq 3/4, \\ 4 - 4x, & 3/4 < x \leq 1. \end{cases} \tag{2.6.1}$$

Once the graph is drawn, the fixed points are found at the intersection of the graph of $\mathbf{T}^2$ and the 45° line (the graph of $y = x$). Using the graphical method, or brute force, we can identify the fixed points of $\mathbf{T}$ as $\{0, 2/3\}$ and fixed points of $\mathbf{T}^2$ as $\{0, 2/5, 2/3, 4/5\}$. Thus, the periodic points of period 2 for the tent map $\mathbf{T}$ are $\{2/5, 4/5\}$. Observe from (2.6.1) that the formula for the second iterate has four pieces: the graph shows two tents (the width of the "bases" of these tents are $1/2$). The graph of $\mathbf{T}^3$ shows four tents, with base width equal to $1/4$.

$$\mathbf{T}^3(x) = \begin{cases} 4\mathbf{T}(x), & 0 \leq \mathbf{T}(x) < 1/4, \\ 2 - 4\mathbf{T}(x), & 1/4 \leq \mathbf{T}(x) < 1/2, \\ 4\mathbf{T}(x) - 2, & 1/2 \leq \mathbf{T}(x) < 3/4, \\ 4 - 4\mathbf{T}(x), & 3/4 \leq \mathbf{T}(x) \leq 1. \end{cases}$$

It has eight fixed points (including the two fixed points of $\mathbf{T}$ of course!):

$$\{0, 2/7, 4/7, 6/7, 2/9, 4/9, 6/9, 8/9\}.$$

Hence, $\mathbf{T}$ has six periodic points of period 3.

$\mathbf{T}$ is chaotic on $[0, 1]$ according to an alternative definition of chaos [Devaney (1986); Hirsch, Smale and Devaney (2013)]. Here are the widely discussed properties characterizing chaos (subinterval of an interval I is a subset of I that is itself an interval).

Definition 2.6.1. A map α from a closed interval $I = [a, b]$ to itself is chaotic if

(1) periodic points of α are dense in I;
(2) (Transitivity) given any two open subintervals U_1 and U_2 in I, there is a point $x_0 \in U_1$ and some positive integer n such that $\alpha^n(x_0) \in U_2$;
(3) (Sensitive Dependence on Initial Conditions) there exists a *sensitivity constant* (denoted by $\beta > 0$) such that for any $x_0 \in I$ and any open interval U around x_0, there is some $y_0 \in U$ and positive integer n such that $|\alpha^n(x_0) - \alpha^n(y_0)| > \beta$.

In our context, condition (3) follows from (1) and (2) (see Banks *et al.* (1992)).

2.6.2   Lotka–Volterra model

The following theorem is a fundamental theorem on stability.

Theorem 2.6.1. *Let X^* be an equilibrium point for $\dot{X} = F(X(t))$. Let $\mathcal{L} : \mathcal{O} \to R$ be a differentiable function defined on an open set $\mathcal{O}$ containing X^*. Suppose further that*

(i) $\mathcal{L}(X^*) = 0$ *and* $\mathcal{L}(X) > 0$ *if* $X \neq X^*$;
(ii) $\dot{\mathcal{L}} \leq 0$ *in* $\mathcal{O} \backslash X^*$.

Then X^* is stable.

Proof. Let $\delta > 0$ be small enough so that the closed ball $B_\delta(X^*)$ around the equilibrium point X^* of radius δ lies entirely in $\mathcal{O}$. Let m be the minimum value of $\mathcal{L}$ on the boundary of $B_\delta(X^*)$, i.e., on $S_\delta(X^*) = \{X \in \mathcal{O} : d(X, X^*) = \delta\}$. By (i), and Weierstrass theorem, $m > 0$. Define $\mathcal{O}_1 = \{X \in B_\delta(X^*) : \mathcal{L}(X) < m\}$. No solution starting in $\mathcal{O}_1$ can meet $S_\delta(X^*)$, since $\mathcal{L}$ is non-increasing on solution curves. Thus, every solution starting in $\mathcal{O}_1$ never leaves $B_\delta(X^*)$. This proves stability. $\qquad\square$

2.7 Supplementary Exercises

1. The logistic equation is often written as

$$\dot{x} = \lambda x(t) - a[x(t)]^2, \tag{2.7.1}$$

where λ and a are positive constants. Show that for every $x(0) = x_0 > 0$, the solution $x(t) \to \lambda/a$ as $t \to \infty$.

Hints: [▼ Review the method for solving differential equations by separation of variables (and the Bernoulli equation), see, e.g., the classic text by Ince (1956, Chapter 2, pp. 16–32), Hirsch, Smale and Devaney (2013), or Brauer and Castillo-Chaver (2001). ▲] The solution to (2.7.1) is given by

$$x(t) = \frac{\lambda x_0}{[ax_0 + (\lambda - ax_0)e^{-\lambda t}]}. \tag{2.7.2}$$

You can directly verify that the solution satisfies (2.7.1) by taking the derivative and substituting the relevant expressions on the two sides of (2.7.1).

2. Suppose that a population governed by a logistic equation with carrying capacity K, intrinsic growth rate r and a "constant effort harvesting" Ex, where $E > 0$, and $x > 0$.
   Solve the initial value problem

$$\dot{x} = rx(t)[1 - (x(t)/K)] - Ex(t), \quad x(0) = K, \tag{2.7.3}$$

analytically. Verify that if $E < r$, then $\lim_{t \to \infty} x(t) = K(1 - E/r)$, while if $E > r$, $\lim_{t \to \infty} x(t) = 0$.

Hints: This is a Bernoulli equation. Verify that the solution to (2.7.3) is given by

$$x(t) = \frac{[K(r - E)]}{[r - Ee^{-(r-E)t}]}. \tag{2.7.4}$$

3. Let $S = R_+$ and

$$\alpha(x) = \theta_1 x e^{-\theta_2 x}, \tag{2.7.5}$$

where θ_1 and θ_2 are positive parameters.

(i) Verify that $\alpha'(x) = \theta_1 e^{-\theta_2 x} - \theta_1 \theta_2 x e^{-\theta_2 x}$.

(ii) Clearly $x = 0$ is a steady state. Since $\alpha'(0) = \theta_1$, the steady state 0 is locally stable if $0 < \theta_1 < 1$ and locally unstable if $\theta_1 > 1$. Analyze the behavior of trajectories when $\theta_1 = 1$ (Hint: Observe that for $x > 0$, $\alpha(x) = xe^{-\theta_2 x} < x$. Hence, for any initial x, the trajectory from x converges to 0).

(iii) If $\theta_1 > 1$, there is a steady state $\hat{x} = (\log \theta_1)/\theta_2$. Analyze the local stability properties of this steady state (Hint: show that $\alpha'(\hat{x}) = 1 - \log \theta_1$).

4. Let $S = R_+$ and

$$\alpha(x) = \theta_1 x (1 + x)^{-\theta_2}, \qquad (2.7.6)$$

where θ_1 and θ_2 are positive parameters.

(i) Verify that $\alpha'(x) = \theta_1 (1 + x)^{-\theta_2} - \theta_1 \theta_2 x (1 + x)^{-\theta_2 - 1}$.

(ii) Clearly, 0 is a steady state. When is it locally stable?

(iii) Analyze the qualitative properties of the trajectories when $\theta_1 = 1$.

(iv) Find conditions on θ_1 and θ_2 for the existence of a positive steady state.

5. Consider a dynamical system with $S = [0, 1]$. Let $\alpha : S \to S$ be defined as

$$\alpha(x) = \begin{cases} 1/2 & \text{for } x = 0, \\ 2x & \text{for } 0 < x \leq 1/2, \\ 0 & \text{for } 1/2 < x \leq 1. \end{cases}$$

(i) Does α have a periodic point of period 3? Does α have a periodic point of period 2?

(ii) Analyze the behavior of all trajectories.

(iii) How is the example related to the Li–Yorke Theorem 2.3.1?

Chapter 3

Optimal Harvesting: Finite Horizon

> Roughly speaking, conservation means saving for the future. The theory of resource conservation can therefore be addressed as a branch of the theory of capital and investment.
>
> — Colin Clark

3.1 Introduction

In this and the following three chapters, we shall address a number of issues related to management of a renewable resource: "optimal harvesting" or "optimal saving", conservation of inheritance, the role of prices in attaining optimality, the possibility of profit maximization triggering extinction and the link between discounting and extinction. The discrete-time dynamic optimization models and the tools of analysis that we use appeared in the development of intertemporal economics in the 1960s, with prominent contributions from David Gale (1967, 1970) and others. A long list of articles, beginning with Ramsey (1928), use a continuous-time framework to develop the theory of capital and investment. We first explore a finite horizon model.

A regeneration (or reproduction) function f [satisfying the assumptions (F.1)–(F.4)] and an initial stock $a > 0$ are given, and a "terminal" or "target" stock $z > 0$ (meeting a feasibility requirement) is set. In the first period $t = 1$, from the gross output $f(a) = y_1$, a part x_1 is saved as an input, and the remaining part $c_1 = f(a) - x_1$ is harvested (consumed). The input x_1 gives rise to the gross output $f(x_1) = y_2$ in the next period $t = 2$, a part of which is saved as an input x_2, and the remaining part $c_2 = f(x_1) - x_2$, is harvested, and the story is repeated. The choice of x_{T-1} generates $y_T = f(x_{T-1}) \geq z$, and one sets $c_T = f(x_{T-1}) - z$. Here, $T(\geq 2)$ is a positive integer. Thus, a planned choice of inputs, an input program $\mathbf{x} = (a, x_1, \ldots, x_{T-1}, z)$ generates a corresponding list of harvests (or, consumptions) $\mathbf{c} = (c_1, \ldots, c_T)$ and outputs $\mathbf{y} = (y_1, \ldots, y_T)$. The set of all possible consumptions (starting from an initial a, ending with a target z in period T) is denoted by $\mathcal{C}(a, z, T)$. Consumption generates immediate return (utility) according to a function u [satisfying (U.1)–(U.2)]. The total return from $\mathbf{c}$ is denoted by $w(\mathbf{c}) = \sum_{t=1}^{T} \delta^{t-1} u(c_t)$, where $\delta \in (0, 1]$ is the discount factor. The resource manager's objective is to maximize $w(\mathbf{c})$ over $\mathcal{C}(a, z, T)$. Given the assumptions on f and u, note that this is an optimization exercise in the "classical" or "convex" environment. Here is an informal summary of the main results. First, there is a unique solution $(\mathbf{x}^*, \mathbf{y}^*, \mathbf{c}^*)_a^z$ to the optimization problem. The optimal program is *positive*, i.e., $\mathbf{c}^* \gg 0$, $\mathbf{x}^* \gg 0$, $\mathbf{y}^* \gg 0$. It should be stressed that the "endpoint" Uzawa–Inada restrictions on the derivatives of the regeneration and reward functions (see (F.4) and (U.2)) play a role in guaranteeing that $\mathbf{c}^* \gg 0$. The connection between discounting and extinction is explored in detail in Chapters 5 and 6, and we do not touch upon this issue in this chapter.

Next, we explore the connection between optimal and competitive (price guided or decentralized) programs and show roughly that a program is optimal if and only if it is competitive.

The optimal program provides a characterization of the best trade-off between consumptions in two adjacent periods: it is shown (see (3.5.7)) to satisfy

$$u'(c_t) = \delta \cdot u'(c_{t+1}) \cdot f'(x_t),$$

$$\text{or} \quad [u'(c_t)/\delta u'(c_{t+1})] = f'(x_t). \tag{3.1.1}$$

The left-hand side of (3.1.1) is interpreted as the marginal rate of substitution between consumptions in periods t and $t+1$, and this must equal the marginal productivity of input.

A comparative dynamics result (Brock (1971)) comes next: suppose the target stock is raised from z to z'. Let $(\mathbf{x}^*(z), \mathbf{c}^*(z))_a$ and $(\mathbf{x}^*(z'), \mathbf{c}^*(z\prime))_a$ be the corresponding optimal programs. Then for all periods $t = 1, \ldots, T-1$, we must have $x_t^*(z') > x_t^*(z)$: the optimal input stock increases in every period ("one shot" adjustment is not enough).

Finally, some qualitative properties of optimal programs are obtained. Of interest is a critical input stock k_δ^* which is obtained as the unique solution to the equation $\delta \cdot f'(x) = 1$. In the literature on optimal growth, this is the "modified golden-rule" input. It is shown that the stationary (or equitable) program from k_δ^* (i.e., the program $a = k_\delta^* = x_t = z$) is optimal among all programs from k_δ^*. Next, it is seen that if $a > k_\delta^*$, $z > k_\delta^*$, then the unique optimal input program $\mathbf{x}^* = (a, x_1^*, \ldots, x_t^*, \ldots, z)$ satisfies $x_t^* > k_\delta^*$ for all $t = 1, \ldots, T-1$. Similarly, if the opposite inequalities hold, i.e., $a < k_\delta^*$, $z < k_\delta^*$, then one has $x_t^* < k_\delta^*$ for all $t = 1, \ldots, T-1$.

3.2 The Model

Consider a renewable resource whose stock or population size is denoted by $x \geq 0$. The *gross natural reproduction function* is denoted by f which satisfies the following assumptions:

(F.1) $f : R_+ \longrightarrow R_+$ *is continuous.*

(F.2) $f(0) = 0$.

(F.3) $f(x)$ *is twice continuously differentiable at* $x > 0$, *with* $f'(x) > 0$ *and* $f''(x) < 0$.

(F.4) (a) $\lim_{x \to 0} f'(x) = \infty$; $\lim_{x \to \infty} f'(x) = \mu < 1$ *and* (b) $f(x) \to \infty$ *as* $x \to \infty$.

In our context, the critical assumption (F.2) emphasizes that if in any period t, $x_t = 0$, the resource becomes extinct.

3.2.1   Feasible programs

A feasible resource management program (briefly, a *program*) from an initial stock $a > 0$, and a terminal stock $z > 0$, denoted by $\mathbf{x} = (x_t)_{t=0}^T$, is a vector of stocks, satisfying

$$x_0 = a, \quad 0 \leq x_{t+1} \leq f(x_t) \quad \text{for } t \geq 0, \ldots, T-1,$$
$$x_T = z. \tag{3.2.1}$$

Here, T is a positive integer, $T \geq 2$.

One interprets x_t as the input stock available for regeneration at the end of period t. It generates the gross "output" of the resource y_{t+1} in the next period according to the reproduction function f:

$$y_{t+1} = f(x_t). \tag{3.2.2}$$

Associated with a program $\mathbf{x} = (x_t)_{t=0}^T$ is a vector of *harvests* (or, consumptions) $\mathbf{c} = (c_t)$ defined by

$$c_t = f(x_{t-1}) - x_t \quad \text{for } t \geq 1. \tag{3.2.3}$$

The initial stock a generates the output $y_1 = f(a)$ in period 1. A part of the stock c_1 is harvested or consumed. The remaining part x_1 generates the stock $f(x_1)$ in period 2 and the story is repeated. A program $\mathbf{x}$ from a to z is often more explicitly denoted by $(\mathbf{x}, \mathbf{y}, \mathbf{c})_a^z$ or simply by $(\mathbf{x}, \mathbf{y}, \mathbf{c})$. We often refer to $(\mathbf{x}, \mathbf{y})_a^z$ as a *production program* (*generating* $\mathbf{c}$). It will be assumed ("free disposal") that if $(\mathbf{x}, \mathbf{y}, \mathbf{c})_a^z$ is a program from a to z, then for any $\acute{\mathbf{c}} = (\acute{c}_t)$, satisfying $0 \leq \acute{c}_t \leq c_t$, $(\mathbf{x}, \mathbf{y}, \acute{\mathbf{c}})$ is also a program from a to z.

A program $(\mathbf{x}, \mathbf{y}, \mathbf{c})_{\tilde{x}}$ is *positive* if $\mathbf{c} \gg \mathbf{0}$. Since $\mathbf{y} \geq \mathbf{c}$, it follows that if $(\mathbf{x}, \mathbf{y}, \mathbf{c})_a^z$ is *positive*, $\mathbf{y} \gg \mathbf{0}$, and by using (F.2), $\mathbf{x} \gg \mathbf{0}$.

Consumption generates *utility* or *immediate return*. The *return function* $u : R_+ \to R_+$ is assumed to satisfy the following conditions:

(U.1) *u is continuous on R_+; $u(0) = 0$;*
(U.2) *$u(c)$ is twice continuously differentiable at $c > 0$, with $u'(c) > 0$, and $u''(c) < 0$. Moreover, $u'(c) \to \infty$ as $c \to 0$.*

3.2.2 Natural growth

For any initial stock $a > 0$, consider the program $\bar{\mathbf{x}}(a)$ defined by

$$\bar{x}_0 = a, \quad \bar{x}_{t+1}(a) = f(\bar{x}_t(a)), \dots, \bar{z}(a) = \bar{x}_T(a) = f(\bar{x}_{T-1}(a)).$$
$$(3.2.4)$$

This is the *natural growth program* from a, which does not allow any harvesting in any period : $\bar{c}_t = 0$, for all $t = 1, \dots, T$.

Lemma 3.2.1. *There is a unique $\bar{k} > 0$ such that*

(i) $f(\bar{k}) = \bar{k}$;
(ii) $f(k) > k$ *for* $0 < k < \bar{k}$;
(iii) $f(k) < k$ *for* $k > \bar{k}$.

Proof. A sketch is given. For $x > 0$, define

$$A(x) = f(x)/x.$$

First, note that $A'(x) < 0$ for all $x > 0$. Clearly, $A'(x) = [f'(x) \cdot x - f(x)]/x^2$. From the Mean Value Theorem,

$$f(x) = f(0) + x \cdot f'(z) \quad \text{where } 0 < z < x$$
$$> x \cdot f'(x).$$

Next, there is some $\hat{x} > 0$ such that $f'(x) > 1$, for all $0 < x < \hat{x}$. Hence,

$$A(x) = (f(x)/x) > f'(x) > 1 \quad \text{for all } 0 < x < \hat{x}. \qquad (3.2.5)$$

Finally, show that by L'Hospital's rule,

$$A(x) = (f(x)/x) < 1 \quad \text{for all } x > 0 \text{ sufficiently large.}$$

By the Intermediate Value Theorem, there is some $\bar{k} > 0$ such that

$$A(\bar{k}) = 1 \quad \text{or} \quad f(\bar{k}) = \bar{k}.$$

Uniqueness follows from the fact that $A'(x) < 0$ for all $x > 0$. $\qquad \square$

We refer to $\bar{k}$ as the *maximum sustainable stock* of the reproduction function f.

A useful property of natural growth programs is noted.

Lemma 3.2.2. *For any* $a \in (0, \bar{k})$, $\bar{x}_t(a) < \bar{x}_{t+1}(a) < \bar{k}$ *for* $t = 1, 2, \ldots, T - 1$,

$$\lim_{T \to \infty} \bar{x}_T(a) = \bar{k}. \tag{3.2.6}$$

Proof.   Since $a \in (0, \bar{k})$, $f(a) > a$, i.e., $\bar{x}_1(a) > a$. Also, $\bar{x}_1(a) = f(a) < f(\bar{k}) = \bar{k}$. Repeat the argument to get

$$0 < \bar{x}_t(a) < \bar{x}_{t+1}(a) < \bar{k} \quad \text{for } t = 1, 2, \ldots, T - 1.$$

Now, as $T \to \infty$, $\bar{x}_T$ converges to some $\hat{x} \le \bar{k}$. Using $\bar{x}_{t+1}(a) = f(\bar{x}_t(a))$, and the continuity of f, one gets $\hat{x} = f(\hat{x})$. Hence, $\hat{x} = \bar{k}$.   □

In the subsequent part of this chapter, we restrict a and z to be in $(0, \bar{k})$.

3.3   Existence of Optimal Programs

Given the choice of T, and an initial $a > 0$, if the planner sets a terminal stock $\tilde{z} > \bar{x}_T(a)$ (see 3.2.4), there is *no* program from a to $\tilde{z}$. In other words, given the initial stock a, and the reproduction function f, a target stock $\tilde{z} > \bar{x}_T(a)$ is too ambitious. To avoid such a triviality, call a pair (a, z) *admissible* if there is at least one program from a to z. In what follows, *we shall always restrict ourselves to admissible pairs* (a, z) (in other words, z is set no larger than the stock $x_T(a)$ achieved by the natural growth program).

Fix $T \ge 2$, and consider an admissible (a, z). Let $\mathcal{C}(a, z, T)$ be the set of *all* consumption vectors in R_+^T corresponding to *all* programs from a to z. It is clear that this set depends on the parameters

(a, z, T). I shall often write $\mathcal{C}$ to simplify notation. The set $\mathcal{C}$ is non-empty. We now prove that it is closed and bounded.

Theorem 3.3.1. $\mathcal{C}(a, z, T)$ *is a non-empty, compact, convex subset of* R_+^T.

Proof. Recall that $\bar{k}$ is the maximum sustainable stock. One can show (▲ *Exercise* ▼) that if $(\mathbf{x}, \mathbf{y}, \mathbf{c})_a^z$ is in $\mathcal{C}(a, z, T)$,

$$0 \leq x_t \leq \bar{k}, \quad 0 \leq c_t \leq \bar{k} \quad \text{for all } t = 1, \ldots, T. \tag{3.3.1}$$

Hence, $\mathcal{C}$ is bounded.

To show that $\mathcal{C}$ is closed, take a sequence $\mathbf{c}^n \in \mathcal{C}$ converging to some $\mathbf{c}$ in R_+^T. We need to prove that $\mathbf{c} \in \mathcal{C}$. For each n, let $\mathbf{x}^n$ be a program generating $\mathbf{c}^n$. Then (recalling the free disposal assumption), we have

$$x_0^n = a, \quad x_t^n + c_t^n \leq f(x_{t-1}^n), \quad \text{for } t = 1, 2, \ldots, T - 1,$$
$$z + c_T^n \leq f(x_{T-1}^n).$$

Now, using (3.3.1), and the Bolzano–Weierstrass theorem, there is a subsequence of $\mathbf{x}^n$ (retain notation) that converges to some $\mathbf{x} \in R_+^{T+1}$. Taking the limit along this subsequence, we get the following:

$$x_0 = a, \quad x_t + c_t \leq f(x_{t-1}), \quad \text{for } t = 1, 2, \ldots, T-1; \ z + c_T \leq f(x_{T-1}).$$

Hence, $\mathbf{x} = (x_t)$ is a program from a to z that generates $\mathbf{c}$. So, $\mathbf{c} \in \mathcal{C}$.

Convexity of $\mathcal{C}$ is verified directly. Take two distinct consumption programs $\mathbf{c}^1 = (c_t^1)_{t=1}^T$ and $\mathbf{c}^2 = (c_t^2)_{t=1}^T$ in $\mathcal{C}$. Let $\lambda \in (0, 1)$. We have to show that

$$\lambda \mathbf{c}^1 + (1 - \lambda) \mathbf{c}^2 = (\lambda c_t^1 + (1 - \lambda) c_t^2)_{t=1}^T \in \mathcal{C}.$$

Corresponding to $\mathbf{c}^1$ and $\mathbf{c}^2$, there are $\mathbf{x}^1 = (a, x_1^1, \ldots, x_{T-1}^1, z)$ and $\mathbf{x}^2 = (a, x_1^2, \ldots, x_{T-1}^2, z)$, such that (recalling the free disposal assumption)

$$x_1^i + c_1^i \leq f(a),$$
$$x_t^i + c_t^i \leq f(x_{t-1}^i), \quad t = 2, \ldots, T-1,$$
$$z + c_T^i \leq f(x_{T-1}^i), \quad i = 1, 2.$$

Observe that $\lambda \cdot (x_1^1 + c_1^1) \leq \lambda \cdot f(a)$ and $(1-\lambda) \cdot (x_1^2 + c_1^2) \leq (1-\lambda) \cdot f(a)$. Reorganizing terms, we get

$$(\lambda \cdot x_1^1 + (1-\lambda) \cdot x_1^2) + (\lambda \cdot c_1^1 + (1-\lambda) \cdot c_1^2) \leq f(a).$$

Similarly, for $t = 2, \ldots, T-1$, we get

$$((\lambda \cdot x_t^1 + (1-\lambda) \cdot x_t^2) + (\lambda \cdot c_t^1 + (1-\lambda) \cdot c_t^2)$$
$$\leq \lambda \cdot f(x_{t-1}^1) + (1-\lambda)f(x_{t-1}^2) \leq f(\lambda \cdot x_{t-1}^1 + (1-\lambda) \cdot x_{t-1}^2).$$

Finally, for $t = T$, we have

$$z + (\lambda \cdot c_T^1 + (1-\lambda) \cdot c_T^2) \leq \lambda \cdot f(x_{T-1}^1) + (1-\lambda)f(x_{T-1}^2)$$
$$\leq f(\lambda \cdot x_{T-1}^1 + (1-\lambda) \cdot x_{T-1}^2).$$

Hence, the program $\lambda \cdot \mathbf{x}^1 + (1-\lambda) \cdot \mathbf{x}^2$ from a to z generates $\lambda \cdot \mathbf{c}^1 + (1-\lambda) \cdot \mathbf{c}^2$. This means that $\lambda \cdot \mathbf{c}^1 + (1-\lambda) \cdot \mathbf{c}^2 \in \mathcal{C}$.   □

We shall now introduce the criterion of optimality. A program $(\mathbf{x}^*, \mathbf{y}^*, \mathbf{c}^*)_a^z$ is *optimal* if

$$\sum_{t=1}^{T} \delta^{t-1} u(c_t^*) \geq \sum_{t=1}^{T} \delta^{t-1} u(c_t) \tag{3.3.2}$$

for all programs $(\mathbf{x}, \mathbf{y}, \mathbf{c})_a^z$. Let us emphasize that given T, we restrict ourselves to admissible (a, z), and the discount factor $\delta \in (0, 1]$.

Since $\mathcal{C}(a, z, T)$ is non-empty, compact, the assumed continuity property of u guarantees that there is an optimal program (Weierstrass' theorem). We have the following theorem.

Theorem 3.3.2. *There is a unique positive optimal consumption program.*

Proof. To get uniqueness, one uses a proof by contradiction. Assume that there are two distinct consumption programs $\mathbf{c}^1 = (c_t^1)_{t=1}^T$ and $\mathbf{c}^2 = (c_t^2)_{t=1}^T$ in $\mathcal{C}$ that are optimal, i.e.,

$$\sum_{t=1}^T \delta^{t-1} u(c_t^1) = \sum_{t=1}^T \delta^{t-1} u(c_t^2) = m.$$

One chooses some $\lambda \in (0,1)$ and proves that $\lambda \cdot \mathbf{c}^1 + (1-\lambda) \cdot \mathbf{c}^2 \in \mathcal{C}$ by using the relevant part of the proof of Theorem 3.3.1. Now, $w(\mathbf{c}) = \sum_{t=1}^T \delta^{t-1} u(c_t)$ is a strictly concave function on $\mathcal{C}$. Hence, $w(\lambda \cdot \mathbf{c}^1 + (1-\lambda) \cdot \mathbf{c}^2) > \lambda \cdot w(\mathbf{c}^1) + (1-\lambda) \cdot w(\mathbf{c}^2) = m$, a contradiction.

To prove that the optimal consumption program is positive, one needs some calculations. I shall consider the special case $T = 2, \delta = 1$, which captures the essence of the proof.

(▲ *Exercise*: Spell out the proof of the general case. ▼)

With $T = 2$, a *program* (x_t, c_t, y_t) from $a > 0$ to z is given by

$$x_1 + c_1 \leq f(a); \quad z + c_2 \leq f(x_1); \quad x_t \geq 0, \ c_t \geq 0, \ t = 1, 2,$$

where $0 < z < f(f(a))$ (and we assume free disposal).

Let (c_1^*, c_2^*) be the optimal consumption program. Since u is increasing, it cannot be that the optimal consumptions $c_1^* = c_2^* = 0$.

Suppose that $c_1^* = 0$. Then $x_1^* = f(a)$, $c_2^* = [f(x_1^*) - z] > 0$.

This implies that $f(x_1^*) > z$. Now, since f is continuous and increasing, we can choose ε_1 positive, but sufficiently small, such that for all $0 < \varepsilon \leq \varepsilon_1$, $f(x_1^* - \varepsilon) > z$.

Consider a program obtained by choosing a positive $\varepsilon < \varepsilon_1$ that satisfies the following:

$$\acute{c}_1 = \varepsilon < x_1^*, \quad \acute{x}_1 = x_1^* - \varepsilon,$$

$$\acute{c}_2 = f(\acute{x}_1) - z. \text{ Then } f(\acute{x}_1) = f(x_1^* - \varepsilon) > z. \text{ So, } \acute{c}_2 > 0.$$

Hence, $(\acute{c}_1, \acute{x}_1; \acute{c}_2, z)$ is a program from a to z.

By the assumed optimality of $(c_1^*, c_2^*) = (0, c_2^*)$, we have

$$u(\acute{c}_1) + u(\acute{c}_2) \le u(0) + u(c_2^*),$$

$$\text{or} \quad [u(\varepsilon) + u(f(x_1^* - \varepsilon) - z)] - [u(0) + u(c_2^*)] \le 0,$$

$$\text{or} \quad [u(\varepsilon) + u(f(x_1^* - \varepsilon) - z)] - u(f(x_1^*) - z) \le 0,$$

$$\text{or} \quad u(\varepsilon) - [u(f(x_1^*) - z) - u(f(x_1^* - \varepsilon) - z)] \le 0.$$

Write, for simplicity of notation, $m^*(\varepsilon) \equiv f(x_1^* - \varepsilon) - z$, and $m^* \equiv f(x_1^*) - z$. Observe that $m^* > m^*(\varepsilon)$. By the Mean Value Theorem,

$$u(m^*) - u(m^*(\varepsilon)) = (m^* - m^*(\varepsilon))u'(q) \quad \text{where } m^*(\varepsilon) < q < m^*$$
$$< (m^* - m^*(\varepsilon))u'(m^*(\varepsilon)).$$

Hence, by substitution,

$$\begin{aligned}
0 \ge u(\varepsilon) &- [u(f(x_1^*) - z) - u(f(x_1^* - \varepsilon) - z)] \\
&= u(\varepsilon) - [u(m^*) - u(m^*(\varepsilon))] \\
&> u(\varepsilon) - (m^* - m^*(\varepsilon))u'(m^*(\varepsilon)) \\
&= u(\varepsilon) - [(f(x_1^*) - z) - (f(x_1^* - \varepsilon) - z)]u'(m^*(\varepsilon)) \\
&= u(\varepsilon) - [(f(x_1^*) - f(x_1^* - \varepsilon)]u'(m^*(\varepsilon)) \\
&= u(\varepsilon) - \varepsilon \cdot [f'(v(\varepsilon)) \cdot u'(m^*(\varepsilon)], \quad \text{where } x_1^* - \varepsilon < v(\varepsilon) < x_1^* \\
&= \varepsilon[u'(j(\varepsilon)) - \{f'(v(\varepsilon)) \cdot u'(m^*(\varepsilon)\}], \quad \text{where } 0 < j(\varepsilon) < \varepsilon.
\end{aligned}$$

Now, as $\varepsilon \to 0$, $f'(v(\varepsilon)) \to f'(x_1^*)$, $u'(m^*(\varepsilon)) \to u'(f(x_1^*) - z)$. On the other hand, $u'(j(\varepsilon)) \to \infty$. So, the last expression is positive for sufficiently small $\varepsilon > 0$, a contradiction.

The case where $c_1^* > 0$ and $c_2^* = 0$ is treated analogously.

Suppose that $c_2^* = 0$. Then $x_1^* = f^{-1}(z)$. Clearly, $c_1^* = f(a) - f^{-1}(z) > 0$. Recall that $f(f(a) > z$ means that there is a positive ε_2 small enough, such that

$$f(f(a)) > z + \varepsilon \quad \text{for all } 0 < \varepsilon < \varepsilon_2,$$

$$\text{or} \quad f(a) > f^{-1}(z + \varepsilon), \quad \text{as } f^{-1} \text{ is increasing.}$$

Consider an alternative feasible program obtained by choosing a positive $\varepsilon < \varepsilon_2$ that satisfies the following:

$$\hat{c}_1 = f(a) - f^{-1}(z + \varepsilon), \quad \hat{x}_1 = f^{-1}(z + \varepsilon),$$
$$\hat{c}_2 = \varepsilon.$$

Clearly,

$$\hat{c}_1 + \hat{x}_1 = f(a), \quad f(\hat{x}_1) = \varepsilon + z.$$

By the assumed optimality of $(c_1^*, c_2^*) = (c_1^*, 0)$, we have

$$u(\hat{c}_1) + u(\hat{c}_2) \leq u(c_1^*) + u(0),$$

or, $u(f(a) - f^{-1}(z + \varepsilon)) + u(\varepsilon) \leq u(f(a) - f^{-1}(z)).$ $\qquad \square$

A clarification is in order: consider the unique positive optimal consumption program $\mathbf{c}^* = (c_1^*, \ldots, c_T^*)$. Since $u'(c) > 0$, if $\mathbf{x}^* = (a, x_1^*, \ldots, x_{T-1}^*, z)$ generates $\mathbf{c}^*$, one must have the equalities:

$$x_1^* + c_1^* = f(a), \ldots, x_t^* + c_t^* = f(x_{t-1}^*), \ldots, z + c_T^* = f(x_{T-1}^*).$$

But this means that $x_1^* = f(a) - c_1^*, x_2^* = f(x_1^*) - c_2^*, \ldots$ are uniquely determined, as are $y_t^* = f(x_{t-1}^*)$. Hence, we refer to $\mathbf{x}^*$ as *the* optimal input program, or to $(\mathbf{x}^*, \mathbf{y}^*, \mathbf{c}^*)_a^z$ as *the* optimal program from a to z.

3.4 Competitive Programs

A program $(\mathbf{x}, \mathbf{y}, \mathbf{c})$ from a to z is *competitive* if there is a non-zero vector $\mathbf{p} = (p_t)_{t=0}^T \in R_+^{T+1}$, such that

(i) $\delta^{t-1} u(c_t) - p_t \cdot c_t \geq \delta^{t-1} u(c) - p_t \cdot c$ for all $c \geq 0$ and for $t = 1, 2, \ldots, T$,

(ii) $p_{t+1} \cdot f(x_t) - p_t \cdot x_t \geq p_{t+1} \cdot f(x) - p_t \cdot x$ for all $x \geq 0$ and for $t = 0, 1, \ldots, T - 1$.

A competitive program will be denoted by $(\mathbf{x}, \mathbf{y}, \mathbf{c}; \mathbf{p})$, where (as a part of the definition), each $p_t \geq 0$, for all $t = 0, 1, \ldots, T$, and for some t, $p_t > 0$. Given our assumptions, more can be concluded.

Suppose that for some period $\tau \geq 1$, $p_\tau = 0$. From (i), we get

$$u(c_\tau) \geq u(c), \qquad (3.4.1)$$

for all $c \geq 0$; but this leads to a contradiction in view of (U.2) (choose $c > c_\tau$). If $p_0 = 0$, we also get a contradiction from condition (ii) in the definition of a competitive program. Hence, if $(\mathbf{x}, \mathbf{y}, \mathbf{c}; \mathbf{p})$ is a *competitive program*, we have $\mathbf{p} \gg 0$ and one can also show that $\mathbf{c} \gg 0$.

(▲ Hint: Suppose that $c_t = 0$ for $t = \tau$.

Then $0 \geq u(c) - p_\tau \cdot c$ for all $c \geq 0$. Note that $u(0) = 0$. Choose $c > 0$ and "sufficiently small" to arrive at a contradiction from

$$0 \geq c[u'(c) - p_\tau]. \ \blacktriangledown)$$

3.5   Competitive and Optimal Programs

We shall study the relation between competitive and optimal programs.

Theorem 3.5.1. *Let $(\mathbf{x}, \mathbf{y}, \mathbf{c}; \mathbf{p})$ be a competitive program from a to z. Then it is optimal from a to z.*

Proof.   Let $(\mathbf{x}', \mathbf{y}', \mathbf{c}')$ be any program from a to z. For $t = 1, 2, \ldots, T$, we get

$$\delta^{t-1} u(c_t) - p_t \cdot c_t \geq \delta^{t-1} u(c_t') - p_t \cdot c_t', \qquad (3.5.1)$$

or

$$[\delta^{t-1} u(c_t) - \delta^{t-1} u(c_t')] \geq p_t \cdot [c_t - c_t']. \qquad (3.5.2)$$

Hence,

$$\sum_{t=1}^{t=T} \delta^{t-1} [u(c_t) - u(c_t')] \geq \sum_{t=1}^{t=T} (p_t \cdot [c_t - c_t']).$$

We can therefore write $c_t = y_t - x_t$ and $c'_t = y'_t - x'_t$ for $t = 1, 2, \ldots, T$. Observe that $x_0 = x'_0 = a$ and $x_T = x'_T = z$. Now,

$$p_t \cdot [c_t - c'_t] = p_t \cdot [(y_t - x_t) - (y'_t - x'_t)]$$
$$= p_t \cdot [(f(x_{t-1}) - x_t) - (f(x'_{t-1}) - x'_t)].$$

Hence,

$$\sum_{t=1}^{t=T} p_t \cdot [c_t - c'_t]\} = \sum_{t=0}^{t=T-1} \{[p_{t+1} \cdot f(x_t) - p_t \cdot x_t] - [p_{t+1} \cdot f(x'_t)$$
$$- p_t \cdot x'_t]\} - (p_T \cdot x_T - p_T \cdot x'_T).$$

Since $x_T = x'_T = z$, we see that the last term on the right side $(p_T \cdot x_T - p_T \cdot x'_T) = 0$. Examine the first term under the summation sign and note that $x_0 = x'_0 = a$, so $-p_0 \cdot x_0 + p_0 \cdot x'_0 = 0$. Use the profit maximization property (see (ii) in the definition) of a competitive program to verify that each term under the summation sign is non-negative. Hence,

$$\sum_{t=1}^{t=T} \delta^{t-1}[u(c_t) - u(c'_t)] \geq \sum_{t=1}^{t=T} p_t \cdot [c_t - c'_t]$$
$$= \sum_{t=0}^{t=T-1} \{[p_{t+1} \cdot f(x_t) - p_t \cdot x_t]$$
$$- [p_{t+1} \cdot f(x'_t) - p_t \cdot x'_t]\} \geq 0.$$

This completes the proof of optimality of $(\mathbf{x}, \mathbf{y}, \mathbf{c}; \mathbf{p})$. □

We now prove a converse. Let $(\mathbf{x}, \mathbf{y}, \mathbf{c})$ be a positive program from a to z, i.e., $\mathbf{c} \gg \mathbf{0}$. Recall that $\mathbf{c} \gg \mathbf{0}$ implies that $\mathbf{x} \gg \mathbf{0}$, $\mathbf{y} \gg \mathbf{0}$.

Theorem 3.5.2. *Let $(\mathbf{x}^*, \mathbf{y}^*, \mathbf{c}^*)$ be the (positive) optimal program from a to z. Then there is a unique strictly positive price vector $\mathbf{p}^* = (p_t^*) \gg 0$, such that*

(i) *for all $t = 1, 2, \ldots, T$, $\delta^{t-1}u(c_t^*) - p_t^* \cdot c_t^* \geq \delta^{t-1}u(c) - p_t^* \cdot c$ for $c \geq 0$;*

(ii) *for all $t = 0, 1, \ldots, T-1$, $p_{t+1}^* \cdot f(x_t^*) - p_t^* \cdot x_t^* \geq p_{t+1}^* \cdot f(x) - p_t^* \cdot x$ for $x \geq 0$.*

Proof.   Optimality of $(\mathbf{x}^*, \mathbf{y}^*, \mathbf{c}^*)$ implies

$$\sum_{t=1}^{T} \delta^{t-1} u(c_t^*) \geq \sum_{t=1}^{T} \delta^{t-1} u(c_t) \qquad (3.5.3)$$

for all programs $(\mathbf{x}, \mathbf{y}, \mathbf{c})$ from a to z. First, rewrite (3.5.4) as follows:

$$\sum_{t=1}^{T} \delta^{t-1} u(f(x_{t-1}^*) - x_t^*) \geq \sum_{t=1}^{T} \delta^{t-1} u(f(x_{t-1}) - x_t) \qquad (3.5.4)$$

for all production programs $(\mathbf{x}, \mathbf{y})$. Observe that, for $t \geq 1$, x_t^* solves the following optimization problem:

$$\max \quad \delta^{t-1}[u(f(x_{t-1}^*) - x)] + \delta^t[u(f(x) - x_{t+1}^*)],$$
$$\text{subject to} \quad 0 \leq x \leq f(x_{t-1}^*) \quad \text{and} \quad f(x) \geq x_{t+1}^*.$$

Since $c_t^* > 0$, $c_{t+1}^* > 0$, and $x_t^* > 0$, we have $0 < x_t^* < f(x_{t-1}^*)$ and $f(x_t^*) > x_{t+1}^*$. The first-order condition for a maximum gives us

$$\frac{d}{dx}\left(\delta^{t-1} u(f(x_{t-1}^*) - x) + \delta^t u(f(x) - x_{t+1}^*)\right) = 0 \qquad (3.5.5)$$

at $x = x_t^*$. This gives us (recall the Chain Rule of differentiation):

$$\delta^{t-1} u'(c_t^*)(-1) + \delta^t u'(c_{t+1}^*) \cdot f'(x_t^*) = 0, \qquad (3.5.6)$$

or, rearranging and canceling,

$$u'(c_t^*) = \delta u'(c_{t+1}^*) \cdot f'(x_t^*). \qquad (3.5.7)$$

Now, for $t = 1, 2, \ldots, T$, define $p_t^* = \delta^{t-1} u'(c_t^*)$ and define $p_0^* = p_1^* \cdot f'(x_0^*)$. From (3.5.7), we get $p_t^* = p_{t+1}^* \cdot f'(x_t^*)$ for all $t = 0, 1, \ldots, T$. Recall the Mean Value theorem and use the assumptions that

$u''(\cdot) < 0$ to get for $t = 1, 2, \ldots, T$,

$$\delta^{t-1}[u(c_t^*) - u(c)] \geq \delta^{t-1}[u'(c_t^*) \cdot (c_t^* - c)] = p_t^* \cdot (c_t^* - c)$$

for $c \geq 0$. By rearranging terms, we are led to

$$\delta^{t-1}u(c_t^*) - p_t^* \cdot c_t^* \geq \delta^{t-1}u(c) - p_t^* \cdot c$$

for $c \geq 0$. Also, using the strict concavity of f, i.e., $f''(\cdot) < 0$, we get, for $t = 1, 2, \ldots, T - 1$,

$$p_{t+1}^* \cdot f(x_t^*) - p_{t+1}^* \cdot f(x) \geq p_{t+1}^* \cdot f'(x_t^*) \cdot [x_t^* - x] = p_t^* \cdot [x_t^* - x] \qquad (3.5.8)$$

for $x \geq 0$. □

The relation (3.5.7) is known as the Ramsey–Euler condition in capital theory. It characterizes the optimal trade-off between consumptions in two adjacent periods.

We now derive a basic comparative dynamic result on a monotonicity property of optimal inputs.

Theorem 3.5.3. *Let $z' > z > 0$. Fix an initial stock a (as usual, both (a, z) and (a, z') are admissible). Assume that $(\mathbf{x}^*(z), \mathbf{y}^*(z), \mathbf{c}^*(z))$ and $(\mathbf{x}^*(z'), \mathbf{y}^*(z'), \mathbf{c}^*(z'))$ are positive optimal programs from a to z and a to z', respectively. Then*

$$x_t^*(z') > x_t^*(z) \quad \text{for all } t = 1, 2, \ldots, T, \qquad (3.5.9)$$

$$c_t^*(z') < c_t^*(z) \quad \text{for all } t = 1, 2, \ldots, T. \qquad (3.5.10)$$

Proof. To simplify typing, we write x_t', x_t, c_t, c_t' instead of $x_t^*(z'), x_t^*(z), c_t^*(z), c_t^*(z')$ in this proof. Suppose, contrary to the claim, that

$$x_1' \leq x_1.$$

Then $c_1' + x_1' = f(a) = c_1 + x_1$ leads to

$$c_1' \geq c_1.$$

Hence, $u'(c_1') \leq u'(c_1)$, and $f'(x_1') \geq f'(x_1)$. Also, $f(x_1') \leq f(x_1)$. From the first inequality, we get

$$1 \leq \frac{u'(c_1)}{u'(c_1')}. \tag{3.5.11}$$

From the Ramsey–Euler condition (3.5.7), we get

$$\delta u'(c_2') \cdot f'(x_1') = u'(c_1') \leq u'(c_1) = \delta u'(c_2) \cdot f'(x_1).$$

Hence,

$$1 \leq \left[\frac{u'(c_1)}{u'(c_1')}\right] = \frac{[u'(c_2) \cdot f'(x_1)]}{[u'(c_2') \cdot f'(x_1')]}$$

$$= \left[\frac{u'(c_2)}{u'(c_2')}\right] \cdot \left[\frac{f'(x_1)}{f'(x_1')}\right] \leq \left[\frac{u'(c_2)}{u'(c_2')}\right].$$

This leads to

$$u'(c_2') \leq u'(c_2),$$

$$\text{or} \quad c_2' \geq c_2.$$

Now,

$$x_2 - x_2' = [f(x_1) - c_2] - [f(x_1') - c_2'] = [f(x_1) - [f(x_1')] + [c_2' - c_2] \geq 0.$$

Summarizing,

$$x_1 \geq x_1' \text{ leads to } x_2 \geq x_2'.$$

Repeating the steps, we get

$$z = x_T \geq x_T' = z' \text{ a contradiction to the assumption } z' > z.$$

To complete the proof, let $\tau \geq 2$ be the *first* period $(1 < \tau \leq T-1)$ such that

$$x_\tau' \leq x_\tau.$$

Then, $x_\tau' + c_\tau' = f(x_{\tau-1}') > f(x_{\tau-1}) = x_\tau + c_\tau$, and the above arguments can be adapted to arrive at a contradiction. Hence,

$$x_t' > x_t \quad \text{for all } t = 1, \ldots, T.$$

Now, $c_1 = f(a) - x_1 > f(a) - x_1' = c_1'$.

Hence,

$$\delta u'(c_2') \cdot f'(x_1') = u'(c_1') > u'(c_1) = \delta u'(c_2) \cdot f'(x_1),$$

$$x_1' > x_1 \text{ implies } f'(x_1') < f'(x_1).$$

Hence,

$$u'(c_2') > u'(c_2), \text{ leading to}$$

$$c_2 > c_2'.$$

Repeating the steps, we get

$$c_t > c_t' \quad \text{for all } t = 1, 2, \ldots, T - 1.$$

This completes the proof. $\qquad\square$

Thus, an *increase* in the target stock from z to z' leads to an *increase* in the *optimal* input in *every period* $t = 1, 2, \ldots, T - 1$ and a *decrease* in the optimal harvest in every period.

Exercise 3.5.1. Let $a' > a > 0$. Fix a target stock $z > 0$, such that both (a, z) and (a', z) are admissible. Assume that $(\mathbf{x}^*(a'), \mathbf{y}^*(a'), \mathbf{c}^*(a'))$ and $(\mathbf{x}^*(a), \mathbf{y}^*(a), \mathbf{c}^*(a))$ are positive optimal programs from a' to z and a to z, respectively. Then

$$x_t^*(a') > x_t^*(a) \quad \text{for } t = 1, 2, \ldots, T - 1,$$

$$c_t^*(a') > c_t^*(a) \quad \text{for } t = 1, 2 \ldots, T.$$

3.6 Optimal Stationary Programs

A program $(\mathbf{x}, \mathbf{y}, \mathbf{c})$ is *stationary* or *equitable* if

$$\begin{cases} x_t = x > 0; & \text{for } t = 1, 2, \ldots, T, \\ y_{t+1} = y = f(x) & \text{for } t = 0, 1, \ldots, T - 1, \\ c_t = c = y - x > 0 & \text{for } t = 1, 2, \ldots, T. \end{cases}$$

Perhaps, I should emphasize that, in terms of our earlier notation, for the stationary program, $a = z = x$. I shall use the notation $\langle x, y, c \rangle$ to stress that we are dealing with a stationary program that specifies the same x for all t. The question that we shall first answer is the following: "Does there exist some $x > 0$ such that the equitable program $\langle x, y, c \rangle$ is, in fact, optimal among *all* programs $(\mathbf{x}, \mathbf{y}, \mathbf{c})$ from x to x?" It turns out that there is a *unique stationary* (*or equitable*) *optimal program*.

For $\delta \leq 1$, let $k_\delta^* > 0$ be the unique solution to the equation

$$\delta f'(x) = 1. \tag{3.6.1}$$

When $\delta = 1$, we shall drop the subscript, if the context leaves no room for misunderstanding. In this case, we refer to $\langle k^*, f(k^*), c^* \rangle$ as the *golden-rule program*. The input k^* is the *golden-rule input*, $f(k^*)$ is the *golden-rule output*, and c^* is the *golden-rule consumption*. When $\delta < 1$, $\langle k_\delta^*, \; f(k_\delta^*), \; c_\delta^* \rangle$ is referred to as the $(\delta-)$ *modified golden-rule program* (with parallel modifications for input, output and stock).

Write $c_\delta^* = f(k_\delta^*) - k_\delta^*$. We shall also write $y_\delta^* = f(k_\delta^*)$.

Theorem 3.6.1. *The stationary program* $\langle k_\delta^*, f(k_\delta^*), c_\delta^* \rangle$ *is optimal among all programs from* k_δ^* *to* k_δ^*. *One has*

$$\delta f(k_\delta^*) - k_\delta^* > \delta f(x) - x \quad \text{for all } x \geq 0, x \neq k_\delta^*. \tag{3.6.2}$$

Proof. First, write $h_\delta(x) = \delta f(x) - x$ for all $x \geq 0$. Then, $h_\delta(0) = 0$. Also, there is some $\theta_1 > 0$, such that $h_\delta(x) > 0$ for $0 < x < \theta_1$. Verify that $h_\delta'(x) = \delta f'(x) - 1$ for all $x > 0$; hence,

$$\left\{ \begin{array}{l} h_\delta'(x) > 0 \text{ if and only if } \delta f'(x) > 1 \\ h_\delta'(x) = 0 \text{ if and only if } \delta f'(x) = 1 \\ h_\delta'(x) < 0 \text{ if and only if } \delta f'(x) < 1 \end{array} \right\},$$

and $h_\delta''(x) = \delta f''(x) < 0$.

Since $h_\delta(x) = \delta f(x) - x < f(x) - x$, we have $h_\delta(x) < 0$ for $x > \bar{k}$. On the closed interval $[0, \bar{k}]$, h_δ attains a global maximum at, say, $\breve{x}_\delta$. Since $h_\delta(x) > 0$ for $0 < x < \theta_1$, $h_\delta(\breve{x}_\delta) > 0$, which means that $\breve{x}_\delta \in (0, \bar{k})$. It follows that $h_\delta'(\breve{x}_\delta) = 0$, or $\delta f'(\breve{x}_\delta) = 1$.

By definition of k_δ^*, we get $k_\delta^* = \breve{x}_\delta$. Hence,

$$h_\delta(k_\delta^*) > 0 \text{ and } h_\delta(k_\delta^*) > h_\delta(x) \quad \text{for all } x \geq 0, \ x \neq k_\delta^*. \quad (3.6.3)$$

We shall now show that the stationary program $\langle k_\delta^*, f(k_\delta^*), c_\delta^* \rangle$ [from k_δ^* to k_δ^*] is competitive; hence, optimal among *all* programs from k_δ^* to k_δ^*. Define $p_t = \delta^{t-1} u'(c_\delta^*)$ for all $t = 1, 2, \ldots, T$ and for $t = 0$, $p_0 = p_1 \cdot f'(k_\delta^*)$.

Now, for $c \geq 0$,

$$\delta^{t-1}[u(c_\delta^*) - u(c)] \geq \delta^{t-1}[c_\delta^* - c] \cdot u'(c^*) = p_t \cdot [c_\delta^* - c].$$

Rearranging, for all $t = 1, 2, \ldots T$,

$$\delta^{t-1} u(c^*) - p_t \cdot c^* \geq \delta^{t-1} u(c) - p_t \cdot c$$

for $c \geq 0$. Also, for all $t = 1, 2, \ldots, T - 1$,

$$\begin{aligned}
p_{t+1} \cdot f(k_\delta^*) - p_t \cdot k_\delta^* &= \delta^t u'(c_\delta^*) \cdot f(k_\delta^*) - \delta^{t-1} u'(c_\delta^*) \cdot k_\delta^* \\
&= \delta^{t-1} u'(c_\delta^*)[\delta f(k_\delta^*) - k_\delta^*] \\
&> \delta^{t-1} u'(c_\delta^*)[\delta f(x) - x] \\
&= p_{t+1} \cdot f(x) - p_t \cdot x \quad \text{for } x \geq 0.
\end{aligned}$$

Exercise 3.6.1. Consider the following maximization problem:

$$\max p_1 \cdot f(x) - p_0 \cdot x,$$

where $p_1 = u'(c_\delta^*)$, $p_0 = p_1 \cdot f'(k_\delta^*)$ and $x \geq 0$.

Show that (using standard calculus methods) $k^{*\delta} > 0$ is the unique solution to this problem. $\qquad\square$

3.7 Qualitative Properties of Optimal Programs

Consider the case when T is finite but "large", we shall obtain some qualitative properties of optimal programs. First, a comparative dynamic result.

Lemma 3.7.1. *Let* $(\mathbf{x}^*, \mathbf{y}^*, \mathbf{c}^*)$ *be a positive optimal program from* a *to* z.

(a) *If* $x_t^* \leq k_\delta^*$ *for some* $t \geq 1$, *then* $c_{t+1}^* \geq c_t^*$ *and* $\delta f'(x_t^*) > 1$.
(b) *If* $x_t^* \geq k_\delta^*$ *for some* $t \geq 1$, *then* $c_{t+1}^* \leq c_t^*$ *and* $\delta f'(x_t^*) < 1$.

Proof.   (a) By equation (3.5.7),

$$u'(c_t^*) = \delta u'(c_{t+1}^*) \cdot f'(x_t^*),$$

or

$$\frac{u'(c_t^*)}{u'(c_{t+1}^*)} = \delta f'(x_t^*) \geq 1,$$

or

$$u'(c_t^*) \geq u'(c_{t+1}^*),$$

leading to $c_{t+1}^* \geq c_t^*$.
    (b) is proved in a similar manner.   □

Lemma 3.7.2.

(a) *If for some* $t \geq 1$, $x_t^* \leq k_\delta^*$ *and* $y_{t+1}^* \leq y_t^*$, *then* $x_{t+1}^* \leq x_t^*$.
(b) *If for some* $t \geq 1$, $x_t^* \geq k_\delta^*$ *and* $y_{t+1}^* \geq y_t^*$, *then* $x_{t+1}^* \geq x_t^*$.

Proof.   Note that $y_{t+1}^* = x_{t+1}^* + c_{t+1}^*$ and $y_t^* = x_t^* + c_t^*$. Hence, $y_{t+1}^* - y_t^* = [x_{t+1}^* - x_t^*] + [c_{t+1}^* - c_t^*]$. Rearranging, $[x_{t+1}^* - x_t^*] = [y_{t+1}^* - y_t^*] - [c_{t+1}^* - c_t^*]$. To prove (a), observe that, by hypothesis, $[y_{t+1}^* - y_t^*] \leq 0$, and, by the earlier lemma, $[c_{t+1}^* - c_t^*] \geq 0$. Similarly, prove (b).   □

It is now possible to provide a description of the properties of optimal programs that are positive.

Let $(\mathbf{x}^*, \mathbf{y}^*, \mathbf{c}^*)$ be a positive optimal program from a to z. There are the following four cases:

Case (i) $x_0^* = a \leq k_\delta^*$, $x_T^* = z \leq k_\delta^*$,
Case (ii) $x_0^* = a \leq k_\delta^*$, $x_T^* = z \geq k_\delta^*$,

Case (iii) $x_0^* = a \geq k_\delta^*,\ x_T^* = z \geq k_\delta^*,$
Case (iv) $x_0^* = a \leq k_\delta^*,\ x_T^* = z \leq k_\delta^*.$

We shall consider the first two cases in detail. The other two are "mirror images" that can be analyzed by similar arguments.

Case (i). $x_0^* = a \leq k_\delta^*,\ x_T^* = z \leq k_\delta^*.$ The first step is to verify $x_t^* \leq k_\delta^*$ for all $t = 0, 1, \ldots, T$. If this claim is false, there is a first period $t_0 \geq 1$, such that $x_{t_0}^* > k_\delta^*$. Then $y_{t_0}^* \leq y_\delta^*$ [since $x_{t_0-1}^* \leq k_\delta^*$].

Now, $y_{t_0+1}^* = f(x_{t_0}^*) > f(k_\delta^*) \equiv y_\delta^* \geq y_{t_0}^*$. By Lemma 3.7.2(b), $x_{t_0+1}^* \geq x_{t_0}^* > k_\delta^*$. Then,

$$y_{t_0+2}^* = f(x_{t_0+1}^*) \geq f(x_{t_0}^*) = y_{t_0+1}^* > \bar{y}^*.$$

Hence, $x_{t_0+2}^* \geq x_{t_0+1}^* \geq x_{t_0}^* > k_\delta^*$, etc., leading to a contradiction since $x_T^* = z \leq k_\delta^*$.

Hence, (c_t^*) must be monotonically non-decreasing. The sequences (x_t^*) [and (y_{t+1}^*)] must be non-decreasing up to a period t_0 and non-increasing thereafter. It is possible that $t_0 = 0$ or T. Let t_0 be the first period such that $x_{t_0+1}^* \leq x_{t_0}^*$. Then $y_{t_0+2}^* \leq y_{t_0+1}^*$ and the fact that $x_t^* \leq k_\delta^*$ for all t lead to $x_{t_0+2}^* \leq x_{t_0+1}^*$, etc.

Case (ii). $x_0^* = a \leq k_\delta^*,\ x_T^* = z \geq k_\delta^*.$ In this case, the sequences (x_t^*) and (y_{t+1}^*) must be monotonically non-decreasing. To see this, let t_0 be the first period, such that $x_{t_0}^* \geq k_\delta^*$. Then, $y_{t_0+1}^* = f(x_{t_0}^*) \geq f(k_\delta^*) = y_\delta^*$. But $x_{t_0-1} < k_\delta^*$ implies $y_{t_0}^* = f(x_{t_0-1}^*) < f(k_\delta^*) = y_\delta^*$. So, $y_{t_0+1}^* > y_{t_0}^*$.

By Lemma 3.7.2(b), $x_{t_0+1}^* \geq x_{t_0}^* \geq k_\delta^*$. Repeating this step to obtain that (x_t^*) is non-decreasing from period t_0 on. On the other hand, if $x_{t'+1}^* < x_{t'}^*$ for some $t' < t_0$, then (since $x_{t'}^* < k_\delta^*$), using Lemma 3.7.2(a), (x_t^*) and (y_{t+1}^*) will be non-increasing thereafter, contradicting $x_T^* \geq k_\delta^*$.

Consumption is non-decreasing up to a period t_0 and non-increasing thereafter. Consumption remains no higher than c_δ^*.

To see this note that $c_{t_0}^*$ is the largest consumption, hence, $c_{t_0}^* \geq c_{t_0-1}^*$ and this means that $x_{t_0-1}^* \leq k_\delta^*$ and $f(x_{t_0-1}^*) = y_{t_0}^* \leq f(k_\delta^*) = y_\delta^*$. So, $c_{t_0}^* = y_{t_0}^* - x_{t_0}^* \leq y_\delta^* - k_\delta^* = c_\delta^*$.

3.8   Conservation of Inheritance

It is perhaps time to step back and reflect upon the limitations of the optimization model in the context of sustainability and conservation. The model does not throw light on the nature of several parameters. We do not know how the "planner" chooses T and, when discounting is present, how δ is arrived at: yet, our exposition stressed the importance of k_δ^* (which is entirely determined by f and δ). The *terminal* stock z becomes the *initial* stock for the *subsequent* planner and is crucial in determining *its* range of options. But we have not characterized the range of values for z that adequately captures our commitment to the economy beyond T. The difficulty of choosing an "acceptable" terminal stock in *any* finite horizon model has been the subject of extended discussion in optimal growth theory (see Koopmans (1967)) and need not be repeated here. In our context, we may introduce a specific constraint: the planner is required to set $a = z$, interpreting it as a "conservation of inheritance" clause. The qualitative behavior of optimal programs can be derived as in Section 3.7 (except that we now have only two cases to consider). We shall explore this clause more fully in Chapter 4 when we introduce planning revision.

3.9   Complements and Details

Some of the results in this chapter are derived in Gale (1970) often without differentiability assumptions.

There is a well-known "Catenary Turnpike Theorem" which will now be sketched. It is of interest when T is "large", and in what follows, we assume that (x, y, c) is the positive optimal program from a to z.

Theorem 3.9.1. *Let $\varepsilon > 0$. There is a positive integer $T_1(\varepsilon)$ such that x_t^* cannot be less than $\bar{k}_\delta^* - \varepsilon$ for more than $T_1(\varepsilon)$ periods.*

Proof.   Consider Case 1, when $x_0^* = a \leq \bar{k}_\delta^*, x_T^* = z \leq \bar{k}_\delta^*$. Then $x_t^* \leq \bar{x}^*$ for all $t = 0, 1, \ldots, T$. There is some constant $\bar{k}$ such that $0 \leq c_t^* \leq \bar{k}$. Now, the positive optimal program satisfies

$$u'(c_1^*) = \delta u'(c_2^*) \cdot f'(x_1^*) = [u'(c_2^*)] \cdot \delta[f'(x_1^*)]. \tag{3.9.1}$$

Repeating, for any finite τ,

$$u'(c_1^*) = [u'(c_\tau^*)] \cdot \Pi_{t=1}^{\tau-1}(\delta f'(x_1^*)). \tag{3.9.2}$$

Now, for all t, $\delta f'(x_t^*) > 1$. Given $\varepsilon > 0$, there is some $\theta > 0$, such that $x \leq \bar{k}_\delta^* - \varepsilon$ implies that $\delta f'(x) \geq 1 + \theta$. So, if $x_t^* \leq \bar{x}^* - \varepsilon$ for P periods, we are led to

$$u'(c_1^*) > [u'(\bar{k})] \cdot (1 + \theta)^P. \tag{3.9.3}$$

Take logarithm to get the bound on P. Modify the argument for Case 2. Let t_0 be the first period such that $x_{t_0}^* \geq \bar{k}_\delta^*$. Then, $x_t^* < \bar{k}_\delta^*$ for all $t = 0, 1, \ldots, t_0 - 1$. Assume that $x_t < \bar{k}_\delta^* - \varepsilon$ for P periods and derive a bound on P. □

Remark 3.9.1. It requires a longer and subtle argument to claim that, given $\varepsilon > 0$, there is a positive integer $T_2(\varepsilon)$ such that x_t^* cannot be more than $\bar{x}^* + \varepsilon$ for more than $T_2(\varepsilon)$ periods (see Gale (1970)).

Chapter 4

Rolling Plans: Efficiency and Long-Run Optimality

The problems associated with the choices and targets for finite plans have been widely discussed and consistent planning is intimately related to an infinite planning horizon... In practice, plans may not be completely executed both for reasons of imperfect foresight or intention. In a rolling plan, the plan is revised at the end of each year and an additional year is appended to the last. The effect of this yearly revision is to always maintain a future planning period of constant length.

— Steven M. Goldman

4.1 Introduction

"Long-run" planning and management have often been explored by using infinite horizon models. But one can also think of the "long-run" plan as a sequence of successive short-run plans in which terminal stocks are chosen and revised in the light of experience. In this and the following two chapters, infinite horizon models are formally introduced. The formal approach in this chapter is a discrete-time version of the idea of continual planning revision process (see Goldman (1968)), with a special emphasis on the conservation of inheritance. Roughly speaking, this process is described as follows: given an initial stock $\tilde{x}$ in $(0, \bar{k})$, consider the undiscounted $(\delta = 1)$ optimal plan over T-periods as in Chapter 3, when the "conservation of inheritance" condition is met. For expository simplicity, think of the case when $T = 2$. The optimal 2-*program* from $\tilde{x}$ to $\tilde{x}$ is a solution to the maximization problem:

$$\text{maximize} \quad u(c_1) + u(c_2),$$

$$\text{subject to} \quad c_1 + x_1 = f(\tilde{x}), \; c_2 + \tilde{x} = f(x_1), \qquad (4.1.1)$$

$$x_1 \geq 0, \; c_t \geq 0, \quad \text{for } t = 1, 2.$$

Here, the regeneration function f and the reward function u satisfy the properties (F.1)–(F.4) and (U.1)–(U.2) listed in Section 3.3.1 of Chapter 3. Denote the optimal 2-*program* by $(\tilde{x}, x_1^*, \tilde{x})$. The optimal first period consumption (say, c_1^*) takes place and the economy moves to $x_1^* = f(\tilde{x}) - c_1^*$. Now, in the next period, another optimization problem (as in (4.1.1) is contemplated with (naturally) x_1^* as the initial stock, and consistent with the conservation of inheritance condition, also, x_1^* as the terminal stock. Again, the optimal consumption takes place, and the story is repeated. The sequence (x_t^*) generated by this procedure is called a "rolling plan". The primary objective of this chapter is to establish that the rolling plan is both intertemporally efficient and maximizes the long-run average return. We begin with some formal definitions involving infinite programs. In Section 4.2.1, we briefly touch upon the problem of identifying intertemporal inefficiency of profit maximizing programs. A few important properties of finite horizon optimal programs are summarized in Section 4.2.2 and a rolling plan is defined in Section 4.3 in terms of a generating function (see (3.3.1)). Some properties of the generating function (monotonicity, continuity, etc.) are derived and welfare properties of a rolling plan are next established.

The rolling plan is an example of a decentralized evolutionary resource allocation mechanism to use the definition of Hurwicz and Weinberger (1990). Decentralization of decision making is achieved by introducing some accounting (Bala–Majumdar–Mitra) prices. On the consumption side, a specific marginal rate of intertemporal substitution and on the production side, the intertemporal profit maximization condition of Malinvaud (1953) are verified. The subtle difference between the B-M-M price scheme and the competitive prices (the "duality theory") of Chapter 3 is spelled out in the final section. Roughly speaking, the difference is due to the fact that even though on the consumption side, the marginal rate of substitution is equated to a price ratio, only the current part of a T-period optimization plan is carried out.

4.2 Infinite Programs

As in Chapter 3, we consider a renewable resource whose stock or population size is denoted by $x \geq 0$. The natural reproduction function $f : R_+ \to R_+$ is assumed to satisfy the properties (F.1)–(F.4) listed in Section 3.2 of Chapter 3. Given an initial stock $x_0 = \tilde{x} > 0$, an *infinite* program from $\tilde{x}$ is a non-negative sequence $\mathbf{x} = (x_t)$, satisfying

$$x_0 = \tilde{x}, \quad 0 \leq x_{t+1} \leq f(x_t) \quad \text{for } t \geq 0. \tag{4.2.1}$$

The (non-negative) consumption sequence $\mathbf{c} = (c_t)$ generated by the (infinite) program $\mathbf{x}$ is defined by

$$c_t = f(x_{t-1}) - x_t \quad \text{for } t \geq 1. \tag{4.2.2}$$

A program $\mathbf{x}$ from $\tilde{x}$ is *positive* if $\mathbf{c} \gg \mathbf{0}$. It follows that for a positive program, $\mathbf{x} \gg \mathbf{0}$. Recall (Lemma 3.2.1) that there is a unique (maximum sustainable stock) $\bar{k} > 0$, such that $f(\bar{k}) = \bar{k}$. Also, recall (Section 3.6 of Chapter 3) that there is a unique $k^* \in (0, \bar{k})$, such that $f'(k^*) = 1$. The (infinite) sequence (x_t) defined by

$$x_t = k^* \quad \text{for } t \geq 0 \tag{4.2.3}$$

is a (stationary or equitable) program (golden-rule program) from k^*. It generates an infinite (constant) sequences $(c^*, c^*, \ldots)$ of consumptions, where $c^* = f(k^*) - k*$. Two criteria for evaluating alternative (infinite) programs from a given initial $\tilde{x}$ will be explored in this chapter. A program $\mathbf{x} = (x_t)$ from $\tilde{x} > 0$ (generating $\mathbf{c} = (c_t)$) is *inefficient* if there is another program $\mathbf{x}' = (x_t')$ from (the same initial)

$\tilde{x} > 0$ (generating $\mathbf{c}'=(c_t')$), such that $c_t' \geq c_t$ for all $t \geq 1$ and $c_t' > c_t$ for some t. It is *efficient* if it not inefficient. It is clear that the program that specifies $c_1 = f(\tilde{x})$, $x_t = c_{t+1} = 0$ for all $t \geq 1$ is efficient. This example shows rather dramatically that the notion of efficiency does not capture any concern for the consumption possibilities of future generations.

Consider the case in which the planner's preferences are described by a return or utility function $u : R_+ \to R_+$ that satisfies the assumptions (U.1) and (U.2) in Section 3.2.1 of Chapter 3. We say that an infinite program $\mathbf{x} = (x_t)$ from $\tilde{x} > 0$ (generating $\mathbf{c} = (c_t)$) maximizes *the long-run average return* if

$$\liminf_{T \to \infty} \left[\sum_{t=1}^{T} u(c_t)/T \right] \geq \liminf_{T \to \infty} \left[\sum_{t=1}^{T} u(c_t')/T \right] \qquad (4.2.4)$$

for every plan $\mathbf{x}'= (x_t')$ from (the same initial) $\tilde{x} > 0$ (generating $\mathbf{c}'=(c_t')$). In this chapter, we say that a program $\mathbf{x}'= (x_t')$ from $\tilde{x} > 0$ (generating $\mathbf{c}'=(c_t')$) is *optimal* if it maximizes the long-run average return and, in addition, efficient.

4.2.1   Intertemporal efficiency and profit maximization

We briefly explore the problem of achieving intertemporal efficiency through profit maximization in infinite horizon models. A price system $\mathbf{p} = (p_t)$ is an infinite sequence of non-negative numbers with $p_0 > 0$. A program $\mathbf{x} = (x_t)$ from $\tilde{x} > 0$ is said to satisfy the condition of *intertemporal profit maximization* at a price system $\mathbf{p} = (p_t)$ if for all $t = 0, 1, \ldots$

$$p_{t+1} \cdot f(x_t) - p_t \cdot x_t \geq p_{t+1} \cdot f(x) - p_t \cdot x \quad \text{for all } x \geq 0. \qquad (4.2.5)$$

Observe that this condition already appeared in the definition of a competitive program with a finite horizon in Section 3.4 of Chapter 3. I shall refer to the pair $(\mathbf{x}, \mathbf{p})$, satisfying (4.2.5) as a profit maximizing program with Malinvaud prices $\mathbf{p}$. If $(\mathbf{x}, \mathbf{p})$ is such a profit maximizing program and $\mathbf{x} \gg \mathbf{0}$, we can show that the Malinvaud prices $\mathbf{p} \gg \mathbf{0}$ and satisfy

$$p_{t+1} = p_t / f'(x_t) \quad \text{for all } t = 0, 1, 2, \ldots. \qquad (4.2.6)$$

It is clear that given the assumption (F.3) if one considers a positive program $\mathbf{x} \gg \mathbf{0}$ and simply *defines* a price system $\mathbf{p}$ by setting $p_0 = 1$, and successive p_t according to (4.2.6), then the pair $(\mathbf{x}, \mathbf{p})$ is a profit maximizing program. For example, consider the *infinite natural growth program* from any $\tilde{x} > 0$ (see Section 3.2.2 of Chapter 3):

$$x_0 = \tilde{x}, \; x_1 = f(\tilde{x}), \ldots, x_{t+1} = f(x_t), \ldots \quad \text{for } t \geq 1. \qquad (4.2.7)$$

As this program generates zero consumption in every period, it is clearly inefficient. Indeed, the example provides a clue to the possibility of inefficiency of competitive programs: overaccumulation (or underharvesting). A sufficient condition for a profit maximizing program to be efficient was proved by Malinvaud (1953) and this is now stated and proved.

Theorem 4.2.1. *Let* $(\mathbf{x}, \mathbf{p})$ *be a profit maximizing program from* $\tilde{x} > 0$ *with* $\mathbf{p} \gg 0$. *If*

$$\lim_{t \to \infty} p_t x_t = 0, \qquad (4.2.8)$$

then $\mathbf{x}$ *is efficient.*

Proof. Let $\mathbf{c}$ be the consumption generated by $\mathbf{x}$. Assume that $\mathbf{x}$ is inefficient. Then, there is some other program, say, $\mathbf{x}' = (x_t')$ from (the same) $\tilde{x}$, which generates a consumption sequence $\mathbf{c}' = (c_t')$, satisfying

$$c_t' \geq c_t \text{ for all } t \geq 1, \text{ and for some } \eta > 0,$$

$$c_\tau' = c_\tau + \eta \text{ for some period } \tau. \qquad (4.2.9)$$

Hence, for all $T > \tau + 1$,

$$\sum_{t=0}^{T-1} [p_{t+1}(c_{t+1}' - c_{t+1})] \geq p_\tau \cdot \eta \equiv \theta \; (say) > 0. \qquad (4.2.10)$$

On the contrary,

$$\sum_{t=0}^{T-1} [p_{t+1}(c'_{t+1} - c_{t+1})] = \sum_{t=0}^{T-1} \{p_{t+1}([f(x'_t) - x'_{t+1}] - [f(x_t) - x_{t+1}])\}$$

$$(4.2.11)$$

$$= \sum_{t=0}^{T-1} \{[p_{t+1}f(x'_t) - p_t x'_t] - [p_{t+1}f(x_t) - p_t x_t]\}$$

$$+ p_0 \tilde{x} - p_0 \tilde{x} - p_T x'_T + p_T x_T$$

$$\leq p_T x_T. \qquad (4.2.12)$$

From (4.2.8), there is some T', such that for all $T > T'$,

$$p_T x_T < \theta. \qquad (4.2.13)$$

Hence, for all $T > T'$,

$$\sum_{t=0}^{T-1} [p_{t+1}(c'_{t+1} - c_{t+1})] < \theta. \qquad (4.2.14)$$

From (4.2.10) and (4.2.14), we arrive at a contradiction. $\qquad\square$

($\blacktriangle$ *Exercise:* Does the proof use (F.1)–(F.4)? $\blacktriangledown$)

The condition (4.2.8) is known as the "transversality condition" in capital theory. It is just a sufficient condition. The issue of identifying inefficiency of profit maximizing programs turned out to be challenging. A definitive treatment of the literature with a unifying criterion can be found in Mitra (1979). A striking result due to Cass (1972) will now be stated and will be used repeatedly. A program $\mathbf{x} = (x_t)$ from $\tilde{x} > 0$ is called an *interior program* if

$$\inf_{t \geq 0} x_t > 0. \qquad (4.2.15)$$

Theorem 4.2.2. *Let* $\mathbf{x} = (x_t)$ *from* $\tilde{x} > 0$ *be an interior program and* $\mathbf{p} = (p_t)$ *be the Malinvaud prices, satisfying* (4.2.6). *It is inefficient* *if and only if*

$$\sum_{t=0}^{\infty} (1/p_t) < \infty. \tag{4.2.16}$$

By using (4.2.16), we see that the (infinite) stationary program $\langle k^*, f(k^*), c^* \rangle$ (see (4.2.3)) from k^* is efficient (but violates (4.2.8)).

4.2.2 Finite horizon: Conservation of inheritance

In this section, we focus entirely on the finite horizon programs (see Section 3.3.1 of Chapter 3) for which the terminal (end of horizon) input stock is the same as the initial input stock, i.e. *those programs that meet the conservation of inheritance property* (Section 3.8 of Chapter 3). Formally, let T be a positive integer greater than one. A *T-program* from $\tilde{x} \in (0, \bar{k})$ is a finite sequence $(x_t)_{t=0}^{t=T}$, satisfying

$$x_0 = x_T = \tilde{x} > 0; \; 0 \le x_{t+1} \le f(x_t) \quad \text{for } t \ge 0. \tag{4.2.17}$$

The (finite) consumption sequence $(c_t)_{t=1}^{T}$ generated by the *T-program* $(x_t)_{t=0}^{t=T}$ is defined by

$$c_t = f(x_{t-1}) - x_t \quad \text{for } 1 \le t \le T. \tag{4.2.18}$$

Two properties, easy to prove, need to be stated:

If $\tilde{x} \in (0, \bar{k})$ there is a T-program from $\tilde{x}$. (▲ *Exercise*: This follows from the property that $f(x) > x$ on $(0, \bar{k})$.▼)
Any T-program from $\tilde{x} \in (0, \bar{k})$ satisfies $(x_t, c_t) \le (\bar{k}, \bar{k})$ for all $t = 1, 2, \ldots, T$ (see 3.3.1).

For any $x \in (0, \bar{k})$, we define (see Section 3.3 of Chapter 3)

$$\mathcal{C}(x) = \{(c_t)_{t=1}^{T} : (c_t) \text{ is a consumption sequence generated by a}$$

$$T\text{-program } (x_t) \text{ from } x\}. \tag{4.2.19}$$

Then $\mathcal{C}(x)$ is a non-empty, closed, bounded, convex subset in R_+^T.

We also define $U : R_+^T \to R$ by

$$U(c_1, c_2, \ldots, c_T) = \sum_{t=1}^{T} u(c_t). \tag{4.2.20}$$

Then U is continuous on R_+^T (therefore on $C(x)$ and strictly concave on $C(x)$).

A *T-program* $(x_t^*)_{t=0}^{t=T}$ from $\tilde{x} \in (0, \bar{k})$ is optimal if

$$\sum_{t=1}^{T} u(c_t^*) \geq \sum_{t=1}^{T} u(c_t) \tag{4.2.21}$$

for every *T-program* $(x_t)_{t=0}^{t=T}$ from the same initial $\tilde{x}$.

Theorem 4.2.3. *If $\tilde{x} \in (0, \bar{k})$, there is a unique optimal T-program $(x_t^*)_{t=0}^{t=T}$ from $\tilde{x}$. Moreover, $(x_t^*, c_t^*) \gg 0$ for all $t = 1, 2, \ldots, T$.*

Proof. (▲ *Exercise* ▼).   □

In view of the repeated use, let us recall the derivation of the Ramsey–Euler condition. Consider for $1 \leq t \leq T - 1$, the following expression

$$V(x) \equiv u(f(x_{t-1}^*) - x) + u(f(x) - x_{t+1}^*). \tag{4.2.22}$$

It must be attaining a maximum at $x = x_t^*$ among all $x \geq 0$ that satisfy $f(x_{t-1}^*) - x \geq 0$, and $f(x) - x_{t+1}^* \geq 0$. Since $(x_t^*, c_t^*) \gg 0$ as well as $c_{t+1}^* > 0$, the first-order condition gives us:

$$V'(x^*) \equiv u'(c_t^*)(-1) + u'(c_{t+1}^*)f'(x_t^*) = 0, \tag{4.2.23}$$

leading to the *Ramsey–Euler condition*

$$u'(c_t^*) = u'(c_{t+1}^*)f'(x_t^*) \quad \text{for } 1 \leq t \leq T - 1. \tag{4.2.24}$$

In Section 4.4, we shall, for the sake of simplicity, consider the case when $T = 2$. A *2-program* from $\tilde{x} \in (0, \bar{k})$ is specified by the input triplet $(\tilde{x}, x_1, \tilde{x})$, generating the consumption pair (c_1, c_2) given by

$c_1 = f(\tilde{x}) - x_1$ and $c_2 = f(x_1) - \tilde{x}$. A convenient characterization of optimal 2-*programs* is noted here for ready reference. □

Proposition 4.2.1. *Let* $(\tilde{x}, x_1, \tilde{x})$ *be a 2-program from* $\tilde{x} \in (0, \bar{k})$, *with* $c_1 \equiv c_1 = f(\tilde{x}) - x_1 > 0$ *and* $c_2 = f(x_1) - \tilde{x} > 0$. *Then* $(\tilde{x}, x_1, \tilde{x})$ *is an optimal 2-program if and only if*

$$u'(f(\tilde{x}) - x_1) = u'(f(x_1) - \tilde{x})f'(x_1). \qquad (4.2.25)$$

Proof. Let $(\tilde{x}, x_1', \tilde{x})$ be any 2-*program* from $\tilde{x} \in (0, \bar{k})$, with associated consumptions $c_1' \equiv c_1 = f(\tilde{x}) - x_1'$ and $c_2' = f(x_1') - \tilde{x}$. Then,

$$
\begin{aligned}
&[u(c_1') + u(c_2')] - [u(c_1) + u(c_2)] \\
&\quad \leq u'(c_1)(c_1' - c_1) + u'(c_2)(c_2' - c_2) \\
&\quad = u'(c_2)[f'(x_1)(c_1' - c_1) + (c_2' - c_2)] \\
&\quad = u'(c_2)[f'(x_1)(x_1 - x_1') + [f(x_1') - f(x_1)] \\
&\quad \leq u'(c_2)[f'(x_1)(x_1 - x_1') + f'(x_1)(x_1' - x_1)] = 0.
\end{aligned}
$$

This establishes that $(\tilde{x}, x_1, \tilde{x})$ is the optimal 2-*program* from $\tilde{x} \in (0, \bar{k})$.

For the other direction, use the argument leading to the derivation of (4.2.24). □

4.3 Rolling Plans

As I stressed, we are looking at *T-programs* satisfying (4.2.17). These generate an infinite *rolling plan* which is now formally defined. From our discussion above, we can conclude that there is a function $\alpha : (0, \bar{k}) \to R_{\div}$ such that if $(x_t)_{t=0}^{t=T}$ *is the unique* optimal *T-program* from x, then $\alpha(x) = x_1$. Observe now that given any $x \in (0, \bar{k})$, $0 < \alpha(x) < f(x) < \bar{k}$. Hence, α is a mapping from $(0, \bar{k})$ into $(0, \bar{k})$.

A *rolling plan* from $\tilde{x} \in (0, \bar{k})$ is an infinite sequence (x_t) defined as

$$x_0 = \tilde{x}, x_{t+1} = \alpha(x_t) \quad \text{for } t \geq 0. \tag{4.3.1}$$

Note that a rolling plan is an infinite program from $\tilde{x}$, according to the definition (4.2.1), since $\alpha(x_t) < f(x_t)$ for $t \geq 0$. It generates a consumption sequence defined by

$$c_t = f(x_{t-1}) - \alpha(x_{t-1}) \quad \text{for } t \geq 1. \tag{4.3.2}$$

In view of Proposition 4.2.1, if $T = 2$, and α is the generating function of a rolling plan, then for every $x \in (0, \bar{k})$,

$$u'(f(x) - \alpha(x)) = u'(f(\alpha(x)) - x)f'(\alpha(x)). \tag{4.3.3}$$

4.3.1   Monotone convergence of rolling plans

In this section, we derive some monotonicity properties of rolling plans. The input sequence of a rolling plan is monotone. If the initial $\tilde{x}$ is below (above) the golden-rule input k^*, then the sequence (x_t) increases (decreases), converging to k^*. We establish the properties by focusing on the generating function of a rolling plan. Specifically, Lemma 4.3.1 shows that the graph of the generating function is above (below) at input stocks below (above) the 45-degree line. Lemma 4.3.2 summarizes the monotonicity, continuity and boundedness properties of the generating function $\alpha(x)$.

Lemma 4.3.1. *If $(x_t^*)_{t=0}^{t=T}$ is the optimal T-program from $\tilde{x} \in (0, \bar{k})$, then (a) $\tilde{x} < k^*$ implies $\tilde{x} < x_1^* < k^*$, (b) $\tilde{x} = k^*$ implies $x_1^* = k^*$, and (c) $\tilde{x} > k^*$ implies $\tilde{x} > x_1^* > k^*$.*

Proof.   We will prove (a), the proofs of (b) and (c) can be worked out analogously. We first establish that $x_1^* < k^*$. Suppose instead that $x_1^* \geq k^*$. Then, $f'(x_1^*) \leq 1$, so by the Ramsey–Euler equation, $u'(c_1^*) = f'(x_1^*)u'(c_2^*) \leq u'(c_2^*)$. Since $u''(c) < 0$ for $c > 0$, we get

$c_1^* \geq c_2^*$. Thus,

$$f(\tilde{x}) - x_1^* \geq f(x_1^*) - x_2^* > f(\tilde{x}) - x_2^*,$$

the last inequality following from $x_1^* \geq k^* > \tilde{x}$, and the fact that f is increasing. Thus, $x_2^* > x_1^* \geq k^*$. We can then repeat the steps to obtain

$$x_T^* > x_{T-1}^* > \cdots > x_2^* > x_1^* > \tilde{x},$$

so that $x_T^* > \tilde{x}$, which contradicts the fact that $x_T^* = \tilde{x}$ by definition of a T-program. Thus, $x_1^* < k^*$.

Next, we establish that $x_1^* > \tilde{x}$. Suppose instead that $x_1^* \leq \tilde{x}$. Since $x_1^* < k^*$, we have $f'(x_1^*) > 1$; so, by the Ramsey–Euler equation, $u'(c_1^*) = f'(x_1^*)u'(c_2^*) > u'(c_2^*)$. Thus, $c_1^* < c_2^*$ and

$$f(\tilde{x}) - x_1^* < f(x_1^*) - x_2^* \leq f(\tilde{x}) - x_2^*.$$

Hence, $x_2^* < x_1^*$. We can then repeat the steps to obtain

$$x_T^* < x_{T-1}^* < \cdots < x_2^* < x_1^* \leq \tilde{x},$$

so that $x_T^* < \tilde{x}$, which contradicts the fact that $x_T^* = \tilde{x}$. Thus, $x_1^* > \tilde{x}$. $\qquad\square$

We note a comparative dynamic property with respect to a change in the initial stock.

Lemma 4.3.2. *If $(x_t^*)_{t=0}^{t=T}$ is the unique optimal T-program from $\tilde{x} \in (0, \bar{k})$ and $(\bar{x}_t)_{t=0}^{T}$ is the unique optimal T-program from $\bar{x} \in (0, \bar{k})$, and $\tilde{x} > \bar{x}$, then*

$$x_1^* > \bar{x}_1. \qquad (4.3.4)$$

Proof. Suppose the hypotheses of Lemma 4.3.2 are valid, but $x_1^* \leq \bar{x}_1$. Then, we have that $c_1^* = f(\tilde{x}) - x_1^* > f(\bar{x}) - \bar{x}_1 = \bar{c}_1$. Thus, we must have $f'(x_1^*) \geq f'(\bar{x}_1)$ and $u'(c_1^*) < u'(\bar{c}_1)$. Using the

Ramsey–Euler equations for the two optimal T-programs, we obtain

$$1 > \frac{u'(c_1^*)}{u'(\bar{c}_1)} = \frac{f'(x_1^*)u'(c_2^*)}{f'(\bar{x}_1)u'(\bar{c}_2)} \geq \frac{u'(c_2^*)}{u'(\bar{c}_2)}.$$

This means $u'(c_2^*) < u'(\bar{c}_2)$, and so, $c_2^* > \bar{c}_2$. Thus, we obtain

$$f(x_1^*) - x_2^* > f(\bar{x}_1) - \bar{x}_2) \geq f(x_1^*) - \bar{x}_2,$$

so that $\bar{x}_2 > x_2^*$. The above argument can then be repeated to get $\bar{x}_t > x_t^*$ for $t = 2, \ldots, T$. Thus, by definition of T-program, we obtain $\tilde{x} = x_T^* < \bar{x}_T = \bar{x}$, which contradicts the hypothesis that $\tilde{x} > \bar{x}$. □

The following result provides a list of properties of the generating function of a rolling plan.

Theorem 4.3.1. *The generating function $\alpha : (0, \bar{k}) \to (0, \bar{k})$ has the following properties:*

(a) *for $x \in (0, \bar{k})$, $0 < \alpha(x) < f(x) < \bar{k}$;*
(b) *for $x \in (0, \bar{k})$, k^* $(>, =, <)$ $\alpha(x)$ according as k^* $(>, =, <)$ x;*
(c) *$\alpha(x)$ is increasing on $(0, \bar{k})$;*
(d) *$\alpha(x)$ is continuous on $(0, \bar{k})$;*
(e) *$\alpha(x) \to 0$ as $x \to 0$; $\alpha(x) \to \bar{k}$ as $x \to \bar{k}$.*

Proof. See Section 4.6.   □

The next result follows immediately from Theorem 4.3.1.

Theorem 4.3.2. *If (x_t) is a rolling pan from $\tilde{x} \in (0, \bar{k})$, then*

(a) *$\tilde{x} < k^*$ implies that x_t monotonically increases to k^* as $t \to \infty$;*
(b) *$\tilde{x} > k^*$ implies that x_t monotonically decreases to k^* as $t \to \infty$;*
(c) *$\tilde{x} = k^*$ implies that $x_t = k^*$ for all $t \geq 1$.*

Proof. We establish (a); the statements (b) and (c) can be proved in a similar manner. If $\tilde{x} < k^*$, then $\tilde{x} < \alpha(\tilde{x}) < k^*$ by Theorem 4.3.1(b), so that

$$\tilde{x} < x_1 < k^*.$$

Using Theorem 4.3.1(b) again, $x_1 < \alpha(x_1) < k^*$, so that

$$x_1 < x_2 < k^*.$$

Repeating this step, we conclude that x_t monotonically increases while remaining bounded above by k^*. Hence, it converges to some $\check{k}$, satisfying

$$0 < \check{k} \leq k^*.$$

But, using the continuity of α (Theorem 4.3.1(d)) and taking the limit as $t \to \infty$,

$$x_{t+1} = \alpha(x_t) \text{ implies that } \check{k} = \alpha(\check{k}).$$

Now, by Theorem 4.3.1(c), $\check{k} = k^*.$ □

We conclude this section by presenting an example where the generating function can be numerically computed.

Example 4.3.1. Let the production function be given by

$$f(x) = 2x^{1/2} \quad \text{for } x \geq 0.$$

Then f satisfies (F.1)–(F.4); the golden-rule input stock $k^* = 1$, and the maximum sustainable input stock $k = 4$. Let the utility function be given by

$$u(c) = c^{1/2} \quad \text{for } c \geq 0$$

Then u satisfies (U.1)–(U.2). Let the planning horizon be fixed at $T = 2$.

A 2-program from $x \in (0, 4)$ is then a vector (x, x_1, x), with $0 \leq x_1 \leq f(x)$ and $x \leq f(x)$. If (x, x_1, x) is an optimal 2-program from $x \in (0,4)$, then using Ramsey–Euler equations, we get

$$2x^{1/2} - x_1 + xx_1 = 2x_1^{3/2}.$$

Denoting $x^{1/2}$ by β, and $x^{1/2}$ by γ, we get

$$2\gamma^3 + (1 - \beta^2)\gamma^2 - 2\beta = 0, \qquad (4.3.5)$$

where $\beta \in (0,2)$.

Given $\beta \in (0,2)$, equation (4.3.5) is a cubic in γ. Since it is of odd degree, with the least coefficient negative $(-2\beta < 0)$ and the first coefficient positive $(2 > 0)$, it has *at least* one positive real root. On the other hand, by Descartes' rule of signs, it has *at most* one positive real root (since regardless of the sign of $(1 - \beta^2)$, there is exactly one change of sign in the equation). Thus, there is exactly one positive real root to equation (4.3.5). If we call this root $\phi(\beta)$, then the generating function, α, is given by

$$\alpha(x) = [\phi(x^{1/2})]^2. \qquad (4.3.6)$$

While our interest is naturally in the unique positive root of equation (4.3.5), we note that *all* the roots of the equation can be found by the standard Cardan–Tartaglia method or the trigonometric method, depending on the sign of the discriminant (see Birkhoff and MacLane (1977, Chapters 4 and 5) for details). □

4.4   Welfare Properties of Rolling Plans

The purpose of this section is to show that rolling plans are optimal according to the definition of optimality we introduced (see Section 4.2). For ease of exposition, in this section, we fix the time horizon of *T-programs* at $T = 2$. The more general case can be handled by invoking the monotonicity properties of optimal *T-programs*.

Thus, we first show that rolling plans are *efficient* (Theorem 4.4.1) and *"good"* (Theorem 4.4.2). Since good programs always maximize the long-run average utility, these two results are combined to establish the optimality of rolling plans (Theorem 4.4.3).

Theorem 4.4.1. *Let (x_t) be a rolling plan from $\tilde{x} \in (0, \bar{k})$. Then (x_t) is efficient.*

Proof. If $\tilde{x} \leq k^*$, then $\tilde{x} < x_t < k^*$ for all $t \geq 0$. Efficiency follows from Theorem 4.1.2. If $\tilde{x} > k^*$, the calculations are spelled out in Section 4.6. $\square$

Following Gale (1967), we define a program (x_t) to be *good if there is a real number* **B**, such that

$$\sum_{t=1}^{\tau} u(c_t) - u(c^*) \geq \mathbf{B} \quad \text{for all } \tau \geq 1. \tag{4.4.1}$$

Theorem 4.4.2. *Suppose that (x_t) is a rolling plan from $\tilde{x} \in (0, \bar{k})$. Then (x_t) is good.*

Proof. See Section 4.6. $\square$

We now come to the main result of this chapter.

Theorem 4.4.3. *Suppose that (x_t) is a rolling plan from $\tilde{x} \in (0, \bar{k})$. Then (x_t) is optimal.*

Proof. Recall that $c^* = f(k^*) - k^*$ is the golden-rule consumption. Denote $u'(c^*)$ by p^*. Let (x'_t) be any program from $\tilde{x}$. Then, for $t \geq 1$,

$$[u(c'_t) - u(c^*)] \leq p^*[c'_t - c^*]$$
$$= p^*[\{f(x'_{t-1}) - x'_t\} - \{f(k^*) - k^*\}]$$
$$= p^*[f(x'_{t-1}) - f(k^*)] - p^*[x'_t - k^*]$$
$$\leq p^*[x'_{t-1} - k^*] - p^*[x'_t - k^*].$$

We use concavity and the fact that $f'(k^*) = 1$ to derive the inequalities above. Now, summing from $t = 1$ to $t = \tau$, we get

$$\sum_{t=1}^{\tau} [u(c'_t) - u(c^*)] \leq p^* \tilde{x}. \tag{4.4.2}$$

Hence, for $\tau \geq 1$,

$$(1/\tau) \sum_{t=1}^{\tau} u(c_t') \leq u(c^*) + (p^* \tilde{x})/\tau. \tag{4.4.3}$$

Taking the inferior limit of both sides in (4.4.3), we get

$$\liminf_{\tau \to \infty} (1/\tau) \sum_{t=1}^{\tau} u(c_t') \leq u(c^*). \tag{4.4.4}$$

Using Theorem 4.4.2, we know that the rolling plan (x_t) is good. Thus, there is some real number **B**, such that for all $\tau \geq 1$,

$$\sum_{t=1}^{\tau} [u(c_t) - u(c^*)] \geq \mathbf{B}. \tag{4.4.5}$$

The inequality (4.4.5) yields, for all $\tau \geq 1$,

$$(1/\tau) \sum_{t=1}^{\tau} u(c_t) \geq u(c^*) + (\mathbf{B}/\tau). \tag{4.4.6}$$

Again, taking the inferior limit of both sides,

$$\liminf_{\tau \to \infty} (1/\tau) \sum_{t=1}^{\tau} u(c_t) \geq u(c^*). \tag{4.4.7}$$

Combining (4.4.4) and (4.4.7),

$$\liminf_{\tau \to \infty} (1/\tau) \sum_{t=1}^{\tau} u(c_t) \geq \liminf_{\tau \to \infty} (1/\tau) \sum_{t=1}^{\tau} u(c_t'). \tag{4.4.8}$$

Using Theorem 4.4.1, the rolling plan (x_t) is also efficient. Hence, the rolling plan (x_t) is optimal. $\qquad\qquad\square$

Remarks:

1. The fact that a good plan maximizes long-run average utility among all plans has already been noted in the literature (see, e.g., Jeanjean (1974)). We have given the proof here for the sake of completeness.

2. A plan which maximizes long-run average utility need not be good. Consider $0 < x < k^*$, satisfying $f(x) > k^*$, and a sequence (x_t) defined by $x_t = x$ for $t = 2^n$ $(n = 0, 1, 2, \ldots)$ and $x_t = k^*$ for $t \neq 2^n$. It can be checked that (x_t) is a plan from x which maximizes long-run average utility and is also efficient, but it is not good. Thus, Theorems 4.4.1 and 4.4.2 actually establish a stronger welfare result about rolling plans than that is reflected in Theorem 4.4.3.

3. The properties of efficiency and "goodness" of a plan are independent of each other. An efficient plan need not be good (see the example in remark 2 above). Similarly, a good plan need not be efficient. Consider $k^* < x < \bar{k}$, and a sequence (x_t) defined by $x_0 = x$, $x_{t+1} = f(x_t) - c^*$ for $t \geq 0$. It can be checked that (x_t) is a plan. It is clearly good, since $c_t = c^*$ for all $t \geq 1$. It is also clearly inefficient since the sequence (x'_t) defined by $x'_0 = x$, $x'_t = k^*$ for $t \geq 1$ is a plan from x with $c_1 > c^*$ and $c_t = c^*$ for all $t \geq 2$.

4.5   Decentralization

It will take us too far to introduce the issues involving decentralization in infinite horizon economies. Informally, the rolling plan that was explored can be obtained through an appropriately designed decentralized resource allocation mechanism (and the mechanism is "evolutionary" in the spirit of Hurwicz and Weinberger (1990)). To show this, Bala, Majumdar and Mitra (1991) constructed a sequence of accounting prices $(\check{p}_t)$ relative to which the intertemporal profit maximization condition holds on the "production side". On the consumption side, there is equality of an appropriate marginal rate of substitution with $\check{p}_t$. But at these prices, the "competitive" condition on the consumption side (see condition (i) in Section 3.4 of Chapter 3) is not met (unless $\tilde{x}$ happens to be k^*). This is because the "second period consumption" of, say, the optimal 2-*plan*, formulated in period t is not carried out in period $t + 1$; instead, the "first period consumption" of the maximal 2-*plan* formulated in period $t + 1$ is carried out in period $t + 1$. Needless to say, the structure of the model and the complete characterization of the optimal 2-*plan* need to be exploited to arrive at this decisive result.

4.6 Complements and Details

This chapter is based entirely on Bala, Majumdar and Mitra (1991). The concept of a decentralized evolutionary mechanism was introduced in Hurwicz and Weinberger (1990). Both papers are available in the collection edited by Majumdar (2016). Maximization of "the long-run average reward" (or criteria that are similar) was studied in the dynamic programming literature (see Howard (1960), Blackwell (1962), Flynn (1976) and Bhattacharya and Majumdar (1989a,b)). Now, we provide the details of some of the proofs.

Proof of Theorem 4.3.1. Clearly, (a) follows from our discussion in Section 4.3. Also, (b) follows from Lemma 4.3.1 and (c) follows from Lemma 4.3.2.

To establish (d), we apply the Maximum Theorem (see Berge (1963, p. 116)). Define $D = (0, \bar{k})$, $\bar{D} = [0, \bar{k}]$; then $\bar{D}^T$ is a compact subset of R^T. Note that $U(c_1, \ldots, c_T)$ is a continuous function from $\bar{D}^T$ to R. Also, $\mathcal{C}(x)$ is a continuous correspondence from D to $\bar{D}^T$. Denoting by $(c_1(x), \ldots, c_T(c))$ the (unique) maximizer of U on $\mathcal{C}(x)$ for each $x \in D$, we note that $(c_1(x), \ldots, c_T(x))$ is a continuous function from D to $\bar{D}_T$. Denoting by $(x_t(x))_{t=0}^T$ the T-program from x, with associated consumption sequence $(c_t(x))_{t=1}^T$, we note that $(x_1(x), \ldots, x_T(x))$ is also a continuous function from D to $\bar{D}$. In particular, then, $\alpha(x) \equiv x_1(x)$ is a continuous function on D.

We can establish (e) as follows. For $0 < x < k^*$, we have, by (a) and (b), $x < \alpha(x) < f(x)$. Thus, as $x \to 0$, $f(x) \to 0$ by continuity of f and $f(0) = 0$, so that $\alpha(x) \to 0$.

If (f) were violated, then there would exist a sequence (x^s), $s = 1, 2, \ldots$, such that $x^s \to \bar{k}$ as $s \to \infty$ and $\alpha(x^s) \to k' < \bar{k}$ (using (c), $k' \geq k^*$). Clearly, $f^{T-1}(k') < k$ (where f^{T-1} is the $(T-1)$ iteration of the function, f), and so we must have, for s large, $f^{T-1}(\alpha(x^s)) < x^s$, which contradicts the definition of α. $\qquad \square$

We first introduce some notation which will ease the writing of our subsequent results and proofs. Recall that $c^* \equiv f(k^*) - k^* > 0$. We write

$$\theta = \min \left[\{c^*/4f'(k^*/2)\}, (k^*/2)\right],$$
$$\hat{\Theta} = (k^* - \theta, k^* + \theta). \tag{4.6.1}$$

It follows from (4.6.1) that $k^* - \theta > 0$, and $0 < c^* - \theta = f(k^*) - (k^* + \theta) < \bar{k} - (k^* + \theta)$; thus, $k^* + \theta < \bar{k}$. Consequently, the set $\hat{\Theta}$ is an open sub-interval of $(0, \bar{k})$. Note that if $(x, z) \in \hat{\Theta}^2$, then $f(x) - z \geq f(k^* - \theta) - f(k^*) + f(k^*) - (k^* + \theta) \geq f'(k^* - \theta)(-\theta) + c^* - \theta > (3c^*/4) - \theta f'(k^*/2) \geq (c^*/2) > 0$.

Lemma 4.6.1. *There is $\delta \in (0, \theta)$ (where θ is given by (4.6.1)), such that the generating function α is continuously differentiable on the set*

$$\Delta \equiv [k^* - \delta, k^* + \delta], \tag{4.6.2}$$

and $0 < \alpha'(x) < 1$ for all $x \in \Delta$.

Proof. Consider the function $\gamma : \hat{\Theta}^2 \to R$ defined by

$$\gamma(x, z) = u'(f(x) - z) - u'(f(z) - x)f'(z).$$

Then γ is continuously differentiable on $\hat{\Theta}^2$. Furthermore, $\gamma(k^*, k^*) = 0$ and $\gamma_z(k^*, k^*) < 0$. So, by the Implicit Function Theorem, there are open neighborhoods N_0 and N_1 of k^* (where both N_0 and N_1 are subsets of $\hat{\Theta}$), and a unique function $L : N_0 \to N_1$, such that $k^* = L(k^*)$, and $\gamma(x, L(x)) = 0$ for all $x \in N_0$. Furthermore, L is continuously differentiable on N_0.

Since N_0 and N_1 are open, we can find $\theta > \delta' > 0$ such that $N' = (k^* - \delta', k^* + \delta')$ is a subset of N_0 and N_1. Since $h(k^*) = k^*$, and $h(\cdot)$ is continuous on $(0, \bar{k})$ by Theorem 4.3.1, we can find $0 < \hat{\delta} < \delta'$, such that for $x \in \hat{N} \equiv (k^* - \hat{\delta}, k^* + \hat{\delta})$, $\alpha(x) \in N'$.

Now, we observe that by the definition of α, $\gamma(x, \alpha(x)) = 0$ for all $x \in \hat{N}$ (see (4.3.3)). By the Implicit Function Theorem (above), this is possible if and only if $\alpha(x) = L(x)$ for all $x \in \hat{N}$. This proves that in the neighborhood $\hat{N}$ of k^*, α is continuously differentiable.

Evaluating the derivative of α on $\hat{N}$,

$$\alpha'(x) = \frac{u''(f(x) - z)f'(x) + u''(f(z) - x)f'(z)}{u''(f(x) - z) + u''(f(z) - x)((f'(z))^2 + u'(f(z) - x)f''(z)},$$

where $z = \alpha(x)$. Thus, evaluating the derivative of α at k^*, we get

$$\alpha'(k^*) = 2u''(c^*)/[2u''(c^*) + u'(c^*)f''(k^*)].$$

Using the facts that $u'(c^*) > 0$, $f''(k^*) < 0$ and $u''(c^*) < 0$, we obtain $0 < \alpha'(k^*) < 1$.

Since α' is continuous on $\hat{N}$, and $0 < \alpha'(k^*) < 1$, we can find $0 < \delta < \hat{\delta}$, such that on $\Delta \equiv [k^* - \delta, k^* + \delta]$, we have $0 < \alpha'(x) < 1$. Clearly, $\delta \in (0, \theta)$.   □

Lemma 4.6.2. *Suppose (x_t) is a rolling plan from $x \in (0, k)$. Then there is $\mathbf{A} > 0$, a positive integer S, and $\rho \in (0, 1)$, such that $x_t \in \Delta$ for $t \geq S$ (where Δ is given by Lemma 4.6.1), and*

$$|x_t - k^*| \leq \mathbf{A}\rho^t \quad \text{for } t \geq S.$$

Proof.   Consider the interval Δ obtained in Lemma 4.6.1. We know that α' is continuous on Δ, and $0 < \alpha'(x) < 1$ for all $x \in \Delta$; then $0 < \rho < 1$.

By Theorem 4.3.2, there is a positive integer S, such that $x_t \in \Delta$ for all $t \geq S$. Since α is continuously differentiable on Δ, we can use the Mean Value Theorem for each $t \geq S$ to obtain

$$|x_{t+1} - k^*| = |\alpha(x_t) - \alpha(k^*)| = |\alpha'(z_t)||x_t - k^*|,$$

where z_t is between x_t and k^*. Since $z_t \in \Delta$, we obtain $0 < \alpha'(z_t) \leq \rho < 1$. Thus, for each $t \geq S$,

$$|x_{t+1} - k^*| \leq \rho|x_t - k^*|.$$

Iterating this inequality, we obtain for $t \geq S$

$$|x_{t+1} - k^*| \leq \rho^{t-S}|x_S - k^*| \leq \rho^{t-S}\delta,$$

where δ is given by Lemma 4.5.1. Defining $\mathbf{A} = (\delta/\rho^S)$, we have $|x_{t+1} - k^*| \leq \mathbf{A}\rho^t$ for $t \geq S$, which proves the lemma.   □

Proof of Theorem 4.4.1. We now consider the case where $\tilde{x} > k^*$. Consider the set $\Delta = [k^* - \delta, k^* + \delta]$ defined in (4.6.2).

Using Lemma 4.6.2, we can find $\mathbf{A} > 0$, a positive integer S, and $\rho \in (0,1)$, such that $x_t \in \Delta$ for $t \geq S$, and

$$|x_t - k^*| \leq \mathbf{A}\rho^t \quad \text{for } t \geq S.$$

Let m be the maximum value of $[-f''(x)]$ on Δ. Choose $s \geq S$, such that $[m\mathbf{A}\rho^S/(1-\rho)] \leq 1/2$. Now, for $t \geq s$, we have $x_t \in \Delta$, and so by the Mean Value Theorem, $f'(x_t) - f'(k^*) = f''(z_t)(x_t - k^*)$, where $x_t \geq z_t \geq k^*$. Since, $z_t \in \Delta$, we have $[-f''(z_t)] \leq m$. Thus, for $t \geq s$, $f'(x_t) \geq 1 - m(x_t - k^*) \geq 1 - m\mathbf{A}\rho^t = 1 - (m\mathbf{A}\rho^s)\rho^{t-s} \geq 1 - [(1-\rho)\rho^{t-s}/2]$. Using the information, we obtain for $t \geq s$,

$$\pi_{t+1} = \prod_{n=0}^{t} f'(x_n) \geq \pi_s \prod_{n=s}^{t} \{1 - [(1-\rho)\rho^{n-s}/2]\}$$

$$\geq \pi_s \left\{1 - \sum_{n=s}^{t}[(1-\rho)\rho^{n-s}/2]\right\}$$

$$\geq \pi_s \left\{1 - \sum_{n=s}^{\infty}[(1-\rho)\rho^{n-s}/2]\right\}$$

$$= \pi_s\{1 - (1/2)\} = \frac{\pi_s}{2}.$$

Thus, $\sum_{n=0}^{\infty} \pi_{n+1}$ is divergent. Since $x_t \geq k^*$ for $t \geq 0$, we can again use Theorem 4.1.2 to conclude that (x_t) is efficient. $\qquad \square$

Proof of Theorem 4.4.2. Using Lemma 4.6.2, there exist $\mathbf{A} > 0$, a positive integer S and $\rho \in (0,1)$, such that $x_t \in \Delta$ (where Δ is given by (4.6.2)) and $|x_t - k^*| \leq \mathbf{A}\rho^t$ for $t \geq S$. Let b be the maximum value of f' on Δ. For $t \geq S$, (x_t, x_{t+1}) is in Δ^2 and so in $\hat{\Theta}^2$, where $\hat{\Theta}$ is given by (4.6.1). Thus, $f(x_t) - x_{t+1} \geq (c^*/2)$ for $t \geq S$.

Consider first the case where $x < k^*$. In this case, by Theorem 4.3.2, $0 \leq (k^* - x_t) \leq \mathbf{A}\rho^t$ for $t \geq S$. Then, for $t \geq S$, we obtain $c^* - c_{t+1} = [f(k^*) - k^*] - [f(x_t) - x_{t+1}] \leq [f(k^*) - f(x_t)] \leq f'(x_t)[k^* - x_t] \leq b\mathbf{A}\rho^t$. This information yields for $t \geq S$, $u(c^*) - u(c_{t+1}) \leq u'(c_{t+1})(c^* - c_{t+1}) \leq u'(c^*/2)b\mathbf{A}\rho^t$. Summing this inequality from S to ∞, we get

$$\sum_{t=S}^{\infty}[u(c^*) - u(c_{t+1}] \leq u'(c^*/2)b\mathbf{A}\rho^S/(1-\rho).$$

It follows immediately that (x_t) is good.

Consider the next case where $x \geq k^*$. By Proposition 4.3.2, $0 \leq (x_t - k^*) \leq \mathbf{A}\rho^t$ for $t \geq S$. Then for $t \geq S$, we obtain $c^* - c_{t+1} = [f(k^*) - k^*] - [f(x_t) - x_{t+1}] \leq (x_{t+1} - k^*) \leq \mathbf{A}\rho^{t+1}$. This yields for $t \geq S$

$$u(c^*) - u(c_{t+1}) \leq u'(c_{t+1}(c^* - c_{t+1}) \leq u'(c^*/2)\mathbf{A}\rho^{t+1}.$$

Then, following the above procedure, (x_t) is good. $\square$

Chapter 5

Infinite Horizon Models: Discounting and Sustainability

> We do not discount later enjoyments in comparison with earlier ones, a practice which is ethically indefensible and arises primarily from the weakness of imagination.
>
> — Frank Ramsey

5.1 Introduction

Infinite horizon models have provided us with intellectually satisfying (and often analytically challenging!) frameworks to deal with "long-run" problems of an economy evolving over time with no predetermined terminal date. In this and the following chapter, we study a class of discrete-time infinite horizon models. In contrast with our exposition in Chapter 3, we allow for biological reproduction functions that may exhibit an initial phase of increasing marginal productivity (a Knightian regeneration function). The optimization problem that we pose takes us beyond "convex analysis" and forces us to exploit the structure of the models in order to derive qualitative properties of optimal harvesting policies. The properties of the regeneration function are first spelled out carefully. Following the tradition of the literature on optimal savings, it is useful to distinguish between the "undiscounted" and "discounted" cases as the total return from a harvesting program is the "sum" of one period returns. In the undiscounted case, the major analytical results deal with the existence of an optimal program (according to the Ramsey criterion) and its convergence to the optimal stationary input ("the golden-rule" input). In the model studied here, "an optimal program" does not lead to extinction.

In the discounted case, we review results on the existence and monotonicity properties of an optimal program. The conflict between impatience and sustainability emerges clearly. We derive two ranges of the discount factors. In the first range, where the values of discount factors are "close to 1" (this captures the intuitive idea of "mild discounting"), the qualitative behavior of any optimal program is similar to that in the undiscounted case, and its consumption approaches a steady state (modified golden-rule consumption). On the contrary, there is a range of values of the discount factors "close to 0" ("heavy discounting") where optimal consumptions, inputs and outputs all converge to 0 (irrespective of initial stocks). This is a case of inevitable extinction. In the intermediate range (studied in detail in Chapter 6), the qualitative behavior depends crucially on the initial stock, and "tipping points" emerge. Corresponding to a discount factor in the intermediate range, there is a critical stock, such that optimal programs from all initial stocks below the critical level lead to extinction, whereas optimal programs from initial stocks above the critical level approach a modified golden-rule program.

5.2   The Model: A Knightian Regeneration Function

We go beyond the convex optimization problems of Chapters 3 and 4. The following assumptions on the regeneration function $f : R_+ \to R_+$ are maintained throughout this chapter and Chapter 6.

(F.1) $f(0) = 0$.
(F.2) $f(x)$ *is strictly increasing in* $x \geq 0$.
(F.3) $f(x)$ *is twice continuously differentiable at* $x \geq 0$.
(F.4) f *satisfies the following endpoint condition:*

$$f'(\infty) < 1 < f'(0) < \infty.$$

(F.5) *There is a (finite) number $k_1 > 0$, such that* (i) $f''(x) = 0$ *for* $x = k_1$, (ii) $f''(x) > 0$ *for* $0 \le x \le k_1$, *and* (iii) $f''(x) < 0$ *for* $x > k_1$.

In contrast to the present model, in the traditional convex framework, we replace (F.5) by:

(F.5′) $f(x)$ *is strictly concave on* R_+, *with* $f''(x) < 0$ *for* $x > 0$ *while preserving* (F.1)–(F.4).

(In some versions, (F.3) and (F.4) are also modified to allow $f'(0) = \infty$ (the Uzawa condition), i.e., we assume that f is twice continuously differentiable at $x > 0$, $\lim_{x \to 0} f'(x) \to \infty$ and $\lim_{x \to \infty} f'(x) < 1$.)

In the discussions to follow, we will find it convenient to refer to a model with assumptions (F.1)–(F.4) and (F.5′) as "classical" and to a model with (F.1)–(F.5) as "non-classical".

We define a function, A (representing the *average product function*), as follows:

$$A(x) = f(x)/x \text{ for all } x > 0; \quad A(0) = \lim_{x \to 0}[f(x)/x] = f'(0). \quad (5.2.1)$$

Under (F.1)–(F.5), it is easily checked that $A(0) = f'(0)$.

Under (F.1)–(F.5), there exist uniquely determined finite numbers $k^*, \bar{k}$ and k_2, satisfying

(i) $0 < k_1 < k_2 < k^* < \bar{k}$;
(ii) $f'(k^*) = 1$;
(iii) $f'(k_2) = A(k_2)$;
(iv) $f(\bar{k}) = \bar{k}$.

Recall that k^* is the golden-rule input and $\bar{k}$ is the maximum sustainable stock.

The stock levels $(k_1, k_2, k^*, \bar{k})$ are of critical importance in the qualitative analysis that follows. Furthermore, observe that

(a) for $0 \le x < k^*$, $f'(x) > 1$, and for $x > k^*$, $f'(x) < 1$;
(b) for $0 < x < \bar{k}$, $x < f(x) < \bar{k}$, and for $x > \bar{k}$, $\bar{k} < f(x) < x$;

(c) for $0 < x < k_2$, $f'(x) > A(x)$, and for $x > k_2$, $f'(x) < A(x)$;

(d) for $0 \leq x < k_2$, $A(x)$ is increasing, and for $x > k_2$, $A(x)$ is decreasing;

(e) for $0 \leq x < k_1$, $f'(x)$ is increasing and for $x > k_1$, $f'(x)$ is decreasing.

The function f together with the numbers k_1, k_2, k^* and $\bar{k}$ may be represented diagrammatically as follows in Fig. 5.1.

5.2.1   Optimality, efficiency and competitive prices

To avoid undue repetitions, we recall only a few basic definitions, appropriately modified to deal with the infinite horizon. The reader is advised to review Section 4.2 of Chapter 4 at this stage. An infinite program from an initial stock $\tilde{x} > 0$ is an infinite sequence $\mathbf{x} = (x_t)$, satisfying

$$x_0 = \tilde{x}, 0 \leq x_{t+1} \leq f(x_t) \quad \text{for } t \geq 0. \tag{5.2.2}$$

It is referred to as an infinite sequence of inputs, generating a corresponding output sequence $\mathbf{y} = (y_t)$ defined as

$$y_t = f(x_{t-1}) \quad \text{for } t \geq 1. \tag{5.2.3}$$

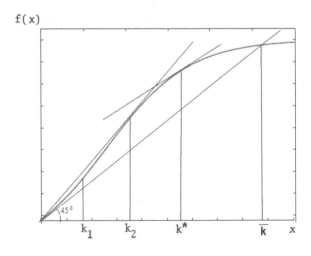

Fig. 5.1. A Knightian production function.

The consumption sequence generated by $\mathbf{x} = (x_t)$ is defined as

$$c_t = f(x_{t-1}) - x_t \quad \text{for } t \geq 1. \tag{5.2.4}$$

We assume "free disposal": if $(\mathbf{x}, \mathbf{y}, \mathbf{c})$ is any (infinite) program from $\tilde{x}$, then for any sequence $\acute{\mathbf{c}} = (\acute{c}_t)$ satisfying $0 \leq \acute{c}_t \leq c_t$, $(\mathbf{x}, \mathbf{y}, \acute{\mathbf{c}})$ is also a program from $\tilde{x}$.

As before, we restrict $\tilde{x} \in (0, \bar{k})$. This means that if $(\mathbf{x}, \mathbf{y}, \mathbf{c})$ is any (infinite) program, we have the following bounds:

$$0 \leq x_t \leq \bar{k},$$
$$0 \leq y_t \leq \bar{k}, \tag{5.2.5}$$
$$0 \leq c_t \leq \bar{k}.$$

The planner or the resource manager has a return function $u : R_+ \to R_+$. As in Chapter 3, the following assumptions on u are maintained:

(U.1) $u(c)$ *is continuous at $c \geq 0$, and twice continuously differentiable with $u'(c) > 0$ and $u''(c) < 0$ at $c > 0$.*
(U.2) $u'(c) \to \infty$ *as $c \to 0$.*

We set $u(0) = 0$.

The discount factor $\delta \in (0, 1]$. We refer to the case $\delta = 1$ as the *undiscounted* case, and the case $0 < \delta < 1$ as the *discounted* case. Observe that in the discounted case, for *any* program $(\mathbf{x}, \mathbf{y}, \mathbf{c})$ from $\tilde{x} \in (0, \bar{k})$,

$$\sum_{t=1}^{\infty} \delta^{t-1} u(c_t) \leq \sum_{t=1}^{\infty} \delta^{t-1} u(\bar{k}) \leq u(\bar{k})[1/(1-\delta)]. \tag{5.2.6}$$

An infinite program $(\mathbf{x}, \mathbf{y}, \mathbf{c})$ from $\tilde{x}$ is *competitive* if there is a (non-zero) non-negative sequence of Gale prices $\mathbf{p} = (p_t)$, such that

(i) $\delta^{t-1} u(c_t) - p_t \cdot c_t \geq \delta^{t-1} u(c) - p_t \cdot c$ for all $c \geq 0$ and for $t \geq 1$;
(ii) $p_{t+1} \cdot f(x_t) - p_t \cdot x_t \geq p_{t+1} \cdot f(x) - p_t \cdot x$ for all $x \geq 0$, and for $t \geq 0$.

A competitive program is denoted by $(\mathbf{x}, \mathbf{y}, \mathbf{c}; \mathbf{p})$. It has two important properties: the sequence of prices is strictly positive: $\mathbf{p} \gg \mathbf{0}$, and (ii) the sequence of consumptions is also strictly positive: $\mathbf{c} \gg \mathbf{0}$ (see Section 3.4 of Chapter 3). In other words, a competitive program $(\mathbf{x}, \mathbf{y}, \mathbf{c}; \mathbf{p})$ is necessarily a positive program.

We now define the criterion for evaluation alternative programs. A program $(\mathbf{x}^*, \mathbf{y}^*, \mathbf{c}^*)$ from $\tilde{x}$ is *optimal* if

$$\limsup_{T \to \infty} \sum_{t=1}^{T} \delta^{t-1} [u(c_t) = u(c_t^*)] \leq 0 \qquad (5.2.7)$$

for every program $(\mathbf{x}, \mathbf{y}, \mathbf{c})$ from $\tilde{x}$.

A positive program $(\mathbf{x}, \mathbf{y}, \mathbf{c})$ from $\tilde{x}$ is an Euler program if

$$u'(c_t) = \delta \cdot f'(x_t) u'(c_{t+1}) \quad \text{for all } t \geq 1. \qquad (5.2.8)$$

An Euler stationary (equitable) program (ESP) from $\tilde{x}$ is a stationary program which is also an Euler program. Recall that an optimal stationary program (OSP) from $\tilde{x}$ is a stationary program from $\tilde{x}$ that is also an optimal program from $\tilde{x}$.

Lemma 5.2.1. (i) *If $(\mathbf{x}^*, \mathbf{y}^*, \mathbf{c}^*)$ is an optimal program from $\tilde{x} > 0$, then it is an Euler program.* (ii) *If $(\mathbf{x}, \mathbf{y}, \mathbf{c}, \mathbf{p})$ is a competitive program from $\tilde{x} > 0$, then it is an Euler program and*

$$[f(x_t)/x_t] \geq f'(x_t) \quad \text{for } t \geq 0. \qquad (5.2.9)$$

Proof. To prove (i), observe that by (U.2), $c_t^* > 0$ for $t \geq 1$, so $(x_t^*, y_{t+1}^*) \gg 0$ for $t \geq 0$. For each $t \geq 1$, the expression,

$$u[f(x_{t-1}^*) - x] + \delta u[f(x) - x_{t+1}^*],$$

is maximized at $x = x_t^*$ among all $x \geq 0$, satisfying $f(x_{t-1}^*) \geq x$ and $f(x) \geq x_{t+1}^*$. Since the maximum is at an interior point,

$$u'(c_t^*) = \delta u'(c_{t+1}^*) \cdot f'(x_t^*) \quad \text{for } t \geq 1.$$

To prove (ii), recall that $\mathbf{p} \gg 0$. Then, using the definition of a competitive program, $p_{t+1} f'(x_t^*) = p_t$ for $t \geq 0$, and

$\delta^{t-1}u'(c_t^*) = p_t$ for $t \geq 1$. Hence, for $t \geq 1$,

$$u'(c_t^*) = \delta u'(c_{t+1}^*) \cdot f'(x_t^*) \quad \text{for } t \geq 1.$$

So, $(\mathbf{x}, \mathbf{y}, \mathbf{c})$ is an Euler program. Also,

$$p_{t+1}f(x_t) - p_{t+1}f'(x_t)x_t \geq p_{t+1}f(x) - p_{t+1}f'(x_t)x \quad \text{for } t \geq 0.$$

So, using $x = 0$ in the above inequality, we get

$$p_{t+1}f(x_t) - p_{t+1}f'(x_t)x_t \geq 0. \qquad \square$$

An important property of competitive programs is noted now.

Theorem 5.2.1. *Let* $(\mathbf{x}, \mathbf{y}, \mathbf{c})_{\tilde{x}}$ *be a competitive program from* $\tilde{x}$. *If*

$$\lim_{T \to \infty} p_T x_T = 0, \qquad (5.2.10)$$

then $(\mathbf{x}, \mathbf{y}, \mathbf{c})_{\tilde{x}}$ *is an optimal program from* $\tilde{x}$.

Proof. Let $(\bar{\mathbf{x}}, \bar{\mathbf{y}}, \mathbf{c})_{\tilde{x}}$ be any other program from $\tilde{x} > 0$. For $t \geq 1$,

$$\delta^{t-1}u(\bar{c}_t) - p_t\bar{c}_t \leq \delta^{t-1}u(c_t) - p_t c_t,$$

or

$$\delta^{t-1}[u(\bar{c}_t) - u(c_t)] \leq p_t\bar{c}_t - p_t c_t.$$

Hence, for any T,

$$\sum_{t=1}^{T}[\delta^{t-1}[u(\bar{c}_t) - u(c_t)]] \leq \sum_{t=1}^{T}(p_t\bar{c}_t - p_t c_t)$$

$$= \sum_{t=1}^{T}\{p_t[(\bar{y}_t - \bar{x}_t) - (y_t - x_t)]\}.$$

Note that $p_1\bar{y}_1 = p_1 y_1 = p_1 f(\tilde{x})$. Hence,

$$\sum_{t=1}^{T}[\delta^{t-1}[u(\bar{c}_t) - u(c_t)] \leq \sum_{t=1}^{T-1}[(p_{t+1}\bar{y}_{t+1} - p_t\bar{x}_t) - (p_{t+1}y_{t+1} - p_t x_t)]$$

$$- p_T\bar{x}_T + p_T x_T$$

$$\leq p_T x_T.$$

Hence,

$$\limsup_{T \to \infty} \sum_{t=1}^{T} \delta^{t-1}[u(\bar{c}_t) - u(c_t)] \leq 0.$$

□

The condition (5.2.10) (already in (4.2.8)) is the transversality condition that was deeply explored in the theory literature on intertemporal decentralization (see Majumdar (1988) and Majumdar and Mitra (1991)).

We shall summarize a few results on characterizing intertemporal efficiency. To this end, it is useful to look at the function $g(x)$ defined by

$$g(x) = \min\left(A(x), f'(x)\right) \quad \text{for } x \geq 0.$$

Associated with any positive program $\mathbf{x} = (x_t)$ is a sequence $\mathbf{q} = (q_t)$ given by

$$q_0 = 1, \quad q_{t+1} = q_t/g(x_t) \quad \text{for } t \geq 0$$

and a sequence $\mathbf{p} = (p_t)$ given by

$$p_0 = 1, \quad p_{t+1} = [p_t/f'(x_t)] \quad \text{for } t \geq 0.$$

Theorem 5.2.2. *If a positive program* $(\bar{\mathbf{x}}, \bar{\mathbf{y}}, \bar{\mathbf{c}})_{\tilde{x}}$ *from* $\tilde{x} \in (0, \bar{k})$ *is inefficient, then*

$$\sum_{t=0}^{\infty}(1/q_t) < \infty.$$

Proof.   See Majumdar and Mitra (1982).   □

A part of Theorem 4.4.2 can be extended to our framework.

Theorem 5.2.3. *An interior program* $(\bar{\mathbf{x}}, \bar{\mathbf{y}}, \bar{\mathbf{c}})_{\tilde{x}}$ *from* $\tilde{x} > 0$ *is inefficient if*

$$\sum_{t=0}^{\infty}(1/p_t) < \infty.$$

Proof.   Follow exactly the method of Cass (1972, pp. 218–220), noting that concavity of f is nowhere required.   □

Suppose that a program $\mathbf{x}$ from any $\tilde{x} \in (0, \bar{k})$ satisfies $\liminf_{t \to \infty} x_t > k^*$. Then it is inefficient (see Phelps (1965)).

We turn to an example. Recall that $k_1 > 0$ is the point with the property that for $0 \le x < k_1$, $f'(x)$ is increasing, for $x > k_1$, $f'(x)$ is decreasing.

Example 5.1. Let $\tilde{x} = k_1$, and consider the stationary program $\langle k_1 \rangle$ defined by $x_t = k_1$ for all t. Then this program $\langle k_1 \rangle$ is efficient. We claim that this is not profit maximizing.

If it were, there is a non-null sequence (p_t) of non-negative prices, such that

$$p_{t+1} f(k_1) - p_t k_1 \ge p_{t+1} f(x) - p_t x \quad \text{for } x \ge 0, \qquad (5.2.11)$$

In fact $p_0 > 0$ (otherwise, $p_t = 0$ for all t ▲ *Exercise* ▼). This means that $p_1 > 0$. Otherwise, choosing $x = 0$ on the right side of (5.2.11), we get

$$-p_0 k_1 \ge 0, \quad \text{a contradiction.}$$

Hence,

$$p_1 f'(k_1) = p_0.$$

So, using $x = 0$ on the right side again,

$$p_1 f(k_1) - p_0 k_1 = p_1 f(k_1) - p_1 f'(k_1)k_1 = p_1[f(k_1) - f'(k_1)k_1] \ge 0,$$

$$\text{or} \quad (f(k_1)/k_1) \ge f'(k_1),$$

a contradiction. $\square$

5.3 Harvesting: The Undiscounted Case

If one wishes to capture the preferences of a social planner (Lange's Central planning Board) who has serious concerns for the "return" or utility of distant generations, one thinks of the "undiscounted" Ramsey criterion ($\delta = 1$) or the case in which δ is "very close" to one. In this section, we focus on the undiscounted case.

Following are the main results: (a) There is a unique Euler stationary program, and this is also the (unique) optimal stationary program, this program is competitive at a stationary price sequence; (b) optimal programs exist from every positive initial input level; they converge monotonically to the optimal stationary program. Thus, in the undiscounted case, optimality is consistent with conservation.

5.3.1   Optimal stationary programs

Consider the set $A = \{c : c = f(x) - x, 0 \le x \le \bar{k}\}$. Clearly, A is compact. Hence, there is a c^* in A, such that $c \le c^*$ for all c in A. Since $0 < x < \bar{k}$ implies $f(x) - x > 0$, then $c^* > 0$. Associated with c^* is x^*, such that $0 < x^* < \bar{k}$ and $f(x^*) - x^* = c^*$. Since x^* maximizes $[f(x) - x]$ over the set $\{x : 0 \le x \le \bar{k}\}$, and the maximum is attained at an interior point,

$$f'(x^*) = 1.$$

Since k^* is unique non-negative solution to $f'(x) = 1$, then $x^* = k^*$, and k^* *is the unique input level, which maximizes* c over the set A.

Consider the program from k^* given by $x_t^* = k^*$, $y_{t+1}^* = f(k^*)$, $c_{t+1}^* = f(k^*) - k^*$, for $t \ge 0$. Then $\langle k^*, f(k^*), c^* = f(k^*) - k^* \rangle$ is a feasible program from k^*. Clearly, it is stationary and positive. Since $f'(k^*) = 1$, then it is an Euler Stationary Program. Since k^* is the unique non-negative solution to $f'(k^*) = 1$, then it is also *the only Euler Stationary Program*.

We show next that $\langle k^*, f(k^*), c^* \rangle$ is an optimal stationary program from k^*. For this, we need two preliminary results.

Lemma 5.3.1. *There is $p^* > 0$, such that*

$$u(c^*) - p^* c^* \ge u(c) - p^* c \quad \text{for } c \ge 0, \tag{5.3.1}$$

$$p^* f(k^*) - p^* k^* \ge p^* f(x) - p^* x \quad \text{for } x \ge 0. \tag{5.3.2}$$

Proof.   Denote $u'(c^*)$ by p^*, then $p^* > 0$. By concavity of u, we have for $c \ge 0$, $u(c) - u(c^*) \le u'(c^*)(c - c^*) = p^*(c - c^*)$. By transposing terms, (5.3.1) is verified.

By definition of k^*, $f(k^*) - k^* \geq f(x) - x$ for $0 \leq x \leq \bar{k}$. For $x > \bar{k}$, $f(k^*) - k^* > 0 > f(x) - x$. So, for all $x \geq 0$, $f(k^*) - k^* \geq f(x) - x$. Multiplying this inequality by $p^* > 0$ yields (5.3.2). $\qquad \square$

For any $c \geq 0$, define the *consumption value loss* at p^* as

$$\alpha(c) \equiv [u(c^*) - p^* c^*] - [u(c) - p^* c].$$

Similarly, define the *loss of intertemporal profit* at p^*

$$\beta(x) \equiv [p^* f(k^*) - p^* k^*] - [p^* f(x) - p^* x].$$

Using (5.3.1) and (5.3.2), we see that $\alpha(c) \geq 0$ for $c \geq 0$ and $\beta(x) \geq 0$ for $x \geq 0$. Of particular interest is the following lemma.

Lemma 5.3.2. *Given any $\theta > 0$, there is $\eta > 0$, such that if $0 \leq x \leq \bar{k}$ and $|k^* - x| \geq \theta$, then $\beta(x) \geq \eta$.*

Proof. Suppose, on the contrary, that there is a sequence (x_n), such that $x_n \geq 0$, $(k^* - x_n) \geq \theta$, for $n = 1, 2, 3, \ldots$, but $[p^* f(k^*) - p^* k^*] - [p^* f(x_n) - p^* x_n] \to 0$ as $n \to \infty$. Clearly, x_n is in $[0, k^*]$ for each n, so consider a subsequence of (x_n) converging to $\hat{x}$. Then $\hat{x}$ is in $[0, k^*]$ and by continuity of f, $[p^* f(k^*) = p^* k^*] - [p^* f(\hat{x}) - p^* \hat{x}]$. Hence, $f(\hat{x}) - \hat{x} = f(k^*) - k^*$. Since $(k^* - x_n) \geq \theta$ for each n, then $(k^* - \hat{x}) \geq \theta$ and $\hat{x} < k^*$. So, $f(k^*) - f(\hat{x}) = f'(z)(k^* - \hat{x})$ where $\hat{x} < z < k^*$. Then $f'(z) > 1$, so $f(k^*) - f(\hat{x}) > k^* - \hat{x}$, a contradiction. This establishes the result. $\qquad \square$

Theorem 5.3.1. *The program $\langle k^*, f(k^*), c^* \rangle$ is an optimal program from k^*.*

Proof. See Section 5.5. $\qquad \square$

Remark 5.1. Note that, by construction, $\langle k^*, f(k^*), c^* \rangle$ is a stationary program from k^*, which has the maximum stationary consumption (hence utility) among *all stationary programs*.

Remark 5.2. The program $\langle k^*, f(k^*), c^* \rangle$ is the only optimal stationary program in this model from a positive initial input. For if there were another, say, $\langle x, y, c \rangle$ from $x > 0$, then it would be a positive program and an Euler program. But $\langle k^*, f(k^*), c^* \rangle$ is the only Euler Stationary Program, so $\langle x, y, c \rangle$ could not be an optimal stationary program.

5.3.2   Non-stationary programs

We will show that:

(a) there is an optimal program from every $\tilde{x} \in (0, \bar{k})$;
(b) every optimal program converges to the optimal stationary program, $\langle k^*, f(k^*), c^* \rangle$.

In the long run, optimal programs approach the optimal stationary program and sustain a positive stock and harvest level. Estimates of the sums of value losses along a program can be obtained from the following useful lemma (although the calculation is routine!).

Lemma 5.3.3. *If* $(\mathbf{x}, \mathbf{y}, \mathbf{c})$ *is a program from* $\tilde{x} > 0$, *then for any finite* $T \geq 1$,

$$
\sum_{t=1}^{T}[u(c_t) - u(c^*)] = \sum_{t=1}^{T} p^*(c_t - c^*) - \alpha(c_t)
$$

$$
= \sum_{t=1}^{T}[p^*(f(x_{t-1}) - x_{t-1}) - p^*(f(k^*) - k^*]
$$

$$
+ p^* x_0 - p^* x_T
$$

$$
= -\left[\sum_{t=1}^{T} \alpha(c_t) + \sum_{t=1}^{T-1} \beta(x_t)\right] + p^* x_0 - p^* x_T.
$$

Recall that (from (4.4.1)) a feasible program $(\mathbf{x}, \mathbf{y}, \mathbf{c})$ is *good* if there exists $\mathbf{B}$ such that

$$
\sum_{t=1}^{T}[u(c_t) - u(c^*)] \geq \mathbf{B} \quad \text{for all } T \geq 1.
$$

It is *bad* if

$$\sum_{t=1}^{T}[u(c_t) - u(c^*)] \to -\infty \quad \text{for all } T \to \infty.$$

Lemma 5.3.4. *There exists a good program from every* $\tilde{x} \in (0, \bar{k})$.

Proof. Consider two cases: (i) $\tilde{x} \geq k^*$; (ii) $\tilde{x} < k^*$. In case (i), the sequence $(\mathbf{x}, \mathbf{y}, \mathbf{c})$ given by $x_0 = \tilde{x}$, $y_1 = f(\tilde{x})$, $c_1 = f(\tilde{x}) - k^*$, $x_t = k^*$, $y_{t+1} = f(k^*)$, $c_{t+1} = f(k^*) - k^*$ for $t \geq 2$ is a feasible program, which is good.

In case (ii), consider that natural growth program $\bar{\mathbf{x}} = (\bar{x}_t)$ from $\tilde{x} > 0$. We know that $\bar{x}_t$ converges to $\bar{k}$ as $t \to \infty$. Since $\bar{k} > k^*$, there is some finite T such that $\bar{x}_t \geq k^*$ for all $t \geq T$. Consider the program $(\mathbf{x}, \mathbf{y}, \mathbf{c})$ defined as $x_0 = \tilde{x}, x_t = \bar{x}_t$ for $t = 1, \ldots, T-1$, $x_t = k^*$ for all $t \geq T$. Then $c_t = c^*$ for all $t > T$. Clearly, this program is good. $\square$

Lemma 5.3.5. *If a program* $(\mathbf{x}, \mathbf{y}, \mathbf{c})$ *from* $\tilde{x} \in (0, \bar{k})$ *is not good, then it is bad.*

Proof. Suppose that a program $(\mathbf{x}, \mathbf{y}, \mathbf{c})_{\tilde{x}}$ is *not good*. Then, given any real number $\mathbf{B}$, there is some T such that

$$\sum_{t=1}^{T}[u(c_t) - u(c^*)] < \mathbf{B}.$$

Using the fact that $x_t \leq \bar{k}$ for all t, and Lemma 5.2.1, we have, for all $\tau > T$,

$$\sum_{t=1}^{\tau}[u(c_t) - u(c^*)] < \mathbf{B} + p^*\bar{k},$$

which implies that it is bad. $\square$

We now come to the celebrated "consumption turnpike theorem".

Theorem 5.3.2. *If a program* $(\mathbf{x}, \mathbf{y}, \mathbf{c})$ *from* $\tilde{x} \in (0, \bar{k})$ *is good, then*

$$(x_t, y_t, c_t) \to (k^*, \ f(k^*), c^*) \quad \text{as } t \to \infty.$$

Proof. Suppose that x_t does not converge to k^*. Then there is some $\theta > 0$ and a subsequence of periods, such that $|k^* - x_t| > \theta$ along this subsequence. By Lemma 5.3.2, there is some $\eta > 0$, such that $\beta(x_t) \geq \eta$ for this subsequence of periods. Now, using Lemma 5.2.1, it follows that $(\mathbf{x}, \mathbf{y}, \mathbf{c})$ is not good. Thus, x_t converges to k^* as $t \to \infty$; consequently, continuity of f gives us $y_t = f(x_{t-1}) \to f(k^*)$ as $t \to \infty$; finally, $c_t = [f(x_{t-1}) - x_t] \to [f(k^*) - k^*] = c^*$ as $t \to \infty$. $\square$

Theorem 5.3.3. *There exists an optimal program from every* $x \in (0, \bar{k})$.

Proof. See Section 5.5. $\square$

Since there exists a good program, then an optimal program is necessarily good. Consequently, by Theorem 5.3.2, every optimal program $(\mathbf{x}, \mathbf{y}, \mathbf{c})$ from $\tilde{x} \in (0, \bar{k})$ has the property that $(x_t, y_t, c_t) \to \langle k^*, f(k^*), c^* \rangle$ as $t \to \infty$. Furthermore, if $\tilde{x} = k^*$, then $\langle k^*, f(k^*), c^* \rangle$ itself is an optimal program by Theorem 5.3.1. One can show that if $\tilde{x} < k^*$, then an optimal program $(\mathbf{x}, \mathbf{y}, \mathbf{c})$ has the property that (x_t, y_{t+1}, c_{t+1}) is *monotonically increasing* for all $t \geq 0$, and $(x_t, y_{t+1}, c_{t+1}) \leq (k^*, f(k^*), c^*)$ for all $t \geq 0$. Similarly, if $\tilde{x} > k^*$, then an optimal program $(\mathbf{x}, \mathbf{y}, \mathbf{c})$ has (x_t, y_{t+1}, c_{t+1}) *monotonically decreasing* for all $t \geq 0$, and $(x_t, y_{t+1}, c_{t+1}) \geq (k^*, f(k^*), c^*)$ for all $t \geq 0$. Roughly speaking, an optimal program sustains c^* in the long run.

Two remarks form the difference between the "classical" and "non-classical" models. First, it is fairly easy to check that for $\tilde{x} \geq k_2$, every optimal program is unique. But, for $\tilde{x} < k_2$, it is not known whether the result is true.

Secondly, optimal programs from $\tilde{x} \geq k_2$ can be shown to be competitive, but we cannot assert the same for $\tilde{x} < k_2$.

5.4   Discounting and Sustainability

We turn to the relation between discounting and sustainability. In this section, we stress two cases formally. First, when the discount factor δ is "sufficiently close" to 1 (formally, $1 > \delta > [\,1/f'(0)]$), the qualitative behavior of optimal programs is similar to that in the undiscounted case: from *any* initial $\tilde{x} \in (0, \bar{k})$, *any* optimal input sequence x_t^* converges to k_δ^*, the (modified) golden-rule input. Thus, in the long run, any optimal program sustains (roughly) the consumption level $c_\delta^*[= f(k_\delta^*) - k_\delta^*]$. Secondly, when the discount factor δ is "sufficiently small" ($\delta < [1/f'(k_1)]$), from *any* initial $\tilde{x} \in (0, \bar{k})$, *any* optimal input sequence x_t^* and its corresponding consumption sequence c_t^* converge to 0. Thus, strong impatience leads to inevitable extinction irrespective of the initial stock, and no positive consumption level is indefinitely sustainable.

In the intermediate range of values of δ, "tipping points" or "critical thresholds" emerge: for each δ in the intermediate range, there is a critical stock $\hat{k}_\delta(>0)$, such that optimal input sequences x_t^* from all initial $\tilde{x}$ below $\hat{k}_\delta$ converge to 0. For all initial $\tilde{x}$ above $\hat{k}_\delta$, any optimal sequence converges to $K^*(\delta)$, the larger root of the equation $\delta f'(x) = 1$. A complete examination of this pattern is too long and technically demanding and is not explored here. In Chapter 6 (when we deal with a linear return function $u(c) = c$)), we provide a self-contained analysis of the appearance of similar threshold stocks in the intermediate range of discount factors.

5.4.1 Existence and monotonicity

Concepts and techniques of dynamic programming have been very useful in treating the discounted case. We begin with two basic results. In the discounted case, a program $(\mathbf{x}^*, \mathbf{y}^*, \mathbf{c}^*)$ from $\tilde{x} \in (0, \bar{k})$ is optimal if

$$\sum_{t=1}^{\infty} \delta^{t-1} u(c_t^*) \geq \sum_{t=1}^{\infty} \delta^{t-1} u(c_t) \qquad (5.4.1)$$

for all programs $(\mathbf{x}, \mathbf{y}, \mathbf{c})$ from $\tilde{x}$.

Theorem 5.4.1. *For any $\tilde{x} \in (0, \bar{k})$, there is an optimal program.*

Proof. See Section 5.5. $\qquad\qquad\square$

Next, we have a monotonicity property.

Theorem 5.4.2. *Let $\tilde{x}$ and $\hat{x}'$ belong to $(0, \bar{k})$ and $\tilde{x} > \hat{x}'$. Let $\mathbf{x}^* = (x_t^*)$ and $\mathbf{x}^{*\prime} = (x_t^{*\prime})$ be optimal input programs from $\tilde{x}$ and $\hat{x}'$, respectively. Then*

$$x_t^* > x_t^{*\prime} \quad \text{for } t \geq 1. \tag{5.4.2}$$

Proof.   See Majumdar and Nermuth (1982) or Majumdar (2006, pp. 186–189) for a dynamic programming argument.   □

Consider now an optimal input program $\mathbf{x}^* = (x_t^*)$ from $\tilde{x}$. Note that for any $T \geq 1$, $(x_t^*)_{t=T}^{\infty}$ is an optimal input program from x_T^*. If $x_1^* > \tilde{x}$, then comparing the sequence $(x_1^*, x_2^*, x_3^*, \ldots)$ as an optimal input program from x_1^* and the sequence $(\tilde{x}, x_1^*, x_2^*, \ldots)$ as an optimal input program from $\tilde{x}$, we see that $x_{t+1}^* > x_t^*$ for all $t \geq 1$. Similarly, if $x_1^* < \tilde{x}$, we get $x_{t+1}^* < x_t^*$ for all $t \geq 1$. This property is particularly useful in what follows.

5.4.2   Mild discounting

We consider the case when δ is "close to one" and establish a convergence property similar to the undiscounted case. More precisely, assume that

$$1 > \delta > [1/f'(0)]. \tag{5.4.3}$$

In this case, there is a unique solution to the equation $\delta f'(x) = 1$, which is denoted by k_δ^*. Surely, $k^* > k_\delta^* > k_2$. Consider the unique Euler stationary program defined by

$$x_t^* = k_\delta^*, \; y_{t+1}^* = f(k_\delta^*), c_{t+1}^* = f(k_\delta^*) - k_\delta^*, \quad \text{for } t \geq 0. \tag{5.4.4}$$

One can show (see Section 5.5) that this stationary program is profit maximizing at prices defined by

$$p_t^* = \delta^{t-1} u'(c_t^*) \quad \text{for } t \geq 1, \tag{5.4.5}$$

$$p_0^* = p_1^* \cdot f'(k_\delta^*).$$

Moreover, note that, by concavity of u, $\delta^{t-1}[u(c) - u(c_t^*)] \leq \delta^{t-1}[u'(c_t^*)(c - c_t^*)] = p_t^*(c - c_t^*)$.

Thus, the stationary program $\langle k_\delta^*, f(k_\delta^*), (f(k_\delta^*) - k_\delta^*) \rangle$ from k_δ^* is competitive at the prices defined by (5.4.5).

We now state the basic result on the existence of an optimal stationary program.

Theorem 5.4.3. *The stationary program $\langle k_\delta^*, f(k_\delta^*), (f(k_\delta^*) - k_\delta^*) \rangle$ from k_δ^* is the unique optimal stationary program.*

Proof. We claim that the stationary program $\langle k_\delta^*, f(k_\delta^*), (f(k_\delta^*) - k_\delta^*) \rangle$ from k_δ^* is the optimal program from k_δ^*. It has been shown to be competitive at prices p_t^* defined by (5.4.5).

Moreover, since $0 < \delta < 1$,

$$\lim_{t \to \infty} p_t^* x_t^* = \lim_{t \to \infty} [\delta^{t-1} u'(c_{t+1}^*) x_t^*]$$

$$= \lim_{t \to \infty} [\delta^{t-1} \cdot u'(f(k_\delta^*) - k_\delta^*) \cdot k_\delta^*] = 0. \qquad (5.4.6)$$

Optimality follows from Theorem 5.2.1. Uniqueness follows from Lemma 5.2.1 (as k_δ^* is the unique solution of $\delta f'(x) = 1$). $\qquad \square$

We now turn to the asymptotic properties of optimal programs in this case. Choose any $0 < \tilde{x} < k_\delta^*$.

Any optimal program $(\mathbf{x}^*, \mathbf{y}^*, \mathbf{c}^*)$ from $\tilde{x}$ must satisfy the following:

(i) $x_t^* < k_\delta^*$ (use (5.4.2);
(ii) (x_t^*) is increasing: $x_{t+1}^* > x_t^*$ for all $t \geq 0$.

To prove (ii), suppose that (x_t^*) is non-increasing: $x_{t+1}^* \leq x_t^*$ for all $t \geq 0$. It must converge to some $\bar{x} \geq 0$. If $\bar{x} > 0$, we have $0 < \bar{x} \leq \tilde{x} < k_\delta^* < \bar{k}$. In the limit, $\bar{c} = f(\bar{x}) - \bar{x} > 0$. Using the Ramsey–Euler condition and taking the limit, we get $\delta f'(\bar{x}) = 1$, and this contradicts the uniqueness property of k_δ^*. On the contrary, if $\bar{x} = 0$, there is some T, such that $\delta f'(x_t^*) > 1$ for all $t \geq T$ (since $\delta f'(0) > 1$, and x_t^* is converging to 0). From the Ramsey–Euler condition,

$$u'(c_t^*) = \delta f'(x_t^*)u'(c_{t+1}^*) \quad \text{for all } t \geq T,$$

$$\text{or} \quad u'(c_t^*) > u'(c_{t+1}^*) \quad \text{for all } t \geq T,$$

$$\text{or} \quad c_t^* < c_{t+1}^* \quad \text{for all } t \geq T. \tag{5.4.7}$$

Hence, if x_t^* is converging to 0, we also have $f(x_t^*)$ converging to 0, and arrive at a contradiction from (5.4.8) for a sufficiently large t. This completes the proof of (ii).

Since the optimal input sequence x_t^* is increasing and bounded above by k_δ^*, it must converge to some $\bar{x} \leq k_\delta^*$. It follows that c_{t+1}^* converges to $\bar{c} = f(\bar{x}) - \bar{x} > 0$. From

$$u'(c_t^*) = \delta f'(x_t^*)u'(c_{t+1}^*) \quad \text{for all } t \geq 1,$$

we take the limit and get $\delta f'(\bar{x}) = 1$. By the uniqueness of k_δ^*, we assert that $\bar{x} = k_\delta^*$.

One can deal with the case $\tilde{x} > k_\delta^*$ in a similar manner. To summarize, we have the following theorem when (5.4.3) holds.

Theorem 5.4.4. *If $0 < \tilde{x} < k_\delta^*$, any optimal input sequence (x_t^*) from $\tilde{x}$ satisfies* (i) $x_{t+1}^* > x_t^*$ *for all $t \geq 0$ and* (ii) $\lim_{t \to \infty} x_t^* = k_\delta^*$.

If $\tilde{x} > k_\delta^*$, any optimal input sequence (x_t^*) from $\tilde{x}$ satisfies (i) $x_{t+1}^* < x_t^*$ *for all $t \geq 0$, and* (ii) $\lim_{t \to \infty} x_t^* = k_\delta^*$.

5.4.3   Heavy discounting: Inevitability of extinction

We turn to the central result on the inevitability of extinction when the discount factor δ is "small"; more precisely, assume that

$$\delta f'(k_1) < 1. \tag{5.4.8}$$

Hence, for all $x \geq 0$, we have

$$\delta f'(x) \leq \delta f'(k_1)) < 1. \tag{5.4.9}$$

Consider an optimal program $(\mathbf{x}^*, \mathbf{y}^*, \mathbf{c}^*)$ from $\tilde{x} \in (0, \bar{k})$. It is an Euler program, so that

$$u'(c_t^*) = \delta f'(x_t^*)u'(c_{t+1}^*) \quad \text{for all } t \geq 1. \tag{5.4.10}$$

From (5.4.10) and (5.4.11), we get

$$u'(c_t^*) < u'(c_{t+1}^*) \quad \text{for all } t \geq 1,$$

$$c_t^* > c_{t+1}^* \quad \text{for all } t \geq 1. \tag{5.4.11}$$

Hence, there is no Euler program that is stationary and no optimal stationary program. Now, the positive decreasing sequence c_t^* converges to some $\bar{c} \geq 0$ and the bounded monotone sequence x_t^* converges to some $\bar{x} \geq 0$. We now show that if $\bar{c} > 0$, we arrive at a contradiction. If $\bar{c} > 0$, we take the limit in (5.4.11) as $t \to \infty$ and get

$$\delta f'(\bar{x}) = 1. \tag{5.4.12}$$

From (5.4.9), we get $\delta f'(\bar{x}) \leq \delta f'(k_1) < 1$, a contradiction. Hence,

$$\lim_{t \to \infty} c_t^* = 0.$$

From

$$c_{t+1}^* = f(x_t^*) - x_{t+1}^*, \tag{5.4.13}$$

we get, by taking the limit as $t \to \infty$,

$$0 = f(\bar{x}) - \bar{x}. \tag{5.4.14}$$

Hence, $\bar{x}$ is either 0 or $\bar{k}$. But if x_t^* converges to $\bar{k}$, it is inefficient by Theorem 5.3.3 (see the remark following the proof), this contradicts the optimality of $(\mathbf{x}^*, \mathbf{y}^*, \mathbf{c}^*)$. Hence,

$$\lim_{t \to \infty} x_t^* = 0, \tag{5.4.15}$$

which implies that

$$\lim_{t \to \infty} y_t^* = 0. \tag{5.4.16}$$

To summarize the discussion on the inevitability of extinction, we have the following theorem when (5.4.9) holds.

Theorem 5.4.5. *Let $\tilde{x} \in (0, \bar{k})$. Any optimal program $(\mathbf{x}^*, \mathbf{y}^*, \mathbf{c}^*)$. from $\tilde{x}$ has the following property:*

The sequences of inputs (x_t^*), consumptions (c_t^*) and outputs (y_t^*) are all monotone decreasing, and

$$\lim_{t\to\infty} x_t^* = 0, \ \lim_{t\to\infty} c_t^* = 0, \ \lim_{t\to\infty} y_t^* = 0.$$

5.5   Complements and Details

A large part of this chapter is based on Majumdar and Mitra (1982) which contains detailed proofs, additional results, and comments on the contrast between the "classical" and "non-convex" optimization problems. Majumdar and Nermuth (1982) allowed for a larger class of non-concave production functions and derived monotonicity properties of optimal programs. A more detailed classification of the domains of attractions of multiple steady states was also presented there. Following Majumdar and Mitra (1982), Dechert and Nishimura (1983) made a notable contribution to the study of tipping points. Extension of the discounted case to incorporate uncertainty is in Majumdar, Mitra and Nyarko (1989). A comprehensive review of the intertemporal allocation models in "convex" environments, the duality theory and the fundamental issues of decentralization in infinite horizon economies is given in Majumdar and Mitra (1991). Many of the duality results in a multi-sector framework (without differentiability assumptions) were presented in Cass and Majumdar (1979).

Some comparative dynamic problems involve studying the sensitivity of optimal programs with respect to (a) variations in the discount factor, (b) variations of parameters in the return/regeneration functions. For a sample of results, see Bhattacharya and Majumdar (1989b), Dutta (1991), Dutta, Majumdar and Sundaram (1994).

Proof of Theorem 5.3.1. Suppose, on the contrary, that there is a feasible program (x, y, c) from k^*, a scaler $\alpha > 0$, and a sequence of periods T_n $(n = 1, 2, 3, \ldots)$, such that

$$\sum_{t=1}^{T_n} [u(c_t) - u(c^*)] \geq \alpha \quad \text{for all } n. \tag{5.5.1}$$

Using Lemma 5.3.1, we have, for $t \geq 1$,

$$
\begin{aligned}
u(c_t) - u(c^*) &\leq p^*(c_t - c^*) \\
&= p^*[f(x_{t-1}) - x_t] - p^*[f(k^*) - k^*] \\
&= [p^* f(x_{t-1}) - p^* x_{t-1}] + [p^* x_{t-1} - p^* x_t] \\
&\quad - p^*[f(k^*) - k^*] \\
&\leq [p^* x_{t-1} - p^* x_t].
\end{aligned}
$$

Hence, for $T \geq 1$, we have

$$
\sum_{t=1}^{T} [u(c_t) - u(c^*)] \leq \sum_{t=1}^{T} [p^* x_{t-1} - p^* x_t] = p^* k^* - p^* x_T. \qquad (5.5.2)
$$

Hence, for all n, we have, using (5.4.1) and (5.4.2),

$$
p^*(k^* - x_{T_N}) \geq \alpha. \qquad (5.5.3)
$$

This means that $(k^* - x_{T_N}) \geq (\alpha/p^*)$ for all n, so by Theorem 5.3.2, there is $\varepsilon > 0$, such that

$$
[p^* f(k^*) - p^* k^*] \geq [p^* f(x_{T_N}) - p^* x_{T_N}] + \varepsilon \quad \text{for all } n. \qquad (5.5.4)
$$

Using Lemma 5.3.1, we have, for $t = T_n + 1$,

$$
\begin{aligned}
u(c_t) - u(c^*) &\leq p^*(c_t - c^*) \\
&= p^*[f(x_{t-1}) - x_t] - p^*[f(k^*) - k^*] \\
&= [p^* f(x_{t-1}) - p^* x_{t-1}] + [p^* x_{t-1} - p^* x_t] \\
&\quad - p^*[f(k^*) - k^*] \\
&\leq [p^* x_{t-1} - p^* x_t] - \varepsilon,
\end{aligned}
$$

and for $t \neq T_n + 1$, we have, by our previous calculations, $u(c_t) - u(c^*) \leq [p^* x_{t-1} - p^* x_t]$. Hence, for all n,

$$
\alpha \leq \sum_{t=1}^{T_n} [u(c_t) - u(c^*)] \leq p^*(k^* - x_{T_N}) - (n-1)\varepsilon
$$

$$
\leq p^* k^* - (n-1)\varepsilon.
$$

For n large, this is a contradiction. Hence, $(k^*, f(k^*), c^*)$ is an OSP.   □

Proof of Theorem 5.3.3. The first part of the proof establishes an important property of good programs. The second part is based on a diagonalization argument.   □

Lemma 5.5.1. *If* $(\mathbf{x}, \mathbf{y}, \mathbf{c})_{\tilde{x}}$ *is a good program from* $\tilde{x} \in (0, \bar{k})$,

$$\lim_{T \to \infty} \sum [u(c_t^*) - u(c^*)] \text{ exists.}$$

Proof.   Since $(\mathbf{x}, \mathbf{y}, \mathbf{c})_{\tilde{x}}$ is good, there is some real number $\mathbf{B}$, such that

$$\sum_{t=1}^{T} [u(c_t) - u(c^*)] \geq \mathbf{B}.$$

Using Lemma 5.2.1, we get

$$- \left[\sum_{t=1}^{T} \alpha(c_t) + \sum_{t=1}^{T-1} \beta(x_t) \right] + p^* x_0 - p^* x_T \geq \mathbf{B},$$

or

$$\left[\sum_{t=1}^{T} \alpha(c_t) + \sum_{t=1}^{T-1} \beta(x_t) \right] \leq p^* x_T - p^* x_0 - \mathbf{B}.$$

Since $x_T \leq \bar{k}$, we have

$$\left[\sum_{t=1}^{T} \lim \alpha(c_t) + \sum_{t=1}^{T-1} \beta(x_t) \right] \leq p^* (\bar{k} - \tilde{x}) - \mathbf{B}.$$

Since $\alpha(c_t) \geq 0$ for all $t \geq 1$, and $\beta(x_t) \geq 0$ for all $t \geq 0$, we see that

$$\mathbf{L}(\mathbf{x}, \mathbf{y}, \mathbf{c}) \equiv \lim_{T \to \infty} \left[\sum_{t=1}^{T} \lim \alpha(c_t) + \sum_{t=1}^{T-1} \lim \beta(x_t) \right]$$

exists. From Lemma 5.2.1, we get

$$\sum_{t=1}^{T} \lim[u(c_t) - u(c^*)] = - \left[\sum_{t=1}^{T} \lim \alpha(c_t) + \sum_{t=1}^{T-1} \beta(x_t) \right] + p^* \tilde{x} - p^* x_T.$$

$$(5.5.5)$$

The right-hand side has a limit (using Theorem 5.2.2). Hence,

$$\lim_{T \to \infty} \left[\sum_{t=1}^{T} [u(c_t) - u(c^*)] \right] = p^*(\tilde{x} - k^*) - \mathbf{L}(\mathbf{x}, \mathbf{y}, \mathbf{c}),$$

and the proof is completed. □

To continue with the proof of the theorem, let us write

$$\mathbf{L}(\tilde{x}) = \inf\{\mathbf{L}(\mathbf{x}, \mathbf{y}, \mathbf{c}) : (\mathbf{x}, \mathbf{y}, \mathbf{c}) \text{ is a good program from } \tilde{x}\}.$$

Take a sequence $(\mathbf{x}^n, \mathbf{y}^n, \mathbf{c}^n)$ of good programs from $\tilde{x}$, such that

$$\mathbf{L}(\mathbf{x}^n, \mathbf{y}^n, \mathbf{c}^n) \leq \mathbf{L}(\tilde{x}) + (1/n).$$

Hence, by a diagonalization argument (see 13.1.5), there is a subsequence (retain notation), such that for each $t \geq 0$,

$$(x_t^n, y_{t+1}^n, c_{t+1}^n) \to (x_t^*, y_{t+1}^*, c_{t+1}^*) \quad \text{as } n \to \infty,$$

where $(\mathbf{x}^*, \mathbf{y}^*, \mathbf{c}^*) = (x_t^*, y_{t+1}^*, c_{t+1}^*)$ is, in fact, a program from $\tilde{x}$. Using Lemma 5.2.1, one can show that $(\mathbf{x}^*, \mathbf{y}^*, \mathbf{c}^*)$ is a good program. We claim that

$$\mathbf{L}(\mathbf{x}^*, \mathbf{y}^*, \mathbf{c}^*) = \mathbf{L}(\tilde{x}).$$

If this claim is false, we can find a positive integer T and some $\varepsilon > 0$, such that

$$\left[\sum_{t=1}^{T} \alpha(c_t^*) + \sum_{t=1}^{T-1} \beta(x_t^*) \right] \geq \mathbf{L}(\tilde{x}) + \varepsilon.$$

But continuity of f and u imply that there is some $\bar{n} > 0$, such that for all $n \geq \bar{n}$,

$$\left[\sum_{t=1}^{T} \alpha(c_t^n) + \sum_{t=1}^{T-1} \beta(x_t^n) \right] \geq \mathbf{L}(\tilde{x}) + \varepsilon/2.$$

But this implies that for all $n \geq \bar{n}$,

$$\mathbf{L}(\mathbf{x}^n, \mathbf{y}^n, \mathbf{c}^n) \geq \mathbf{L}(\tilde{x}) + \varepsilon/2.$$

This leads to a contradiction for $n > \max(\bar{n}, 2/\varepsilon)$. This establishes the claim and optimality of $(\mathbf{x}^*, \mathbf{y}^*, \mathbf{c}^*)$. $\qquad\square$

Proof of Theorem 5.4.1. Let $(\mathbf{x}, \mathbf{y}, \mathbf{c})$ be any program from $\tilde{x}$. From (5.2.6), we get

$$\sum_{t=1}^{\infty} \delta^{t-1} u(c_t) \leq u(\bar{k})[1/(1-\delta)]. \qquad (5.5.6)$$

Define $\mathfrak{a} = \sup\{\sum_{t=1}^{\infty} \delta^{t-1}\delta^{t-1} u(c_t) : \tilde{x}\}$ as a program from $\tilde{x}$. Now, choose a sequence of programs from $\tilde{x}$, denoted by $(\mathbf{x}^n, \mathbf{y}^n, \mathbf{c}^n)$, such that

$$\sum_{t=1}^{\infty} \delta^{t-1} u(c_t^n) \geq \mathfrak{a} - (1/n) \quad \text{for } n = 1, 2, 3, \ldots. \qquad (5.5.7)$$

Using (5.2.5), continuity of f and the Cantor diagonal method, there is a program $(\bar{\mathbf{x}}, \bar{\mathbf{y}}, \bar{\mathbf{c}})$ from $\tilde{x}$, and some subsequence of $(\mathbf{x}^n, \mathbf{y}^n, \mathbf{c}^n)$ (retain notation) such that for all $t \geq 1$,

$$(x_t^n, y_t^n, c_t^n) \to (\bar{x}_t, \bar{y}_t, \bar{c}_t) \quad \text{as } n \to \infty. \qquad (5.5.8)$$

We claim that, in fact, $\sum_{t=1}^{\infty} \delta^{t-1} u(\bar{c}_t) = \mathfrak{a}$; hence, the program $(\bar{\mathbf{x}}, \bar{\mathbf{y}}, \bar{\mathbf{c}})$ from $\tilde{x}$ is indeed optimal. If this claim is false, there is some

$\varepsilon > 0$, such that

$$\sum_{t=1}^{\infty} \delta^{t-1} u(\bar{c}_t) < \mathfrak{a} - \varepsilon. \qquad (5.5.9)$$

Pick T such that $\delta^{T+1}[u(\bar{k})/(1-\delta)] < (\varepsilon/3)$. Using (5.5.7) and the continuity of u, one can find $\bar{n}$, such that for all $n \geq \bar{n}$,

$$\sum_{t=1}^{T} \delta^{t-1} u(c_t^n) < \sum_{t=1}^{T} \delta^{t-1} u(\bar{c}_t) + (\varepsilon/3). \qquad (5.5.10)$$

Thus, for all $n \geq \bar{n}$, we get

$$\sum_{t=1}^{T} \delta^{t-1} u(c_t^n) \leq \sum_{t=1}^{T} \delta^{t-1} u(c_t^n) + \delta^{T+1}[u(\bar{k})/(1-\delta)]$$

$$< \sum_{t=1}^{T} \delta^{t-1} u(c_t^n) + (\varepsilon/3)$$

$$< \sum_{t=1}^{T} \delta^{t-1} u(\bar{c}_t) + (2\varepsilon/3)$$

$$< \mathfrak{a} - (\varepsilon/3).$$

But this leads to a contradiction to (5.5.7) for sufficiently large n. □

Proof of Theorem 5.4.3. Following is an informal sketch of the claim that the stationary program,

$$x_t^* = k_\delta^*, \ y_{t+1}^* = f(k_\delta^*), \ c_{t+1}^* = f(k_\delta^*) - k_\delta^*, \quad \text{for } t \geq 0, \qquad (5.5.11)$$

is profit maximizing at prices:

$$p_t^* = \delta^{t-1} u'(c_t^*) \quad \text{for } t \geq 1, \qquad (5.5.12)$$

$$p_0^* = p_1^* \cdot f'(k_\delta^*).$$

Define $\phi(x) = [f(k_2)/k_2] \cdot x$ for $0 \leq x \leq k_2$ and $\phi(x) = f(x)$ for $x \geq k_2$. Then $\phi(x)$ is an increasing, concave and differentiable function for $x \geq 0$, and $\phi(x) \geq f(x)$. Hence, for $x \geq 0$, we have

$$
\begin{aligned}
p^*_{t+1}[f(x) - f(x^*_t)] &\leq p^*_{t+1}[\phi(x) - \phi(x^*_t)] \\
&\leq p^*_{t+1}p^*_{t+1}\phi'(x^*_t)[x - x^*_t][x - x^*_t] \\
&= p^*_{t+1}f'(x^*_t)[x - x^*_t], \quad\quad (5.5.13)
\end{aligned}
$$

since $\phi'(x^*_t) = \phi'(k^*_\delta) = f'(k^*_\delta) = f'(x^*_t)$ for $t \geq 0$. Hence,

$$
p^*_{t+1} \cdot f(x) - p^*_{t+1} \cdot f'(x^*_t)x \leq p^*_{t+1} \cdot f(x^*_t) - p^*_{t+1} \cdot f'(x^*_t)x^*_t. \quad (5.5.14)
$$

Using the relations

$$
\begin{aligned}
p^*_{t+1} \cdot f'(x^*_t) &= \delta^t u'(c^*_{t+1}) \cdot f'(x^*_t) \\
&= (\delta^{t-1}u'(c^*_{t+1})) \cdot (\delta f'(x^*_t) \\
&= (\delta^{t-1}u'(c^*_t)) \cdot (\delta f'(k^*_\delta)) = p^*_t \quad \text{for } t \geq 1, \quad (5.5.15)
\end{aligned}
$$

and $p^*_1 \cdot f'(x^*_0) = p^*_1 \cdot f'(k^*_\delta) = p^*_0$, we have, from (5.5.14), upon substitution, the profit maximizing condition:

$$
p^*_{t+1} \cdot f(x) - p^*_t x \leq p^*_{t+1} \cdot f(x^*_t) - p^*_t x^*_t \quad \text{for } t \geq 0. \quad\quad \square
$$

5.5.1   Exercises

1. Consider an infinite horizon economy in which the regeneration function f satisfies (F.1)–(F.4) listed in Section 3.2 of Chapter 3. The one period return function u satisfies (U.1) and (U.2) of Section 3.2.1 of Chapter 3 (set $u(0) = 0$). Define an infinite program $(\mathbf{x}, \mathbf{y}, \mathbf{c})$ from $\tilde{x} \in (0, \bar{k})$ as in Section 4.2 of Chapter 4. Finally, an infinite program $(\mathbf{x}^*, \mathbf{y}^*, \mathbf{c}^*)$ from $\tilde{x} > 0$ is optimal if

$$
\limsup_{T \to \infty} \sum_{t=1}^{T} \delta^{t-1}[u(c_t) - u(c^*_t)] \leq 0,
$$

where δ, the discount factor, satisfies $0 < \delta \leq 1$.

(a) Show that there is a unique optimal program.

[▲ Hint: Deal with the discounted and undiscounted cases separately. Also, establish convexity properties of appropriate sets for uniqueness. ▼]

2. A program $(\mathbf{x}, \mathbf{y}, \mathbf{c})$ from $\tilde{x} \in (0, \bar{k})$ is competitive if there is a non-negative, non-zero sequence of Gale prices $\mathbf{p} = (p_t)$, such that

   (i) $\delta^{t-1} u(c_t) - p_t \cdot c_t \geq \delta^{t-1} u(c) - p_t \cdot c$ for all $c \geq 0$, and for $t \geq 1$;
   (ii) $p_{t+1} \cdot f(x_t) - p_t \cdot x_t \geq p_{t+1} \cdot f(x) - p_t \cdot x$ for all $x \geq 0$, and for $t \geq 0$.

   Show that a competitive program $(\mathbf{x}, \mathbf{y}, \mathbf{c}; \mathbf{p})$ is optimal if and only if

   $$\lim_{t \to \infty} p_t x_t = 0 \quad \text{when } 0 < \delta < 1,$$

   ($p_t x_t$ is a bounded sequence ($\delta = 1$)). $\hspace{2cm}$ (5.5.16)

3. Let $(\mathbf{x}^*, \mathbf{y}^*, \mathbf{c}^*)$ be a program from $\tilde{x} > 0$. There is a positive price sequence $\mathbf{p} \gg \mathbf{0}$, satisfying (i) and (ii) of Exercise 2 if and only if

   (i) $x_t^* > 0$, $y_{t+1}^* > 0$, $c_{t+1.}^* > 0$ for all $t \geq 0$;
   (ii) $u'(c_t^*) = \delta f'(x_t^*) u'(c_{t+1}^*)$ for all $t \geq 1$ (Ramsey–Euler condition).

4. Let $(\mathbf{x}^*, \mathbf{y}^*, \mathbf{c}^*)$ be the optimal program from $\tilde{x} > 0$. Then there is a positive sequence of Gale prices $\mathbf{p} = (p_t) \gg \mathbf{0}$, satisfying (i) and (ii) and

   $$\lim_{t \to \infty} p_t x_t = 0 \quad \text{when } 0 < \delta < 1,$$

   ($p_t x_t$ is a bounded sequence ($\delta = 1$)). $\hspace{2cm}$ (5.5.17)

   [▲ Hint: The proof is spelled out in Majumdar and Mitra (1991) ▼]

Chapter 6

Profit Maximization and Extinction

The hidden hand will always do its work; but it may work by strangulation.

— Joan Robinson

Does species extinction really matter, given that one might be able to compensate, through chemical inventiveness, for whatever chemical gems might be lost through extinction? As regards development of new medicinal products, I would say that it does indeed matter... Nature is the ally of organic chemists and a frequent source of their inspiration. What we risk losing in chemical knowledge through extinction is doubtless in large measure irreplaceable.

— Thomas Eisner

6.1 Introduction

The exploitation of natural resources based on *economic motives* (such as profit maximization) is often considered to be in conflict with *conservation* of such resources. This conflict is at the heart of a voluminous literature. This chapter provides a systematic treatment of this theme in a simple framework. While the analytical methods are applicable in a variety of contexts, we find it convenient as a matter of interpretation to focus on a particular renewable resource, a fishery.

The gross reproduction (regeneration) function (called the "stock recruitment function") of our fishing firm is S-shaped, consisting of an initial convex segment, followed by a concave segment (see Clark (2010, Chapter 7)). The economic objective of the management of the fishery is to maximize the discounted net revenues (profits) from "harvests", given a (constant) net revenue per unit $p > 0$, a constant interest rate $i > 0$. One can interpret $p = q - k > 0$, where q (≥ 0) is the return (price) per unit of harvesting, and k (≥ 0) is the constant unit cost of harvesting. A particular case $k = 0$ (costless harvesting) is perhaps the most convenient pedagogical tool.

Following is an informal summary of our results:

(1) If the rate of interest i is "sufficiently high" relative to the productivity of the fishery, then an *extinction* program is optimal for every initial stock (see Theorem 6.3.1); this is the most alarming scenario: extinction is inevitable.

(2) If the rate of interest is i "sufficiently low" relative to the productivity, then a "sustained yield program" is optimal for every initial stock, i.e., it is optimal to attain (in a finite number of periods) and remain at a uniquely defined stationary stock (see Theorems 6.3.2 and 6.3.3). Here, the relatively high productivity ensures that the market outcome is consistent with conservation objective.

(3) In the "intermediate" range of values of i, we assert the existence of a critical positive stock $\hat{k}_\delta$, such that if the initial stock of the fishery $\tilde{x}$ is less than $\hat{k}_\delta$, the extinction program is optimal. If the initial stock $\tilde{x}$ is greater than $\hat{k}_\delta$, then the extinction program is *not* optimal, and an optimal program is a "sustained yield program" of the type discussed in (2) (see case (3), Theorems 6.3.4 and 6.3.5). Thus, the long-run behavior of optimal programs may critically depend on the initial stock.

This critical stock, $\hat{k}_\delta$, *is the minimum safe standard of conservation.* A policy that allows "economic" exploitation of a fishery only when the stock of the fishery exceeds $\hat{k}_\delta$ will ensure that the stock of the resource can be sustained: in other words, an aggressive "initial" intervention will mean that the "economic motives" will not be in conflict with sustainability. It is clear that such an intervention will not work in case (1) and is not needed in case (2).

A model of the fishery in which the reproduction function is concave has also been extensively discussed in the literature. We note that in such a model, results like those discussed in cases (1) and (2) will hold (with the terms "sufficiently low" and "sufficiently high" appropriately redefined) and for the same economic reasons. But the phenomenon in case (3) cannot occur in such a model. Thus, it is the behavior in (3), a *sensitivity to the initial stock*, which really distinguishes the "non-convex" (or, in Hurwicz's words, "non-classical") model from the "convex" ("classical") model extensively used in earlier studies by many authors.

6.2   The Model

We consider a fishery with a Knightian reproduction (regeneration) function $f : R_+ \to R_+$. The assumption on f and the definitions of the important stock levels $(k, k_2, k^*, \bar{k})$ are all spelled out in Section 5.2.1 of Chapter 5. The graph of the Knightian regeneration function is reproduced in Fig. 6.1.

As usual, an infinite program from an initial stock $\tilde{x} > 0$ is a sequence of stocks $\mathbf{x} = (x_t)$, satisfying

$$x_0 = \tilde{x}, \ 0 \leq x_{t+1} \leq f(x_t) \quad \text{for } t \geq 0. \tag{6.2.1}$$

Associated with a program $\mathbf{x}$ from $\tilde{x} > 0$ is a sequence of *harvests* $\mathbf{c} = (c_t)$ defined by

$$c_t = f(x_{t-1}) - x_t \quad \text{for } t \geq 1. \tag{6.2.2}$$

Thus, the initial $\tilde{x}$ gives rise to a total fish population $f(\tilde{x})$ in period $t = 1$. A part of the stock is harvested in period 1, and this is denoted by $c_1 \geq 0$, and the remaining part of the population $x_1 [= f(\tilde{x}) - c_1]$ generates the total stock $f(x_1)$ in period $t = 2$, and the story is repeated period after period. We shall assume that if

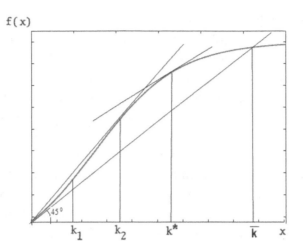

Fig. 6.1. A Knightian regeneration function.

$c' = (c_t)$ is generated by $\mathbf{x}$, so is $0 \leq \mathbf{c}' \leq c_t$ for any sequence $\mathbf{c}' = c'_t$ (free disposal).

In the simple case we consider, we assume that the net return from harvesting a unit of fish $p > 0$ is given and remains constant over time. Now, assume either that there is no cost of harvesting, or that $p = q - k > 0$ where q is the price per unit and k is the constant cost of harvesting. Similarly, the rate of interest $i > 0$ is also constant over time. The firm's objective is the maximization of the discounted sum of net revenues (or payoffs). Thus, a program $\mathbf{x}^*$ from $\tilde{x} > 0$ is *optimal* if

$$\sum_{t=1}^{\infty} [(pc_t^*)/(1+i)^{t-1}] \geq \sum_{t=1}^{\infty} [(pc_t)/(1+i)^{t-1}] \qquad (6.2.3)$$

for every program $\mathbf{x}$ from $\tilde{x}$. We denote $1/(1+i)$ by δ and refer to δ as the *discount factor* corresponding to the interest rate i. Clearly, $0 < \delta < 1$. We see from (6.2.3) that a program $\mathbf{x}^*$ from $\tilde{x} > 0$ is *optimal* if

$$\sum_{t=1}^{\infty} \delta^{t-1} c_t^* \geq \sum_{t=1}^{\infty} \delta^{t-1} c_t \qquad (6.2.4)$$

for every program $\mathbf{x}$ from $\tilde{x} > 0$.

We note that by adapting the proof of Theorem 5.4.1, we get the following theorem.

Theorem 6.2.1. *There is an optimal program from every $\tilde{x} > 0$.*

In the analysis of the qualitative behavior of optimal programs, a critical role is played by the solutions to the equation

$$\delta f'(x) = 1, \text{ or equivalently,}$$

$$f'(x) = 1 + i.$$

This equation might not have any (real) solution. If it has a non-negative real solution, we call it K_δ^*. If it has two distinct non-negative real solutions, we denote the smaller one by k_δ^* and the larger by K_δ^*.

A program $\hat{\mathbf{x}}$ from $\tilde{x}$ is called the *extinction program* from $\tilde{x}$ if

$$c_1 = f(\tilde{x}), \ x_t = 0 \quad \text{for } t \geq 1.$$

In the extinction program, the entire stock $f(\tilde{x})$ generated by the initial $\tilde{x}$ is harvested in period 1.

A program $\mathbf{x}$ from $\tilde{x}$ is a *sustained yield program* if there are some $\tilde{k}$ satisfying $0 < \tilde{k} < \bar{k}$, and a positive integer T, such that

$$x_t = \tilde{k} \quad \text{for } t \geq T.$$

Finally, a program $\mathbf{x}$ from $\tilde{x} < K_\delta^*$ is a *regeneration program* if there is some $T' \geq 1$, such that $x_t > x_{t-1}$ for $1 \leq t \leq T'$ and $x_t = K_\delta^*$ for $t \geq T'$.

6.3   Characterization of Optimal Programs

In this section we describe the qualitative properties of optimal programs from alternative initial stocks. Recall that given the interest rate i, we denote by δ, the discount factor, defined as $\delta \equiv 1/(1+i)$.

For convenience of exposition, we distinguish among three cases:

Case (1). $(1+i) \geq f'(k_2)$ [i.e., $\delta f'(k_2) \leq 1$].
Case (2). $f'(0) \geq (1+i)$ [i.e., $\delta f'(0) \geq 1$].
Case (3). $f'(0) < (1+i) < f'(k_2)$ [i.e., $\delta f'(0) < 1 < \delta f'(k_2)$].

In case (1), the extinction program is an optimal program. Excepting the borderline case of $(1 + i) = f'(k_2)$, the extinction program is the *unique* optimal program. This is a situation in which the interest rate which captures the rate of return on alternative viable investments is "too large" relative to what the fishery delivers given its productivity. We can interpret case (2) in a similar manner. Case (3) brings out *the conflict between revenue maximization and sustainability sharply and suggests a simple intervention policy.* In this case, there is a critical stock $\hat{k}_\delta > 0$, such that if the initial stock $\tilde{x} < \hat{k}_\delta$, the extinction program is optimal, whereas if $\tilde{x} > \hat{k}_\delta$, a sustained yield program is optimal. The behavior of optimal programs even in the long run depends critically on the initial stock.

Case (1). $(1 + i) \geq f'(k_2)$ [i.e., $\delta f'(k_2) \leq 1$].

In this case, we show that the extinction program is optimal for every initial stock $\tilde{x} > 0$.

The proof is based on two key steps.

Lemma 6.3.1.

(i) $[\delta f(x) - x] \leq 0$ *for all* $x \geq 0$.
(ii) *If* $\delta f'(k_2) < 1$, *then* $[\delta f(x) - x] < 0$ *for all* $x > 0$.

Proof.   To prove (i), suppose to the contrary that there is some $x' \geq 0$, such that

$$[\delta f(x') - x'] > 0.$$

Clearly, $x' > 0$, and

$$(\delta f(x'))/x' > 1.$$

But, $(\delta f(x'))/x' \leq (\delta f(k_2))/k_2 = \delta f'(k_2) \leq 1$, a contradiction.

To prove (ii), suppose to the contrary that there is some $x' > 0$, such that

$$[\delta f(x') - x'] \geq 0.$$

Then,

$$(\delta f(x'))/x' \geq 1.$$

But, $(\delta f(x'))/x' \leq (\delta f(k_2))/k_2 = \delta f'(k_2) < 1$, a contradiction. $\square$

The following lemma provides an upper bound on the present value of any feasible harvest c (multiply both sides of (6.3.1) by p).

Lemma 6.3.2. *If* **x** *is a program from* $\tilde{x}$, *then*

$$\sum_{t=1}^{\infty} \delta^{t-1} c_t \equiv \sum_{t=1}^{\infty} c_t/(1+i)^{t-1} \leq f(\tilde{x}). \qquad (6.3.1)$$

Proof. For $t \geq 2$, denote $\{-\delta^{t-2}[\delta f(x_{t-1}) - x_{t-1}]\}$ by θ_t. By Lemma 6.3.1, $\theta_t \geq 0$ for $t \geq 2$.

Also, for $t \geq 2$, we have

$$\begin{aligned}
\delta^{t-1} c_t &= \delta^{t-1}[f(x_{t-1}) - x_t] \\
&= \delta^{t-2}[\delta f(x_{t-1}) - x_{t-1}] + [\delta^{t-2} x_{t-1} - \delta^{t-1} x_t] \\
&= [\delta^{t-2} x_{t-1} - \delta^{t-1} x_t] - \theta_t. \qquad (6.3.2)
\end{aligned}$$

For $t = 1$,

$$\delta^{t-1} c_t = f(\tilde{x}) - x_1. \qquad (6.3.3)$$

Hence, for $T \geq 2$,

$$\sum_{t=1}^{T} \delta^{t-1} c_t = f(\tilde{x}) - \delta^{T-1} x_T - \sum_{t=2}^{T} \theta_t \leq f(\tilde{x}). \qquad (6.3.4)$$

Thus (see 13.1.6),

$$\sum_{t=1}^{\infty} \delta^{t-1} c_t \leq f(\tilde{x}). \qquad (6.3.5)$$

$\square$

The next result captures the inevitability of extinction in this case.

Theorem 6.3.1. *If $(1 + i) > f'(k_2)$, then the extinction program $\hat{\mathbf{x}}$ is the unique optimal program from any $\tilde{x} > 0$.*

Proof.   For any $\tilde{x} > 0$, the extinction program $\hat{\mathbf{x}}$ from $\tilde{x}$ satisfies

$$\sum_{t=1}^{\infty} \delta^{t-1} c_t^* = f(\tilde{x}).\tag{6.3.6}$$

Hence, by Lemma 6.3.2, $\hat{\mathbf{x}}$ is optimal.

To prove uniqueness, consider *any* feasible program $\mathbf{x}$ from $\tilde{x}$, which is *distinct from* $\hat{\mathbf{x}}$. Then there is some period $s \geq 1$, such that $x_s > 0$. Then, by Lemma 6.3.1, $\theta_{s+1} > 0$. Consequently, by using (6.3.4), we get, for $T \geq s + 1$,

$$\sum_{t=1}^{T} \delta^{t-1} c_t \leq f(x') - \theta_{s+1}.\tag{6.3.7}$$

Hence, one obtains

$$\sum_{t=1}^{\infty} \delta^{t-1} c_t \leq f(x') - \theta_{s+1} < f(x') = \sum_{t=1}^{\infty} \delta^{t-1} c_t^*.\tag{6.3.8}$$

$\square$

Remark 6.3.1. When $f''(k_2) = 1 + i$, the extinction program is still an optimal program from x'; it need not be unique.

Proof.   See Section 6.4. $\square$

The classical model

Consider now the classical model, satisfying (A.1)–(A.4) and (A.5′). Then the methods developed above can be used to prove the following corollary.

Corollary 6.3.1. *If $f'(0) \leq 1 + i$ (i.e., $\delta f'(0) \leq 1$), then the extinction program is the unique optimal program from every $\tilde{x} > 0$.*

Proof. The key step in the proof is to establish

$$[\delta f(x) - x] < 0 \quad \text{for } x > 0.$$

Suppose, to the contrary, that there is some $x' > 0$, with $[\delta f(x') - x'] \geq 0$; then $[\delta f(x')/x'] \geq 1$. By the Mean Value Theorem, we have

$$f(x') = f(x') - f(0) = f'(z)x', \quad \text{where } 0 < z < x'. \tag{6.3.9}$$

Hence,

$$\delta f'(z) = \delta f(x')/x' \geq 1. \tag{6.3.10}$$

But then,

$$f'(0) - f'(z) = f''(\zeta)(-z) > 0, \quad \text{where } 0 < \zeta < z. \tag{6.3.11}$$

Hence,

$$1 \geq \delta f'(0) > \delta f'(z) = \delta f(x')/x' \geq 1, \tag{6.3.12}$$

a contradiction. $\square$

Exercise 6.3.1. ▲ Complete the rest of the proof. ▼

Case (2). $f'(0) \geq 1 + i$ (i.e., $\delta f'(0) \geq 1$).

Recall that the positive real roots of the equation $\delta f'(x) = 1$ are denoted by $k_\delta^* < K_\delta^*$. In this case, $K_\delta^* > k_2$. If k_δ^* exists, then $k_1 > k_\delta^* \geq 0$. If $\tilde{x} < K_\delta^*$, let $\hat{T}$ be the smallest integer, such that $\bar{x}_{\hat{T}} \geq K_\delta^*$ (where $(\bar{x}_t)$ is the natural growth program from $\tilde{x}$ defined by $\bar{x}_{t+1} = f(\bar{x}_t)$, see Lemma 3.2.2).

A precise description of the optimal program in this case is given by Theorems 6.3.2 and 6.3.3. Informally, the optimal program "attains K_δ^* as soon as possible" and stays at K_δ^* in the subsequent periods.

Lemma 6.3.3.

(i) $\delta f(K_\delta^*) - K_\delta^* \geq \delta f(x) - x$ *for all* $x \geq 0$.
(ii) $\delta f(K_\delta^*) - K_\delta^* > \delta f(x) - x$ *for* $x \neq K_\delta^*$.

Proof.   To prove (i), define

$$g(x) = \begin{cases} [f(k_2)/k_2]x & \text{for } 0 \le x \le k_2, \\ f(x) & \text{for } x \ge k_2. \end{cases} \quad (6.3.13)$$

Then, $g(0) = 0$, and g is an increasing, concave, differentiable function of $x \ge 0$. Also, $g(x) \ge f(x)$ for all $x \ge 0$, and, since $K_\delta^* > k_2$, $g(K_\delta^*) = f(K_\delta^*)$. Hence,

$$f(x) - f(K_\delta^*) \le g(x) - g(K_\delta^*). \quad (6.3.14)$$

Now,

$$g(x) - g(K_\delta^*) \le g'(K_\delta^*)(x - K_\delta^*) = f'(K_\delta^*)(x - K_\delta^*), \quad (6.3.15)$$

or

$$f(x) - f(K_\delta^*) \le f'(K_\delta^*)(x - K_\delta^*). \quad (6.3.16)$$

Hence,

$$\delta[f(x) - f(K_\delta^*)] \le \delta[f'(K_\delta^*)(x - K_\delta^*)] = (x - K_\delta^*). \quad (6.3.17)$$

Transposing terms, we get

$$\delta f(K_\delta^*) - K_\delta^* \ge \delta f(x) - x, \quad (6.3.18)$$

completing the proof of (i).

To prove (ii), suppose on the contrary that there is some $x' \ne K_\delta^*$, such that

$$\delta f(x') - x' \ge \delta f(K_\delta^*) - K_\delta^*.$$

First, note that $x' \ne 0$. For $\delta f(K_\delta^*) - K_\delta^* = f(K_\delta^*)[\delta - (K_\delta^*/f(K_\delta^*))] > f(K_\delta^*)[\delta - (1/f'(K_\delta^*))] = 0$.

So, $\delta f(x') - x' > 0$, and this implies that $x' \ne 0$. Now, either (a) $x' \ge k_2$ or (b) $x' < k_2$.

In case (a), by the Mean Value Theorem,

$$f(x) - f(K_\delta^*) = f'(K_\delta^*)(x - K_\delta^*) + (1/2) \, f''(\xi)(x - K_\delta^*)^2$$

$$\text{where } x < \xi < K_\delta^*. \tag{6.3.19}$$

Since $f''(\xi) < 0$, we get

$$f(x) - f(K_\delta^*) < f'(K_\delta^*)(x - K_\delta^*). \tag{6.3.20}$$

Multiplying by δ,

$$\delta f(x) - \delta f(K_\delta^*) < \delta f'(K_\delta^*)(x - K_\delta^*) = x - K_\delta^*, \tag{6.3.21}$$

a contradiction.

In case (b) (using the function g defined above (6.3.13)),

$$f(x) - f(K_\delta^*) < g(x) - f(K_\delta^*) = g(x) - g(K_\delta^*)$$

$$\le g'(K_\delta^*)(x - K_\delta^*) = f'(K_\delta^*)(x - K_\delta^*). \tag{6.3.22}$$

Multiplying by δ,

$$\delta f(x) - \delta f(K_\delta^*) < \delta f'(K_\delta^*)(x - K_\delta^*) = x - K_\delta^*, \tag{6.3.23}$$

a contradiction. $\qquad \square$

Theorem 6.3.2. *If $\tilde{x} \ge K_\delta^*$, the program $\mathbf{x}^*$ from $\tilde{x}$ defined by $x_0^* = \tilde{x}$, $x_t^* = K_\delta^*$ for $t \ge 1$, is the unique optimal program from $\tilde{x}$.*

Proof. Let $\mathbf{x}$ be any program from $\tilde{x}$, distinct from $\mathbf{x}^*$. Then there is some period $s \ge 1$, such that $x_s \ne x_s^*$.

For $t = 1$, we have

$$\delta^{t-1} c_t - \delta^{t-1} c_t^* = [f(\tilde{x}) - x_t] - [[f(\tilde{x}) - x_t^*] = [x_t^* - x_t]. \tag{6.3.24}$$

For $t \ge 2$, we have

$$\delta^{t-1} c_t - \delta^{t-1} c_t^* = \delta^{t-1}[f(x_{t-1}) - x_t] - \delta^{t-1}[f(x_{t-1}^*) - x_t^*]$$

$$= \delta^{t-2}[\delta f(x_{t-1}) - x_{t-1}] + [\delta^{t-2} x_{t-1} - \delta^{t-1} x_t]$$

$$- \delta^{t-2}[\delta f(x_{t-1}^*) - x_{t-1}^*] - [\delta^{t-2} x_{t-1}^* - \delta^{t-1} x_t^*]$$

$$= [\delta^{t-2} x_{t-1} - \delta^{t-1} x_t] - [\delta^{t-2} x_{t-1}^* - \delta^{t-1} x_t^*] - \eta_t, \tag{6.3.25}$$

where

$$\eta_t = \delta^{t-2}[\delta f(x_{t-1}^*) - x_{t-1}^*] - \delta^{t-2}[\delta f(x_{t-1}) - x_{t-1}]. \tag{6.3.26}$$

Hence, for $T \geq 2$, we have

$$\sum_{t=1}^{T}[\delta^{t-1}c_t - \delta^{t-1}c_t^*] = \delta^{T-1}(x_T^* - x_T) - \sum_{t=2}^{T}\eta_t. \tag{6.3.27}$$

Using the fact that $x_t^* = K_\delta^*$ for $t \geq 1$, from Lemma 6.3.3(i), we note that $\eta_t \geq 0$ for $t \geq 2$, and from Lemma 6.3.3(ii), we note that $\eta_{s+1} > 0$. Hence, for $T \geq s + 1$,

$$\sum_{t=1}^{T}[\delta^{t-1}(c_t - c_t^*)] \leq \delta^{T-1}x_T^* - \eta_{s+1}. \tag{6.3.28}$$

Taking limits,

$$\sum_{t=1}^{\infty}[\delta^{t-1}(c_t - c_t^*)] \leq -\eta_{s+1} < 0. \tag{6.3.29}$$

Hence, $\mathbf{x}^*$ is the unique optimal program.   □

Theorem 6.3.3. *If $\tilde{x} < K_\delta^*$, the program $\mathbf{x}^*$ from $\tilde{x}$ defined by $x_0^* = \tilde{x}$, $x_t^* = \bar{x}_t$ for $t = 1, 2, \dots, \hat{T} - 1$, $x_t^* = K_\delta^*$, for $t \geq \hat{T}$, is the unique optimal program from $\tilde{x}$.*

Proof.   Let $\mathbf{x}$ be any feasible program from $\tilde{x}$, distinct from $\mathbf{x}^*$. Then there is some period $s \geq 1$, such that $x_s \neq x_s^*$.

As in the proof of Theorem 6.3.2, for $t = 1$, we have

$$\delta^{t-1}c_t - \delta^{t-1}c_t^* = [f(\tilde{x}) - x_t] - [f(\tilde{x}) - x_t^*] = [x_t^* - x_t]. \tag{6.3.30}$$

For $t \geq 2$, we have

$$\begin{aligned}
\delta^{t-1}c_t - \delta^{t-1}c_t^* &= \delta^{t-1}[f(x_{t-1}) - x_t] - \delta^{t-1}[f(x_{t-1}^*) - x_t^*] \\
&= \delta^{t-2}[\delta f(x_{t-1}) - x_{t-1}] + [\delta^{t-2}x_{t-1} - \delta^{t-1}x_t] \\
&\quad - \delta^{t-2}[\delta f(x_{t-1}^*) - x_{t-1}^*] - [\delta^{t-2}x_{t-1}^* - \delta^{t-1}x_t^*] \\
&= [\delta^{t-2}x_{t-1} - \delta^{t-1}x_t] - [\delta^{t-2}x_{t-1}^* - \delta^{t-1}x_t^*] - \eta_t,
\end{aligned} \tag{6.3.31}$$

where

$$\eta_t = \delta^{t-2}[\delta f(x_{t-1}^*) - x_{t-1}^*] - \delta^{t-2}[\delta f(x_{t-1}) - x_{t-1}]. \qquad (6.3.32)$$

For all $t = 2, 3, \ldots, \hat{T}$, $x_{t-1} \le x_{t-1}^* = \bar{x}_{t-1}$. If in some period $\hat{t} \in \{2, 3, \ldots, \hat{T}\}$, $x_{\hat{t}-1} = x_{\hat{t}-1}^*$ holds, then $\eta_{\hat{t}} = 0$. Otherwise, $x_{t-1} < x_{t-1}^* < K_\delta^*$, and, by using the Mean Value Theorem,

$$[\delta f(x_{t-1}^*) - x_{t-1}^*] - [\delta f(x_{t-1}) - x_{t-1}]$$
$$= \delta f'(\xi_{t-1})(x_{t-1}^* - x_{t-1}) - (x_{t-1}^* - x_{t-1}),$$
$$\text{where } x_{t-1} < \xi_{t-1} < x_{t-1}^* < K_\delta^*. \qquad (6.3.33)$$

Since $\delta f'(\xi_{t-1}) > 1$, $\eta_t > 0$ in all such periods. For all $t > \hat{T}$, using Lemma 6.3.3, as in the proof of Theorem 6.3.2, we have $\eta_t \ge 0$, and if for some $\hat{t} > \hat{T}$, $x_{\hat{t}-1}^* \ne x_{\hat{t}-1}$, then $\eta_{\hat{t}} > 0$. To summarize,

$$\eta_t \ge 0 \text{ for all } t \ge 2, \text{ and } \eta_{s+1} > 0. \qquad (6.3.34)$$

Hence, taking $T \ge s + 1$,

$$\sum_{t=1}^{T}[\delta^{t-1}(c_t - c_t^*)] = \delta^{T-1}(x_T^* - x_T) - \sum_{t=2}^{T}\eta_t \le \delta^{T-1}x_T^* - \eta_{s+1}. \quad (6.3.35)$$

Taking limits,

$$\sum_{t=1}^{\infty}[\delta^{t-1}(c_t - c_t^*)] \le -\eta_{s+1} < 0. \qquad (6.3.36)$$

Hence, $\mathbf{x}^*$ is the unique optimal program from $\tilde{x}$. $\qquad \square$

Remark 6.3.2. The program $\mathbf{x} = \langle K_\delta^* \rangle$ from K_δ^*, given by $x_t = K_\delta^*$ for $t \ge 0$, is the unique optimal stationary program.

The classical model. Consider, once again, the classical model. If $\delta f'(0) > 1$, there is a unique positive solution, call it $k_\delta^* > 0$, of the equation

$$\delta f'(x) = 1.$$

One can show, by using the methods developed above, the following corollaries.

Corollary 6.3.2. *If $\tilde{x} \geq k_\delta^*$, the program $\mathbf{x}^*$ defined by $x_0^* = \tilde{x}$, $x_t^* = k_\delta^*$, for $t \geq 1$, is the unique optimal program from $\tilde{x}$.*

Corollary 6.3.3. *If $\tilde{x} < k_\delta^*$, the program $\mathbf{x}^*$ defined by $x_0^* = \tilde{x}$, $x_t^* = \bar{x}_t'$, for $t = 1, 2, \ldots, M' - 1$, $x_t^* = k_\delta^*$, for $t \geq M'$, is the unique optimal program from $\tilde{x}$. Here, to be sure, $\bar{\mathbf{x}}'$ is the natural growth program from $\tilde{x}$ in the classical model, and M' is the smallest integer, such that $\bar{x}_{M'}' \geq k_\delta^*$.*

Case (3). $f'(0) < 1 + i < f'(k_2)$ (i.e., $\delta f'(0) < 1 < \delta f'(k_2)$).

Here, $0 < k_\delta^* < k_1 < k_2 < K_\delta^* < k^*$. Also, there is some $d > 0$, such that $k_\delta^* < d < k_2$, and $\delta[f(d)/d] = 1$.

The case $\tilde{x} \geq K_\delta^*$ can be dispensed with easily by using the methods of case (2). More precisely, we have the following lemma.

Lemma 6.3.4.

(i) $\delta f(K_\delta^*) - K_\delta^* \geq \delta f(x) - x$ *for all* $x \geq 0$.
(ii) $\delta f(K_\delta^*) - K_\delta^* > \delta f(x) - x$ *for* $x \neq K_\delta^*$.

Proof.   Follow exactly the proof of Lemma 6.3.3.   □

Theorem 6.3.4. *If $\tilde{x} \geq K_\delta^*$, the program $\mathbf{x}^*$ from $\tilde{x}$ defined by $x_0^* = \tilde{x}$, $x_t^* = K_\delta^*$ for $t \geq 1$, is the unique optimal program from $\tilde{x}$.*

Proof.   Follow exactly the proof of Theorem 6.3.2.   □

Thus, we are left with the case in which $\tilde{x} < K_\delta^*$. This difficult case requires a careful, extensive analysis for a precise characterization of optimal programs. We shall leave some steps for Section 6.4.

Theorem 6.3.5. *There is a unique optimal stationary program $\mathbf{x}^* = \langle K_\delta^* \rangle$ given by $x_t^* = K_\delta^*$ for all $t \geq 0$.*

Proof.   Before getting into details, note that we proceed in two steps. First, $\mathbf{x}^* = \langle K_\delta^* \rangle$ is stationary, and using Theorem 6.3.4, it is optimal from K_δ^*.

Next, let $\mathbf{x} = \langle \tilde{x}' \rangle$ be a stationary program from some $\tilde{x}' > 0$. For $\mathbf{x}$ to be an *optimal* program from $\tilde{x}'$, one must have $[f(\tilde{x}') - \tilde{x}'] > 0$, and if $\mathbf{x}$ is an *optimal stationary* program, the expression,

$$W(x) = [f(\tilde{x}') - x] + \delta[f(x) - \tilde{x}'], \qquad (6.3.37)$$

is maximized at $x = \tilde{x}'$, among all $x \geq 0$ satisfying $f(\tilde{x}') \geq x$, $f(x) \geq \tilde{x}'$. Since the maximum is at an interior point $\tilde{x}'$, the first-order condition leads to $\delta f'(\tilde{x}') = 1$. So, $\tilde{x}' = k_\delta^*$ or K_δ^*. Since $W(x)$ is *maximized* at $\tilde{x}'$, $W''(\tilde{x}') = \delta f''(\tilde{x}') \leq 0$. Since $f''(k_\delta^*) > 0$, $\tilde{x}' = K_\delta^*$. Hence, $\mathbf{x}^* = \langle K_\delta^* \rangle$ is the unique optimal program. $\qquad \square$

Lemma 6.3.5. *If $\tilde{x} < K_\delta^*$, and $\mathbf{x}$ is an optimal program from $\tilde{x}$, then $x_t \leq K_\delta^*$ for all $t \geq 0$.*

Proof. See Section 6.4. $\qquad \square$

Lemma 6.3.6. *If the extinction program from $\tilde{x}$ is optimal, then the extinction program is optimal from all $\tilde{x}'$, satisfying $0 < \tilde{x}' \leq \tilde{x}$.*

Proof. Suppose the extinction program $\hat{\mathbf{x}}$ is optimal from $\tilde{x}$, but, contrary to the claim, there is some $0 < \tilde{x}' < \tilde{x}$, such that the extinction program is not optimal from $\tilde{x}'$. Then there is a feasible program $\langle x_t' \rangle$ from $\tilde{x}'$, such that $\sum_{t=1}^{\infty} \delta^{t-1} c_t' > f(\tilde{x}')$. Consider the sequence $\langle x \rangle$ given by $x_0 = \tilde{x}$, $x_t = x_t'$ for $t \geq 1$. Then $\langle x \rangle$ is a feasible program from $\tilde{x}$, since $c_1 = f(\tilde{x}) - x_1 = f(\tilde{x}) - f(\tilde{x}') + f(\tilde{x}') - x_1' = f(x) - f(\tilde{x}') + c_1' > c_1' \geq 0$, and $c_t = c_t' \geq 0$ for $t \geq 2$.

Now,

$$\sum_{t=1}^{\infty} \delta^{t-1} c_t = c_1 + \sum_{t=2}^{\infty} \delta^{t-1} c_t'$$

$$= f(\tilde{x}) - f(\tilde{x}') + c_1' + \sum_{t=2}^{\infty} \delta^{t-1} c_t'$$

$$= f(\tilde{x}) - f(\tilde{x}') + \sum_{t=1}^{\infty} \delta^{t-1} c_t' > f(\tilde{x}) - f(\tilde{x}') + f(\tilde{x}') = f(\tilde{x}).$$

So, the extinction program in not optimal from $\tilde{x}$, a contradiction. $\qquad \square$

Lemma 6.3.7. *If (x_t) is an optimal program from $\tilde{x}$, then for each $T \geq 0$, the sequence (x'_t) given by $x'_t = x_{t+T}$ for $t \geq 0$ is an optimal program from x_T.*

Proof.   See Section 6.4.   □

Lemma 6.3.8. *If (x_t) is an optimal program from $\tilde{x} < K^*_\delta$, and $0 < x_{t+1} \leq x_t$ for some $t = s \geq 0$, then $x_{s+1} = K^*_\delta$.*

Proof.   See Section 6.4.   □

Lemma 6.3.9. *If the extinction program is not optimal from $\tilde{x} < K^*_\delta$, and $\mathbf{x}$ is an optimal program from $\tilde{x}$, and if $\tilde{x} \leq x_{t-1} < K^*_\delta$ for some $t = s \geq 1$, then $x_s > x_{s-1}$.*

Proof.   Suppose the conditions of the lemma hold, but $x_s \leq x_{s-1}$. Then $x_s < K^*_\delta$. We will show that $x_s > 0$. If $x_s = 0$, then the sequence (x'_t) given by $x'_0 = x_{s-1}$, $x'_t = 0$ for $t \geq 1$ is a feasible program from x_{s-1}. Also, $x'_t = x_{t+s-1}$ for $t \geq 0$. So, by Lemma 6.3.7, $\langle x' \rangle$ is optimal from x_{s-1}. Since $\langle x' \rangle$ is the extinction program from x_{s-1}, the extinction program is optimal from x_{s-1}. But, then, since $\tilde{x} \leq x_{s-1}$, the extinction program is optimal from $\tilde{x}$, by Lemma 6.3.7, a contradiction. So, $x_s > 0$.

Thus, we have $0 < x_s \leq x_{s-1}$, for some $s \geq 1$, and $x_s \neq K^*_\delta$. This violates Lemma 6.3.8, and proves that $x_s > x_{s-1}$.   □

Lemma 6.3.10. *If the extinction program is not optimal from $\tilde{x} < K^*_\delta$, and (x_t) is any optimal program from $\tilde{x}$, then (x_t) is a regeneration program.*

Proof.   See Section 6.4.   □

Lemma 6.3.11. *There is some $\tilde{x} > 0$, such that the extinction program from $\tilde{x}$ is an optimal program from $\tilde{x}$.*

Proof.   See Section 6.4.   □

On the contrary (recall that there is $d > 0$ such that $k^*_\delta < d < k_2$, and $\delta[f(d)/d] = 1$), we have the following lemma.

Lemma 6.3.12. *If $\tilde{x}$ satisfies $d < \tilde{x} < K_\delta^*$, then the extinction program is not an optimal program from $\tilde{x}$.*

Proof. Suppose, on the contrary, there is some $\tilde{x}$, satisfying $d < \tilde{x} < K_\delta^*$, such that the extinction program $\hat{\mathbf{x}}$ from $\tilde{x}$ is optimal from $\tilde{x}$. Then

$$\sum_{t=1}^{\infty} \delta^{t-1} c_t^* = f(\tilde{x}).$$

Consider the sequence x_t given by $x_t = \tilde{x}$ for $t \geq 0$. Then $\langle \tilde{x} \rangle$ is a feasible program from $\tilde{x}$, and

$$\sum_{t=1}^{\infty} \delta^{t-1} c_t = \frac{f(\tilde{x}) - \tilde{x}}{(1 - \delta)} = f(\tilde{x}) \frac{[1 - \{\tilde{x}/f(\tilde{x})\}]}{(1 - \delta)}.$$

Now, $f(\tilde{x})/\tilde{x} > f(K_\delta^*)/K_\delta^* > f'(K_\delta^*) > 1$. So, $\tilde{x}/f(\tilde{x}) < K_\delta^*/f(K_\delta^*) < 1/f'(K_\delta^*) = \delta < 1$. Hence, $[1 - \{\tilde{x}/f(\tilde{x})\}] > [1 - \delta]$, and $[f(\tilde{x}) - \tilde{x}]/(1 - \delta) > f(\tilde{x})$. Hence, $\hat{\mathbf{x}}$ cannot be optimal from $\tilde{x}$, a contradiction. This establishes the lemma. $\qquad\square$

In view of the above results, define

$$\mathcal{A} = \{\tilde{x} \in (0, K_\delta^*) : \text{the extinction program from } \tilde{x} \text{ is an}$$
$$\text{optimal program from } \tilde{x}\}.$$

Observe that (using Lemma 6.3.11), $\mathcal{A}$ is non-empty. From Lemma 6.3.12, if $\tilde{x} \in A$, $\tilde{x} \leq d$. Thus, $\mathcal{A}$ is a non-empty set of reals that is bounded above, define

$$\hat{k}_\delta = \sup\{x : x \in \mathcal{A}\}.$$

Then, $0 < \hat{k}_\delta \leq d$. We can finally state the following main result in case (3).

Theorem 6.3.6.

(i) *If $0 < \tilde{x} < \hat{k}_\delta$, then the extinction program from $\tilde{x}$ is an optimal program from $\tilde{x}$.*

(ii) *If $\hat{k}_\delta < \tilde{x} < K_\delta^*$, then the extinction program is not optimal from $\tilde{x}$, and any optimal program is a regeneration program.*

Proof.   To prove (i), pick any $\tilde{x}$ satisfying $0 < \tilde{x} < \hat{k}_\delta$, then there is some $\tilde{x}'$ satisfying $\tilde{x} < \tilde{x}' < \hat{k}_\delta$ such that $\tilde{x}' \in \mathcal{A}$ (otherwise, $\hat{k}_\delta$ is not $\sup\{x : x \in \mathcal{A}\}$. By Lemma 6.3.6, the extinction program is optimal from $\tilde{x}$.

To prove (ii), pick any $\tilde{x}$ satisfying $\hat{k}_\delta < \tilde{x} < K_\delta^*$. Then, this $\tilde{x}$ does not belong to $\mathcal{A}$ (otherwise, the definition of $\hat{k}_\delta$ is contradicted). Hence, the extinction program from this $\tilde{x}$ is not an optimal program. By Lemma 6.3.10, any optimal program is a regeneration program.   □

Remark 6.3.3.

(i) It can be checked that if $0 < \tilde{x} < \hat{k}_\delta$, and $f(\tilde{x}) < \hat{k}_\delta$, then the extinction program from $\tilde{x}$ is the unique optimal program from $\tilde{x}$.

(ii) If $\tilde{x} = \hat{k}_\delta$, then the optimal program from $\tilde{x}$ is not unique.

6.4   Complements and Details

This chapter is entirely based on Majumdar and Mitra (1980, 1983), inspired by an earlier paper by Clark (1972) which left case (3) as an open problem. Using continuous-time models, Clark (2010) has an expanded treatment of the problem of renewable resource management. Next, we provide the details of some proofs.

Proof of Remark 6.3.1. For the extinction program $\hat{\mathbf{x}}$ from $\tilde{x}$, we clearly have

$$\sum_{t=1}^{\infty} \delta^{t-1} c_t = f(\tilde{x}). \tag{6.4.1}$$

Hence, by Lemma 6.3.2, $\hat{\mathbf{x}}$ is optimal. Now, if $\delta f'(k_2) = 1$, then an optimal program need not be unique. To confirm this, consider the case where $\tilde{x} = k_2$. Consider the stationary program $\langle \tilde{x} \rangle$ from $\tilde{x}$, given by $x_t = \tilde{x}$ for $t \geq 0$. Then $\langle\tilde{x}\rangle$ is a feasible program from $\tilde{x}$, and

$$\sum_{t=1}^{\infty} \delta^{t-1} c_t = \frac{[f(k_2) - k_2]}{(1 - \delta)} = \frac{f(k_2)[1 - (k_2/f(k_2))]}{(1 - \delta)}$$

$$= \frac{f(k_2)[1 - \{1/f'(k_2)\}]}{(1 - \delta)} = \frac{f(k_2)[1 - \delta]}{(1 - \delta)} = f(x).$$

Hence, $\langle\tilde{x}\rangle$ is an optimal program from $\tilde{x}$. By Theorem 6.3.2, the extinction program $\hat{x}$ (which is clearly distinct from $\langle x \rangle$) is also an optimal program from $\tilde{x}$. Thus, an optimal program is not unique in this case.

In fact, here, there is an infinite number of optimal programs. Consider the programs (x_t^n), (for $n = 1, 2, \ldots$) given by $(x_t^n) = (\tilde{x}_n)$ for $t = 0, 1, \ldots, n$; $x_t^n = 0$ for $t > n$. Then each program (x_t^n) is a feasible program from $\tilde{x}$. Furthermore, for each n, we have

$$\sum_{t=1}^{\infty} \delta^{t-1} c_t = \sum_{t=1}^{n} \delta^{t-1} c_t + \delta^n f x = \frac{[f(k_2) - k_2][1 - \delta^n]}{[1 - \delta]} + \delta^n f(k_2)$$

$$= f(k_2)[1 - \delta^n] + \delta^n f(k_2) = f(k_2) = f(\tilde{x}).$$

Hence, by Lemma 6.3.2, each program (x_t^n) is optimal from $\tilde{x}$. $\qquad\square$

Proof of Lemma 6.3.5. Suppose, on the contrary, there is some $t \geq 1$, for which $x_t > K_\delta^*$. Let s be the first period this happens. Define a sequence (x_t') by $x_t' = x_t$ for $t = 0, \ldots, s - 1$; $x_t' = K_\delta^*$ for $t \geq s$. Clearly, (x_t') is a feasible program. Now, for $t \geq s + 1$, we have

$$\delta^{t-1} c_t - \delta^{t-1} c_t' = \delta^{t-1}[f(x_{t-1}) - x_t] - \delta^{t-1}[f(x_{t-1}') - x_t']$$

$$= \delta^{t-2}[\delta f(x_{t-1}) - x_{t-1}] + [\delta^{t-2} x_{t-1} - \delta^{t-1} x_t]$$

$$\quad - \delta^{t-2}[\delta f(x_{t-1}) - x_{t-1}'] - [\delta^{t-2} x_{t-1}' - \delta^{t-1} x_t']$$

$$= [\delta^{t-1} x_{t-1} - \delta^{t-1} x_t] - [\delta^{t-2} x_{t-1}' - \delta^{t-1} x_t'] - \eta_t,$$

where $\eta_t = \delta^{t-2}[\delta f(x_{t-1}) - x_{t-1}] - \delta^{t-2}[\delta f(x_{t-1}') - x_{t-1}']$.

Also, for $t = s$,

$$
\begin{aligned}
\delta^{t-1} c_t - \delta^{t-1} c_t' &= \delta^{t-1}[f(x_{t-1}) - x_t] - \delta^{t-1}[f(x_{t-1}') - x_t'] \\
&= \delta^{t-1}[f(x_{t-1}) - f(x_{t-1}')] - \delta^{t-1}[x_t - x_t'] \\
&= -\delta^{s-1}[x_s - x_s'].
\end{aligned}
$$

Hence, for $T \geq s + 1$,

$$
\sum_{t=s}^{T} \delta^{t-1}(c_t - c_t') = \delta^{T-1}[x_T' - x_T] - \sum_{t=s+1}^{T} \eta_t.
$$

Also, we have

$$
\sum_{t=1}^{s-1} \delta^{t-1}(c_t - c_t') = 0.
$$

So, for $T \geq s + 1$,

$$
\sum_{t=1}^{T} \delta^{t-1}(c_t - c_t') = \delta^{T-1}[x_T' - x_T] - \sum_{t=s+1}^{T} \eta_t.
$$

Now, for $t \geq s + 1$, $\eta_t \geq 0$ by Lemma 6.3.4(i) and $\eta_{s+1} > 0$ by Lemma 6.3.4(ii). So, for $T \geq s + 1$,

$$
\sum_{t=1}^{T} \delta^{t-1}(c_t - c_t') \leq \delta^{T-1} z - \eta_{s+1},
$$

and

$$
\sum_{t=1}^{\infty} \delta^{t-1}(c_t - c_t') \leq -\eta_{s+1} < 0.
$$

This proves that $\mathbf{x}$ is not optimal from $\tilde{x}$, a contradiction. □

Proof of Lemma 6.3.7. Suppose, on the contrary, there is some $T \geq 0$, such that the sequence $\langle x' \rangle$ given by $x_t' = x_{t+T}$ for $t \geq 0$ is not an optimal program from x_T. Clearly, $\langle x' \rangle$ is a feasible program from x_T. So, there is a feasible program $\langle x'' \rangle$ from x_T, such that

$$\sum_{t=1}^{\infty} \delta^{t-1} c_t'' > \sum_{t=1}^{\infty} \delta^{t-1} c_t'.$$

Construct the sequence $\langle \tilde{x} \rangle$ given by $\tilde{x}_t = x_t$ for $t = 0, \ldots, T$ and $\tilde{x}_t = x_{t-T}''$ for $t > T$. Then, clearly, $\langle \tilde{x} \rangle$ is a feasible program from $\tilde{x}$. Also,

$$\sum_{t=1}^{\infty} \delta^{t-1} \tilde{c}_t = \sum_{t=1}^{T} \delta^{t-1} \tilde{c}_t + \sum_{t=T+1}^{\infty} \delta^{t-1} \tilde{c}_t$$

$$= \sum_{t=1}^{T} \delta^{t-1} c_t + \sum_{t=T+1}^{\infty} \delta^{t-1} c_{t-T}''$$

$$= \sum_{t=1}^{T} \delta^{t-1} c_t + \left[\sum_{t=1}^{\infty} \delta^{t-1} c_t'' \right] \delta^T$$

$$> \sum_{t=1}^{T} \delta^{t-1} c_t + \delta^T \sum_{t=1}^{\infty} \delta^{t-1} c_t'$$

$$= \sum_{t=1}^{T} \delta^{t-1} c_t + \sum_{T+1}^{\infty} \delta^{t-1} c_{t-T}'$$

$$= \sum_{t=1}^{T} \delta^{t-1} c_t + \sum_{T+1}^{\infty} \delta^{t-1} c_t$$

$$= \sum_{t=1}^{\infty} \delta^{t-1} c_t.$$

This contradicts the fact that $\mathbf{x}$ is optimal from $\tilde{x}$ and establishes the lemma. $\qquad \square$

Proof of Lemma 6.3.8. By Lemma 6.3.7, the sequence (x_t') given by $x_t' = x_{t+s}$ for $t \geq 0$ is an optimal program from x_s and the sequence given by $x_t'' = x_{t+s+1}$ for $t \geq 0$ is an optimal program from x_{s+1}.

Then,

$$\sum_{t=1}^{\infty} \delta^{t-1} c'_t = c'_1 + \sum_{t=2}^{\infty} \delta^{t-1} c'_t$$

$$= [f(x_s) - x_{s+1}] + \delta \sum_{t=1}^{\infty} \delta^{t-1} c'_{t+1}$$

$$= [f(x_s) - x_{s+1}] + \delta \sum_{t=1}^{\infty} \delta^{t-1} c''_t. \qquad (6.4.2)$$

Consider the sequence $(\tilde{x}_t)$ given by $\tilde{x}_0 = x_s$, $\tilde{x}_t = x''_t$ for $t \geq 1$. Then,

$$\tilde{c}_1 = f(x_s) - \tilde{x}_1 = f(x_s) - x''_1 = f(x_s) - x_{s+2}$$

$$= f(x_s) - f(x_{s+1}) + f(x_{s+1}) - x_{s+2}$$

$$= f(x_s) - f(x_{s+1}) + c_{s+2} \geq 0.$$

Also, for $t \geq 2$,

$$\tilde{c}_t = f(\tilde{x}_{t-1}) - \tilde{x}_t = f(x''_{t-1}) - x''_t = c''_t \geq 0.$$

Hence, $\langle \tilde{x} \rangle$ is feasible from x_s.

Now, we certainly have

$$\sum_{t=1}^{\infty} \delta^{t-1} c'_t \geq \sum_{t=1}^{\infty} \delta^{t-1} \tilde{c}_t,$$

so that

$$\sum_{t=1}^{\infty} \delta^{t-1} c'_t \geq f(x_s) - f(x_{s+1}) + c_{s+2} + \sum_{t=2}^{\infty} \delta^{t-1} c''_t$$

$$= f(x_s) - f(x_{s+1}) + \sum_{t=1}^{\infty} \delta^{t-1} c''_t. \qquad (6.4.3)$$

From (6.4.2) and (6.4.3), we have

$$\sum_{t=1}^{\infty} \delta^{t-1} c''_t + f(x_s) - f(x_{s+1}) \leq f(x_s) - x_{s+1} + \delta \sum_{t=1}^{\infty} \delta^{t-1} c''_t.$$

So,

$$\left[\sum_{t=1}^{\infty} \delta^{t-1} c_t'' \right] \leq \frac{[f(x_{s+1}) - x_{s+1}]}{(1 - \delta)}. \tag{6.4.4}$$

Now, consider the sequence $\langle \hat{x} \rangle$ given by $\hat{x}_t = x_{s+1}$ for $t \geq 0$. Clearly, this is a feasible program from x_{s+1}. Also,

$$\sum_{t=1}^{\infty} \delta^{t-1} \hat{c}_t = \frac{[f(x_{s+1}) - x_{s+1}]}{(1 - \delta)}. \tag{6.4.5}$$

Hence, by (6.4.4) and (6.4.5), $\langle \hat{x} \rangle$ is an optimal program from x_{s+1}. Also, $\langle \hat{x} \rangle$ is a stationary program, so $\langle \hat{x} \rangle$ is an OSP. By Theorem 6.3.5, $x_{s+1} = K_\delta^*$. $\square$

Proof of Lemma 6.3.10. By Lemma 6.3.6, either (a) $x_t < K_\delta^*$ for $t \geq 0$, or (b) $x_t = K_\delta^*$ for some $t \geq 1$.

If (a) holds, then we claim that $x_{t+1} > x_t$ for $t \geq 0$. For $t = 0$, since $\tilde{x} = x_0 \leq K_\delta^*$, $x_1 > x_0$ by Lemma 6.3.9. Suppose our claim is true for $t = 0, \ldots, N$. Then $x_{N+1} > x_N > \cdots > x_0 = \tilde{x}$, so $\tilde{x} \leq x_{N+1} < K_\delta^*$, and by Lemma 6.3.9, $x_{N+2} > x_{N+1}$. Hence, the claim is true for $t = N + 1$. This completes the induction step and establishes that $x_{t+1} > x_t$ for $t \geq 0$.

Since $x_t < K_\delta^*$ for $t \geq 0$, x_t converges to some $\hat{x}$; clearly, $\tilde{x} < \hat{x} \leq K_\delta^*$. So, c_t converges to $[f(\hat{x}) - \hat{x}] > 0$. Hence, we can find T large enough, so that for $t \geq T$, $c_t > 0$. Since (x_t) is optimal, for each $t \geq T$, the expression

$$W(x) = [f(x_{t-1}) - x] + \delta f(x - x_{t+1})$$

is maximized at $x = x_t$, among all x satisfying $f(x_{t-1}) \geq x$, $f(x) \geq x_{t+1}$. Since the maximum is at interior point, $W'(x_t) = 0$, $W''(x_t) \leq 0$ for $t \geq T$. This means $\delta f'(x_t) = 1$ for $t \geq T$, and $\delta f''(x_t) \leq 0$ for $t \geq T$. Hence, $x_t = K_\delta^*$ for $t \geq T$. But $x_{t+1} > x_t$ for all t, a contradiction. So, (a) cannot occur. Hence, (b) must occur.

Let M be the smallest integer for which $x_M = K_\delta^*$. Then $M \geq 1$ and $x_t < K_\delta^*$ for $t = 0, \ldots, M - 1$. We claim that $x_{t+1} > x_t$ for $t = 0, \ldots, M-1$. For $t = 0$, $\tilde{x} \leq x_0 < K_\delta^*$, so $x_1 > x_0$ by Lemma 6.3.9.

Suppose our claim is true for $t = 0, \ldots, n$, where $n < M - 1$. Then $x_{n+1} > x_n > \cdots > x_0 = \tilde{x}$, so $\tilde{x} \leq x_{n+1} < K_\delta^*$ and by Lemma 6.3.10, $x_{n+2} > x_{n+1}$. Hence, the claim is true for $t = n + 1$. This completes the induction step and establishes that $x_{t+1} > x_t$ for $t = 0, \ldots, M-1$.

Now, since (x_t) is optimal from $\tilde{x}$, the sequence (x_t') given by $x_t' = x_{t+M}$ for $t \geq 0$ is optimal from x_M by Lemma 6.3.7. Since $x_M = K_\delta^*$, the sequence $\langle x^* \rangle$ given by $x_t^* = K_\delta^*$ for $t \geq 0$ is the unique optimal program from $K_\delta^* = x_M$, by Theorem 6.3.5. Hence, $x_t' = K_\delta^*$ for $t \geq 0$ and $x_{t+M} = K_\delta^*$ for $t \geq 0$.

To summarize, there is an integer $M \geq 1$, such that $x_t > x_{t-1}$ for $1 \leq t \leq M$, and $x_t = K_\delta^*$ for $t \geq M$. Hence, (x_t) is a regeneration program. □

Proof of Lemma 6.3.11. Denote $[1/\delta h(k_\delta^*)]$ by m. Since $\delta h(k_\delta^*) = \delta f(k_\delta^*)/k_\delta^* < 1$, so $m > 1$. Thus, $\delta < \delta m$, and $\delta m = [1/h(k_\delta^*)] < 1$. Denote $[m - 1]/m$ by $\hat{m}$; then $\hat{m} > 0$.

Choose a positive integer M, such that $(\delta m)^M < k_\delta^*$. Since $0 < \delta m < 1$, this can be done. Choose $T > M$, such that

$$(\delta m)^{T+M} \geq \frac{2\delta^T \bar{k}}{\hat{m}}, \tag{6.4.6}$$

and

$$\sum_{t=1}^{T} \delta^{t-1} \geq \frac{1}{2(1 - \delta)}. \tag{6.4.7}$$

Since $(\delta m) > \delta$, so (6.4.6) can be satisfied; (6.4.7) can clearly be satisfied, since

$$\sum_{t=1}^{\infty} \delta^{t-1} = \frac{1}{(1 - \delta)}.$$

Choose $\tilde{x} = [\delta m]^{T+M}$.

Consider the pure accumulation program $\langle \bar{x} \rangle$ from $\tilde{x}$. Then we claim that $\bar{x}_t < k_\delta^*$ for $t = 0, 1, \ldots, T$. This is clearly true for $t = 0$, since $\tilde{x} = [\delta m]^{T+M} > [\delta m]^M < k_\delta^*$. Suppose the claim is true for

$t = 0, \ldots, n$, where $n < T$. Then, we have

$$\bar{x}_{n+1} = f(\bar{x}_n) = \left[\frac{f(\bar{x}_n)}{\bar{x}_n}\right]\bar{x}_n$$

$$\leq \left[\frac{f(k_\delta^*)}{k_\delta^*}\right]\bar{x}_n = [\delta h(k_\delta^*)]\left[\frac{\bar{x}_n}{\delta}\right] = \frac{\bar{x}_n}{(\delta m)}.$$

Iterating on this relationship,

$$\bar{x}_{n+1} \leq \frac{\bar{x}_0}{(\delta m)^{n+1}} = \frac{(\delta m)^{T+M}}{(\delta m)^{n+1}}$$

$$\leq \frac{(\delta m)^{T+M}}{(\delta m)^T} = (\delta m)^M < k_\delta^*.$$

This completes the induction step and establishes our claim that $\bar{x}_t < k_\delta^*$ for $t = 0, 1, \ldots, T$.

Suppose the extinction program is not optimal from $\tilde{x}$. Then, by Lemma 6.3.10, if (x_t) is an optimal program from $\tilde{x}$, (x_t) is a regeneration program. Then $x_t > \tilde{x}$ for $t \geq 0$. Also, $x_t \leq \bar{x}_t$ for $t \geq 0$. In particular, $x_t > \tilde{x}$ for $t = 0, 1, \ldots, T$ and $x_t \leq \bar{x}_t < k_\delta^*$ for $t = 0, 1, \ldots, T$.

For $2 \leq t \leq T + 1$, we then have

$$\delta^{t-1}c_t = \delta^{t-1}f(x_{t-1}) - \delta^{t-1}x_t$$

$$= [\delta^{t-2}x_{t-1} - \delta^{t-1}x_t] - \delta^{t-2}x_{t-1}\left[1 - \frac{\delta f(x_{t-1})}{x_{t-1}}\right].$$

Now, for $2 \leq t \leq T + 1$, we have

$$\frac{\delta f(x_{t-1})}{x_{t-1}} \leq \frac{\delta f(k_\delta^*)}{k_\delta^*} < 1,$$

so

$$\delta^{t-2}x_{t-1}\left[1 - \frac{\delta f(x_{t-1})}{x_{t-1}}\right] \geq \delta^{t-2}x_{t-1}\left[1 - \frac{\delta f(k_\delta^*)}{k_\delta^*}\right] \geq \delta^{t-1}\tilde{x}\hat{m}.$$

Hence, for $2 \leq t \leq T + 1$, we have

$$\delta^{t-1}c_t \leq [\delta^{t-1}x_{t-1} - \delta^{t-1}x_t] - \delta^{t-2}\tilde{x}\hat{m},$$

and

$$\sum_{t=2}^{T+1} \delta^{t-1} c_t \leq x_1 - \delta^T x_{T+1} - \tilde{x}\hat{m} \sum_{t=2}^{T+1} \delta^{t-2}.$$

So,

$$\sum_{t=1}^{T+1} \delta^{t-1} c_t \leq f(\tilde{x}) - \delta^T x_{T+1} - \tilde{x}\hat{m} \sum_{t=2}^{T+1} \delta^{t-2}$$

$$\leq f(\tilde{x}) - \tilde{x}\hat{m} \sum_{t=1}^{T} \delta^{t-1}$$

$$\leq f(\tilde{x}) - \left[\frac{\tilde{x}\hat{m}}{2(1-\delta)} \right].$$

Also,

$$\sum_{t=T+2}^{\infty} \delta^{t-1} c_t \leq \bar{k} \sum_{t=T+2}^{\infty} \delta^{t-1} = \bar{k}\delta^{T+1} \sum_{t=1}^{\infty} \delta^{t-1} = \frac{\bar{k}\delta^{T+1}}{(1-\delta)}.$$

Hence,

$$\sum_{t=1}^{\infty} \delta^{t-1} c_t \leq f(\tilde{x}) - \frac{\tilde{x}\hat{m}}{2(1-\delta)} + \frac{\bar{k}\delta^{T+1}}{(1-\delta)}.$$

Using (6.4.6), we have

$$\tilde{x} = (\delta m)^{T+M} \geq \frac{2\delta^T \bar{k}}{\hat{m}}.$$

So,

$$\frac{\tilde{x}\hat{m}}{2(1-\delta)} \geq \left[\frac{2\delta^T \bar{k}}{\hat{m}} \right] \left[\frac{\hat{m}}{2(1-\delta)} \right] = \frac{\delta^T \bar{k}}{(1-\delta)} > \frac{\delta^{T+1} \bar{k}}{(1-\delta)}.$$

So,

$$\sum_{t=1}^{\infty} \delta^{t-1} c_t < f(\tilde{x}).$$

If $\hat{\mathbf{x}}$ is the extinction program from $\tilde{x}$, then clearly

$$\sum_{t=1}^{\infty} \delta^{t-1} c_t = f(\tilde{x}).$$

Hence, (x_t) is not optimal from $\tilde{x}$. This contradiction establishes that the extinction program must be optimal from $\tilde{x}$. $\qquad\square$

Chapter 7

Utilization of an Exhaustible Resource: A Partial Equilibrium Approach

> OPEC's defenders seem to have the notion that somehow market forces have never properly recognized the value of oil, that its price should always have been higher. This tosses rational economic analysis out of the window.
>
> — *Time* magazine (October 14, 1974, p. 35)

7.1 Introduction

This chapter is an introduction to the literature that grew out of the seminal contribution of Hotelling (1931) on the optimal extraction and consumption of an exhaustible resource. Consider a piece of cake (that remains freshly baked forever!) of size $\underline{S}$ (>0). If c_t is the quantity consumed in period t, it generates utility $u(c_t)$ (*or return*) according to a function u that satisfies the usual continuity, monotonicity and strict concavity properties (see (U.1)–(U.3)). If we plan for T-periods ($T \geq 2$ is a finite positive integer), the problem of optimal consumption over T-periods can be formally stated as

$$\text{maximize} \quad \sum_{t=0}^{T} u(c_t)$$

$$\text{subject to} \quad \sum_{t=0}^{t=T} c_t \leq \underline{S}, \; c_t \geq 0, \quad \text{for all } t = 0, 1, \ldots, T.$$

It can be shown that there is a unique optimal program $(c_t^*)^T$ that is *equitable* in the sense that $c_0^{*T} = c_1^{*T} = \cdots = c_t^{*T} = \cdots = c_T^{*T} = \underline{S}/(T+1)$. The exact value of the optimal c_t^{*T} naturally depends on both the initial stock and the length T of the planning periods. Observe that for *every* period $t = 0, 1, \ldots$, the optimal consumption $c_t^{*T} \to 0$ as $T \to \infty$. Indeed, it can be shown that if we start with an infinite horizon and choose not to introduce discounting, there is *no* optimal program according to the overtaking criterion (see Example 7.2.1 for an elaboration). Thus, we are led to an examination of a simple infinite horizon model with discounting. Let δ satisfy $0 < \delta < 1$, and consider the following problem:

$$\text{maximize} \quad \sum_{t=0}^{\infty} \delta^t u(c_t)$$

$$\text{subject to} \quad \sum_{t=0}^{\infty} c_t \leq \underline{S}, \; c_t \geq 0, \quad \text{for all } t = 0, 1, \ldots.$$

We can prove that *a unique optimal program exists*. With the appropriate differentiability assumption (see (U.4)) on u, one can throw light on the structure of optimal programs. Two cases need to be distinguished:

(i) $\lim_{c \to 0} u'(c) \to \infty$ and (ii) $\lim_{c \to 0} u'(c) \leq \chi$ for some $\chi > 0$.

In the first case (i), the optimal program (c_t^*) is positive in the sense that $c_t^* > 0$. In other words, *the resource is never exhausted* *in finite time*. We can derive the fundamental result that along the optimal program, *the discounted marginal utility or return remains constant* (see Theorem 7.2.3). It also follows that c_t^* goes to 0 monotonically. Two sensitivity results can be derived: an increase in the initial stock (size of the cake!) leads to an increase in the optimal consumption in *every* period. Secondly, a smaller δ raises the optimal consumption in the initial period.

In the second case (ii), we note that there is a finite T such that the resource is *completely exhausted at the end of period T*. Depending on the size of the initial stock $\underline{S}$ and the discount factor δ, it is possible that the resource is used up (the cake is eaten!) in the initial period. The optimal program has the structure that "$c_0^* = \underline{S}$ and $c_t^* = 0$ for all $t \geq 1$" if and only if $u'(\underline{S}) \geq \delta u'(0)$.

7.2   A Model of Resource Extraction

A resource is *exhaustible* if its stock cannot be augmented given the known technology. Let $\underline{S} > 0$ be the initial stock of such an exhaustible resource. An *extraction* (or *consumption*) *program* (briefly, a program) from $\underline{S}$ is a *non-negative* sequence $\mathbf{c} = (c_t)_{t=0}^{\infty}$, satisfying

$$\sum_{t=0}^{\infty} c_t \leq \underline{S}. \qquad (7.2.1)$$

Let $\mathcal{C}$ be the set of all programs from a fixed $\underline{S} > 0$. The set $\mathcal{C}$ is non-empty and convex. We often write a program as $\mathbf{c} = (c_t)$.

A program $\mathbf{c} = (c_t)_{t=0}^{\infty}$ from $\underline{S} > 0$ is *intertemporally efficient* if there does *not* exist another program $\mathbf{c}' = (c_t')_{t=0}^{\infty}$ from (the same) $\underline{S}$, such that

$$c_t' \geq c_t \text{ for all } t, \text{ and } c_t' > c_t \text{ for some } t. \qquad (7.2.2)$$

The following characterization of efficient programs (a special case of Majumdar (1974)) is useful.

Theorem 7.2.1. *A program* $\mathbf{c} = (c_t)_{t=0}^{\infty}$ *from* $\underline{S}$ *is intertemporally efficient if and only if*

$$\sum_{t=0}^{\infty} c_t = \underline{S}. \tag{7.2.3}$$

We shall use the relation (7.2.3) repeatedly.

Extraction (or consumption) of the resource generates a return (or utility). Formally, the *return* function $u : R_+ \to R$ is assumed to satisfy

(U.1) *u is continuous;* $u(0) = 0$.

(U.2) *u is increasing:* $c > c'$ *implies* $u(c) > u(c')$.

(U.3) *u is strictly concave:* $u\left(\theta c + (1-\theta)c'\right) > \theta u(c) + (1-\theta)u(c')$
       for all $0 < \theta < 1$ *and* $c \neq c'$.

We shall introduce additional assumptions as we move on. Alternative programs are evaluated according to the "sum" of returns.

First, we recall Gale's (1967) famous example in the "undiscounted" case, indicating the possibility of non-existence of an optimal program in an infinite horizon model.

Example 7.2.1. Fix an initial stock $\underline{S} > 0$. A *program* $\mathbf{c} = (c_t)_{t=0}^{\infty}$ *from* $\underline{S}$ is said to *overtake* another program $\mathbf{c}' = (c_t')_{t=0}^{\infty}$ from $\underline{S}$ if there exists some T, such that

$$\sum_{t=0}^{\tau} [u(c_t) - u(c_t')] \geq 0 \quad \text{for all } \tau \geq T.$$

A program $\mathbf{c} = (c_t)_{t=0}^{\infty}$ *from* $\underline{S}$ *is optimal* (*according to the "overtaking" criterion, see Weizsacker (1965)*) *if it overtakes all programs from* $\underline{S}$. Clearly, the program of zero consumption (i.e., $c_t = 0$ for all t) cannot be optimal. In any other program $\mathbf{c} = (c_t)$, there must be two values c_T and c_{T+1} that are not equal. But in this case, we can construct a program $c' = (c'_t)$, such that

$$c'_t = c_t \quad \text{for } t \neq T, \ T+1,$$

$$c'_T = c'_{T+1} = \frac{c_T + c_{T+1}}{2}.$$

By strict concavity of u, we have

$$u(c_T) + u(c_{T+1}) < 2u\left(\frac{c_T + c_{T+1}}{2}\right) = u(c'_T) + u(c'_{T+1}).$$

So, $\mathbf{c'} = (c'_t)$ overtakes $\mathbf{c} = (c_t)$. Hence, there is no optimal program according to the "overtaking" criterion. $\qquad\square$

In view of this example, we focus on the "discounted" case: a discount factor δ is given, $0 < \delta < 1$. The *total return* $w(\mathbf{c})$ of a program $\mathbf{c} = (c_t)$ is defined by

$$w(\mathbf{c}) = \sum_{t=0}^{\infty} \delta^t u(c_t). \qquad (7.2.4)$$

Observe that for *any* program $\mathbf{c} = (c_t)_{t=0}^{\infty}$ *from* $\underline{S} > 0$, *the discounted sum of returns is bounded* since we have

$$w(\mathbf{c}) = \sum_{t=0}^{\infty} \delta^t u(c_t) \leq \sum_{t=0}^{\infty} \delta^t u(S) \equiv \frac{u(\underline{S})}{1 - \delta}. \qquad (7.2.5)$$

A program $\mathbf{c}^* = (c_t^*)_{t=0}^{\infty}$ from $\underline{S} > 0$ is *optimal* if

$$w(\mathbf{c}^*) = \sum_{t=0}^{\infty} \delta^t u(c_t^*) \geq \sum_{t=0}^{\infty} \delta^t u(c_t) \qquad (7.2.6)$$

for *all* programs $\mathbf{c} = (c_t)_{t=0}^{\infty}$ from (the same initial stock) $\underline{S} > 0$.

7.2.1   Optimal programs

We begin by proving the existence and uniqueness of an optimal program.

Theorem 7.2.2. *Under* (U.1)–(U.3), *there is a unique optimal program.*

Proof.   We provide a sketch, since the details can be adapted from Theorem 5.4.1. Let $\mathbf{c} = (c_t)_{t=0}^{\infty}$ be a program from $\underline{S} > 0$.
Let

$$\alpha = \sup\{w(\mathbf{c}) : \mathbf{c} \in \mathcal{C}\}. \tag{7.2.7}$$

Choose a sequence $\mathbf{c}^n = (c_t^n)$ in $\mathcal{C}$, such that

$$w(\mathbf{c}^n) \geq \alpha - \frac{1}{n}.$$

By a diagonalization argument (see 13.1.5), there is a subsequence of $\mathbf{c}^n$ (retain notation), such that for all $t \geq 0$,

$$c_t^n \to c_t^* \quad \text{as} \quad n \to \infty,$$

where $\mathbf{c}^* = (c_t^*)_{t=0}^{\infty}$ is some non-negative sequence. The proof is completed by verifying that (i) $\mathbf{c}^* = (c_t^*)_{t=0}^{\infty}$ is a program from $\underline{S}$, and (ii) $w(\mathbf{c}^*) = \alpha$, so that it is indeed an optimal program.
    Uniqueness is a consequence of strict concavity, (U.3). $\square$

    From the assumption (U.2), it follows that the optimal program is intertemporally efficient, i.e., $\sum_{t=0}^{\infty} c_t^* = \underline{S}$.
    For simplicity of exposition, we make the following assumption on u:

(U.4) *u is twice continuously differentiable on R_{++}; $u'(c) > 0$ and $u''(c) < 0$ at $c > 0$.*

    We now study the qualitative properties of the optimal program. We consider two cases: first with an unbounded marginal utility and then with a bounded marginal utility, separately.

Case (i). Here, we make the assumption:

(U.5) (Inada) $u'(c) \to \infty$ *as* $c \to 0$

Under this assumption, the unique optimal program $\mathbf{c}^* = (c_t^*)_{t=0}^{\infty}$ satisfies $c_t^* > 0$.

Lemma 7.2.1. *Under* (U.1), (U.4), (U.5), *the optimal program* $\mathbf{c}^* = (c_t^*)_{t=0}^{\infty}$ *must satisfy* $c_t^* > 0$ *for all* $t > 0$.

Proof. Since u is increasing, $c_t^* > 0$ for some period, say, $t = t_1$. Suppose that $c_t^* = 0$, for some period, say, $t = t_2$. Consider first some $\varepsilon > 0$ small enough, such that $c_{t_1}^* - \varepsilon > 0$. Consider the program $\mathbf{c} = (c_t)$ defined by $c_t = c_t^*$ for $t \neq t_1, t_2$, $c_{t_1} = c_{t_1}^* - \varepsilon$, $c_{t_2} = \varepsilon$. As $\mathbf{c}^*$ is assumed to be optimal,

$$
\begin{aligned}
0 &\geq \left[\sum_{t=0}^{\infty} \delta^t u(c_t) - \sum_{t=0}^{\infty} \delta^t u(c_t^*) \right] \\
&= (\delta^{t_1} u(c_{t_1}) + \delta^{t_2} u(c_{t_2})) - (\delta^{t_1} u(c_{t_1}^*) + \delta^{t_2} u(c_{t_2}^*)) \\
&= (\delta^{t_1} u(c_{t_1}^* - \varepsilon) + \delta^{t_2} u(\varepsilon)) - (\delta^{t_1} u(c_{t_1}^*) + \delta^{t_2} u(0)) \\
&= \delta^{t_2} u(\varepsilon) - \delta^{t_1} (u(c_{t_1}^*) - u(c_{t_1}^* - \varepsilon)) \\
&= \delta^{t_2} u(\varepsilon) - \delta^{t_1} (\varepsilon u'(z_\varepsilon)) \ (\text{where } c_{t_1}^* - \varepsilon < z_\varepsilon < c_{t_1}^*) \\
&= \varepsilon [\delta^{t_2} (u(\varepsilon)/\varepsilon) - \delta^{t_1} u'(z_\varepsilon)] \\
&\geq \varepsilon [\delta^{t_2} (u(\varepsilon)/\varepsilon) - \delta^{t_1} u'(c_{t_1}^* - \varepsilon)].
\end{aligned}
$$

Recall that, by (U.5), as $\varepsilon \to 0$, $(u(\varepsilon)/\varepsilon) \to \infty$, and by continuity of u', $u'(c_{t_1}^* - \varepsilon) \to u'(c_{t_1}^*)$. So, for sufficiently small positive ε, the last expression is positive, a contradiction. $\qquad \square$

Theorem 7.2.3. *Under* (U.1), (U.2), (U.4), *and* (U.5), *if* $\mathbf{c}^* = (c_t^*)_{t=0}^{\infty}$ *is the optimal program from* $\underline{S}$, *then there is a constant* $\lambda^* > 0$, *such that*

$$
\delta^t u'(c_t^*) = \lambda^* \quad \text{for all } t \geq 0. \tag{7.2.8}
$$

Proof. To simplify writing, we write S for $\underline{S}$ in this proof. Consider the optimal $\mathbf{c}^* = (c_t^*)$. Now, for each $T = 1, 2, \ldots$, write $S_T^* = \sum_{t=T+1}^{\infty} c_t^* > 0$. Observe that $(c_t^*)_{t=0}^{T}$ is the unique solution to the following maximization problem:

$$\text{maximize} \quad \sum_{t=0}^{T} \delta^t u(c_t)$$

$$\text{subject to} \quad \sum_{t=0}^{T} c_t = S - S_T^*, \ c_t \geq 0, \ t = 0, 1, \dots, T.$$

Then there is some Lagrangian multiplier λ_T, such that $(c_t^*)_{t=0}^{T}$ maximizes

$$\mathcal{L}\{(c_t), \lambda_T\} \equiv \sum_{t=0}^{T} \delta^t u(c_t) - \lambda_T \left[\sum_{t=0}^{T} c_t - (\underline{S} - S_T^*) \right]. \qquad (7.2.9)$$

Since $c_t^* > 0$, the first-order condition gives us

$$\delta^t u'(c_t^*) = \lambda_T, \quad t = 0, 1, \dots, T. \qquad (7.2.10)$$

But the left side does not involve T, so $\lambda_T = \lambda_{T+1} = \cdots = \lambda^*$, *say*, which leads to (7.2.8). □

Now, to expand the meaning of the condition,

$$\delta^t u'(c_t^*) = \lambda^*, \qquad (7.2.11)$$

we see that the left side is the discounted marginal return (utility) from optimal extraction (consumption) in period t, and λ^*, the "multiplier", is the present value price of the resource (which remains constant over time). Write $\delta \equiv \frac{1}{1+r}$, where r is the *discount rate*.

Then $p_t = \frac{\lambda^*}{\delta^t}$ is the *current price in period t*. As

$$[p_{t+1} - p_t]/p_t = \left[\frac{\lambda^*}{\delta^{t+1}} \Big/ \frac{\lambda^*}{\delta^t} \right] - 1$$

$$= \frac{1}{\delta} - 1 = r,$$

we see that the current price goes up at the discount rate.

We note a converse of Theorem 7.2.3. For the sake of completeness, we sketch a proof.

Theorem 7.2.4. *Under* (U.1)–(U.5), *let a strictly positive sequence* $\mathbf{c}^* = (c_t^*)_{t=0}^{\infty}$ *from* $\underline{S}$ *be an intertemporally efficient program, such that for some* $\lambda^* > 0$, *the condition* (7.2.11) *holds. Then* $\mathbf{c}^* = (c_t^*)_{t=0}^{\infty}$ *is the optimal program from* $\underline{S}$.

Proof. To simplify writing, we write S for $\underline{S}$ in this proof. Take any program $\mathbf{c} = (c_t)_{t=0}^{\infty}$ from $\underline{S} > 0$. Note that for any finite T,

$$\sum_{t=0}^{T} \delta^t u(c_t) - \sum_{t=0}^{T} \delta^t u(c_t^*)$$

$$\leq \sum_{t=0}^{T} [\delta^t (c_t - c_t^*) u'(c_t^*)] \text{ [use } u''(c) < 0 \text{ at } c > 0]$$

$$= \sum_{t=0}^{T} (c_t - c_t^*) \lambda^* \leq \lambda^* \left(\sum_{t=0}^{T} c_t\right) - \lambda^* \left(\sum_{t=0}^{T} c_t^*\right).$$

Looking at the last expression, one notes that $\sum_{t=0}^{\infty} c_t \leq S$ and $\sum_{t=0}^{\infty} c_t^* = S$. Now, taking the limit as $T \to \infty$, one gets (use (7.2.3))

$$\sum_{t=0}^{\infty} \delta^t u(c_t) - \sum_{t=0}^{\infty} \delta^t u(c_t^*) \leq 0,$$

which completes the proof of optimality. □

One immediately concludes from (7.2.11) that

$$\delta u'(c_{t+1}^*) = u'(c_t^*) \quad \text{for all } t \geq 0. \tag{7.2.12}$$

It follows that $c_t^* > c_{t+1}^*$, i.e., along *the optimal program extraction declines over time*.

Two examples where the optimal c_t^* can be explicitly computed are given.

Example 7.2.2. Let $u(c) = \sqrt{c}$. Then the optimal program is given by $c_t^* = \underline{S}(1 - \delta^2)\delta^{2t}$.

Sketch of the Proof. Recall that

$$\delta^t u'(c_t^*) = \lambda^* \quad \text{for all } t \geq 0,$$

$$\text{or} \quad c_t^* = \delta^{2t}/4(\lambda^*)^2.$$

But we also need to satisfy

$$\sum_{t=0}^{\infty} c_t^* = \underline{S},$$

$$\text{or} \quad \sum_{t=0}^{\infty} [\delta^{2t}/(4(\lambda^*)^2)] = \underline{S},$$

$$\text{leading to} \quad \left(\sum_{t=0}^{\infty} \delta^{2t} \right) / \underline{S} = 4(\lambda^*)^2. \qquad \square$$

Example 7.2.3. Let $u(c) = \log c$. Show that $c_t^* = \delta^t(1 - \delta)\underline{S}$.

We shall state two sensitivity results. First, we show that when the stock of the resource is higher, the optimal program involves higher consumption in *every* period. Second, we show that if the consumer is more patient, he consumes less in the *initial* period.

Theorem 7.2.5. *Under* (U.1)–(U.5), *we have the following properties:*

(i) *Let* $\underline{S} > \underline{S}' > 0$ *and* $\mathbf{c}^*(\underline{S}) = (c_t^*(\underline{S}))$ *and* $\mathbf{c}^*(\underline{S}') = (c_t^*(\underline{S}'))$ *be the optimal programs from* $\underline{S}$ *and* $\underline{S}'$, *respectively. Then,*

$$c_t^*(\underline{S}) > c_t^*(\underline{S}') \quad \text{for all } t \geq 0.$$

(ii) *Let* $0 < \delta < \delta' < 1$. *Fix* $\underline{S} > 0$. *Let* $\mathbf{c}^*(\delta) = (c_t^*(\delta))$ *and* $\mathbf{c}^*(\delta') = (c_t^*(\delta'))$ *be the optimal programs from* $\underline{S}$, *corresponding to* δ *and* δ', *respectively. Then,*

$$c_0^*(\delta) > c_0^*(\delta'). \tag{7.2.13}$$

Proof. To simplify writing, we write S, S' for $\underline{S}, \underline{S}'$ in this proof.
(i) First, we show that

$$c_0^*(S) > c_0^*(S').\tag{7.2.14}$$

Suppose, to the contrary, that

$$c_0^*(S) \le c_0^*(S').\tag{7.2.15}$$

We shall arrive at a contradiction. From (7.2.15), we get

$$u'(c_0^*(S)) \ge u'(c_0^*(S')).$$

We also know that

$$u'(c_0^*(S)) = \delta u'(c_1^*(S)) \quad \text{and} \quad u'(c_0^*(S')) = \delta u'(c_1^*(S')).\tag{7.2.16}$$

Upon substitution and cancellation, we get

$$u'(c_1^*(S)) \ge u'(c_1^*(S')),$$
$$\text{or} \quad c_1^*(S) \le c_1^*(S').$$

Repeating the step, we get that (7.2.15) implies

$$c_t^*(S) \le c_t^*(S') \quad \text{for all } t \ge 0.$$

Hence,

$$S = \sum_{t=0}^{\infty} c_t^*(S) \le \sum_{t=0}^{\infty} c_t^*(S') = S',$$

which contradicts $S > S'$ and establishes (ii). Now, we show that

$$c_0^*(S) > c_0^*(S') \text{ implies } c_t^*(S) > c_t^*(S') \quad \text{for all } t \ge 1.\tag{7.2.17}$$

Clearly, $c_0^*(S) > c_0^*(S')$ implies $u'(c_0^*(S)) < u'(c_0^*(S'))$; hence,

$$\delta u'(c_1^*(S)) = u'(c_0^*(S)) < u'(c_0^*(S')) = \delta u'(c_1^*(S')),$$
$$\text{or} \quad c_1^*(S) > c_1^*(S').$$

Repeating the step, we get (7.2.17); hence, we get (i). $\qquad\square$

(ii) Assume, to the contrary, that

$$c_0^*(\delta) \leq c_0^*(\delta'). \tag{7.2.18}$$

This leads to

$$u'(c_0^*(\delta)) \geq u'(c_0^*(\delta')). \tag{7.2.19}$$

By using (7.2.12) and (7.2.19),

$$\delta u'(c_1^*(\delta)) = u'(c_0^*(\delta)) \geq u'(c_0^*(\delta')) = \delta' u'(c_1^*(\delta')). \tag{7.2.20}$$

Hence, the assumption that $\delta < \delta'$ and (7.2.20) lead to

$$u'(c_1^*(\delta)) > u'(c_1^*(\delta')),$$

which means

$$c_1^*(\delta) < c_1^*(\delta').$$

Repeating the step, we get

$$c_t^*(\delta) < c_t^*(\delta') \quad \text{for all } t \geq 1.$$

Hence,

$$S = \sum_{t=0}^{\infty} c_t^*(\delta) < \sum_{t=0}^{\infty} c_t^*(\delta') = S.$$

This contradiction means that (7.2.18) must be false, establishing the claim (7.2.13). □

Case (ii). Here, we make "finite steepness" assumption:

(U.6) $\lim_{c \to 0} u'(c) \leq \chi$ *for some positive (finite)* χ.

We write $u'(0) = \lim_{c=0} u'(c)$ [assumed to exist].

Theorem 7.2.6. *Consider the optimal program* $\mathbf{c}^* = (c_t^*)_{t=0}^{\infty}$. *Under* (U.1)–(U.4) *and* (U.6), *if for some finite* T, $c_T^* = 0$, *then* $c_{T+t}^* = 0$ *for all* $t \geq 1$.

Proof.   Suppose that for some finite T, $c_T^* = 0$, and there is some $\tau > T$, such that $c_\tau^* > 0$. Then, by (U.2) (monotonicity),

$$u(c_\tau^*) > u(c_T^*) = u(0).$$

Consider the program $\mathbf{c} = (c_t)$ defined as $c_t = c_t^*$ for all $t \neq T, \tau$, $c_T = c_\tau^* > 0$ and $c_\tau = c_T^* = 0$. Then, $\mathbf{c}$ is also a program from $\underline{S}$ because $\sum_{t=0}^{\infty} c_t = \underline{S}$. The difference between the sum of utilities of the two programs is

$$\sum_{t=0}^{\infty} \delta^t u(c_t) - \sum_{t=0}^{\infty} \delta^t u(c_t^*) = [\delta^T u(c_T) + \delta^\tau u(c_\tau)] - [\delta^T u(c_T^*) + \delta^\tau u(c_\tau^*)]$$

$$= [\delta^T u(c_\tau^*) + \delta^\tau u(c_T^*)] - [\delta^T u(c_T^*) + \delta^\tau u(c_\tau^*)]$$

$$= (\delta^T - \delta^\tau)[u(c_\tau^*) - u(0)] > 0,$$

which contradicts the optimality of $\mathbf{c}^* = (c_t^*)_{t=0}^{\infty}$.   □

Theorem 7.2.7. *Let* $\mathbf{c}^* = (c_t^*)_{t=0}^{\infty}$ *be the optimal program. Under* (U.1)–(U.4) *and* (U.6), *there is a finite* T, *such that* $c_t^* = 0$ *for all* $t > T$.

Proof.   To simplify writing, we write S for $\underline{S}$ in this proof. If the claim is not true, $c_t^* > 0$ for all $t \geq 0$. We first argue that (7.2.8) also holds under the assumption of this proposition, i.e., for some $\lambda^* > 0$

$$\delta^t u'(c_t^*) = \lambda^* \quad \text{for all } t \geq 0.$$

Fix any finite $T \geq 1$. Write $s_T^* = S - \sum_{t=0}^{t=T} c_t^*$. Then $s_T^* > 0$ for all T. Now, consider all non-negative $c = (c_0, c_1, \ldots, c_T)$, such that

$$\sum_{t=0}^{t=T} c_t \leq S - s_T^*.$$

By the optimality principle, $\sum_{t=0}^{t=T} \delta^t u(c_t^*) \geq \sum_{t=0}^{t=T} \delta^t u(c_t)$. By the standard Lagrangian method, there is some λ_T^*, such that

$$\delta^t u'(c_t^*) = \lambda_T^*.$$

But choosing the horizon $T, T+1, \ldots$, we get

$$\delta^t u'(c_t^*) = \lambda_T^* = \lambda_{T+1}^* = \cdots = \lambda^*, \text{ say.}$$

But $u'(c_t^*) \leq \chi$ and $\delta^t \to 0$, so we get a contradiction.   □

Theorem 7.2.8. *Under* (U.1)–(U.4) *and* (U.6), *the optimal program is* $c_0^* = S$ *and* $c_t^* = 0$ *for all* $t > 0$ *if and only if* $u'(\underline{S}) \geq \delta u'(0)$.

Proof.   Again, to simplify writing, we write S for $\underline{S}$ in this proof. If $c_0^* = S$ is an optimal program, then it must be that $u'(S) \geq \delta u'(0)$, else if $u'(S) < \delta u'(0)$ transferring a small enough ε from period 0 to period 1 would increase the value of the program. Conversely, suppose that $u'(S) \geq \delta u'(0)$, but $c_0^* = S$ is *not* an optimal program, then the optimal consumption is positive in at least the first two periods. In other words, $S > c_0^*$, and $c_0^* > 0$ and $c_1^* > 0$. Hence, by a "transfer argument", we derive

$$u'(c_0^*) = \delta u'(c_1^*). \tag{7.2.21}$$

This leads to

$$u'(S) < u'(c_0^*) = \delta u'(c_1^*) < \delta u'(0). \tag{7.2.22}$$

This contradicts the assumption that $u'(S) \geq \delta u'(0)$.

To derive (7.2.21), let $\mathbf{c}^* = (c_t^*)$ be the optimal program. Choose $\varepsilon > 0$ sufficiently small, so that $c_0^* - \varepsilon > 0$. Consider the program $\mathbf{c} = (c_t)$ defined as

$$c_0 = c_0^* - \varepsilon, c_1 = c_1^* + \varepsilon, c_t = c_t^* \quad \text{for all } t \geq 2. \tag{7.2.23}$$

Then, by the optimality of $\mathbf{c}^* = (c_t^*)$, we have $\sum_{t=0}^{\infty} \delta^t u(c_t^*) \geq \sum_{t=0}^{\infty} \delta^t u(c_t)$ which leads to

$$u(c_0^*) + \delta u(c_1^*) \geq u(c_0) + \delta u(c_1)$$

$$\text{or} \quad u(c_0^*) + \delta u(c_1^*) \geq u(c_0^* - \varepsilon) + \delta u(c_1^* + \varepsilon)$$

$$\text{or} \quad u(c_0^*) - u(c_0^* - \varepsilon) \geq \delta[u(c_1^* + \varepsilon) - u(c_1^*)]$$

$$\text{or} \quad [u(c_0^*) - u(c_0^* - \varepsilon)]/\varepsilon \geq \delta[u(c_1^* + \varepsilon) - u(c_1^*)]/\varepsilon$$

$$\text{Taking limits,} \quad u'(c_0^*) \geq \delta u'(c_1^*).$$

The inequality in the other direction is derived by choosing $\varepsilon > 0$ sufficiently small, so that $c_1^* - \varepsilon > 0$ and considering the program $\mathbf{c} = (c_t)$ defined as

$$c_0 = c_0^* + \varepsilon, c_1 = c_1^* - \varepsilon, c_t = c_t^* \quad \text{for all } t \geq 2. \qquad \square$$

Exercise 7.2.1. Consider the T-period problem introduced in Section 7.1 (T is a finite positive integer):

$$\text{maximize} \quad \sum_{t=0}^{T} u(c_t)$$

$$\text{subject to} \quad \sum_{t=0}^{t=T} c_t \leq \underline{S}, \ c_t \geq 0 \quad \text{for all } t = 0, 1, \ldots, T.$$

Show that under (U.1)–(U.3), there is a unique optimal program:

$$c_t^{*T} = \underline{S}/(T+1).$$

[▼ *Hints*: Use the Weierstrass theorem to prove the existence of an optimal program. Observe that (U.2) implies that any optimal program (c_t^{*T}) must satisfy

$$\sum_{t=0}^{T} c_t^{*T} = \underline{S}.$$

Now, use strict concavity (U.3) to show that $c_0^{*T} = c_1^{*T} = \cdots = c_T^{*T}$. ▲]

7.3 Complements and Details

This chapter is based primarily on parts of Majumdar and Bar (2013). Hotelling's (1931) pioneering paper has led to a huge literature. Dasgupta and Heal (1979) provides a comprehensive introduction (Chapters 10 and 11). Of particular interest as supplementary reading are Solow (1974a) and Solow and Wan (1976). Sweeney (1993) is a review of the many extensions and contains a comprehensive list of references.

Chapter 8

Production with an Exhaustible Resource: Efficiency and Intergenerational Equity

We stand now where two roads diverge. But unlike the roads in Robert Frost's familiar poem, they are not equally fair. The road we have long been traveling is deceptively easy, a smooth superhighway on which we progress with great speed, but at its end lies disaster. The other fork of the road — the one less traveled by — offers our last, our only chance to reach a destination that assures the preservation of the earth.

— Rachel Carson

8.1 Introduction

Declining reserves of natural resources that play an essential role in producing consumption goods pose a challenge to sustainability. In this and the following chapter, we examine an infinite horizon model of intertemporal allocation in which an exhaustible resource is an essential input. To keep the exposition simple, it is assumed that there is a single "producible" good, which can be used for consumption or as an input (labeled "capital") for further production of itself. There is a primary factor of production ("labor"), the supply of which in every period is exogenously given. Moreover, there is an initial stock of a non-augmentable, freely storable natural resource that can be used only as an input. The technology is described by a production function $G : R_+^3 \to R_+$, the arguments of which are capital (k), labor (L) and the resource (r). The stock of the producible good in period $t + 1$, denoted by y_{t+1}, is

$$y_{t+1} = G(k_t, r_t, L_t) + k_t, \quad \text{for } t \geq 0. \tag{8.1.1}$$

It is assumed that "capital" does not depreciate. This stock can be divided into (non-negative) consumption and input:

$$y_{t+1} = c_{t+1} + k_{t+1}, \quad \text{for } t \geq 0. \tag{8.1.2}$$

Given the anthill stocks $(\underline{k}, \underline{S}) \geq 0$, and the exogenous supply of labor L_t (≥ 0) in every period, a program specifies the (non-negative) quantities of all inputs in every period, satisfying, in addition to (8.1.1) and (8.1.2), the requirements $k_0 = \underline{k}$ and

$$\sum_{t=0}^{\infty} r_t \leq \underline{S}. \tag{8.1.3}$$

The last inequality (8.1.3) simply asserts that the total quantity of the resource used over all the periods as an input cannot exceed the available initial stock.

It has been emphasized that the concept of "sustainability" ought to encompass notions of intertemporal efficiency and equity. Of course, as noted earlier, an efficient program can be extremely "selfish" and completely disregard the consumption prospects of the future. Recall that (see Section 1.5 of Chapter 1) intergenerational equity was formally introduced as a requirement that the consumption is the *same* in every period (i.e., that the consumption sequence $\mathbf{c} = (c_{t+1})$ satisfies $c_1 = c_2 = \cdots = c_t = \cdots = c > 0$). A central result (Theorem 8.4.2) identifies sufficient conditions under which there *is* an equitable program which is *also* efficient. Furthermore, it is noted (Corollary 8.4.1) that a program is efficient and equitable if and only if it is a non-trivial maximin (see (8.2.9)) program. In proving this result, it is assumed (Condition E in Section 8.4) that there is some (non-zero) equitable program. A general framework in which such a Condition E holds is explored in Cass and Mitra (1991) and this issue is revisited in Chapter 9 in a Cobb–Douglas economy.

The model of production that we study in this chapter can be interpreted as a special case of the study by Malinvaud (1953). The exhaustible resource is a capital good with a distinctive property (see (8.2.5)). In the Malinvaud–Gale tradition, we naturally turn to "duality" or existence of "competitive" prices, supporting an efficient and equitable program. A fundamental result (Theorem 8.3.3) asserts that for an interior competitive program, the (present value) price of the resource must remain constant over time. The basic duality result (Theorem 8.5.1) that a program is efficient *and* equitable if and only if there is a sequence of strictly positive prices such that (i) the program maximizes intertemporal profit computed at these prices period after period (8.5.1) and (ii) the sum of the present values of capital and the stock of the resource goes to zero in the long run (the transversality condition (8.5.2)). When the resource is sufficiently "important" in production, an interior program is efficient if and only if it satisfies the intertemporal profit maximization condition (8.5.1) and the "resource exhaustion" condition (8.1.3) with an equality.

8.2   Model

Consider an economy with a technology given by a production function, G from R_+^3 to R_+. The production possibilities involve a capital input, k, an exhaustible resource input, r, labor input, L, and current (net) output, $Q = G(k, r, L)$, for $(k, r, L) \geq 0$.

A *total (gross) output function*, F, is defined by

$$F(k, r, L) = G(k, r, L) + k \quad \text{for } (k, r, L) \geq 0. \tag{8.2.1}$$

The production function, G, is assumed to satisfy

(A.1) $G(k, r, L)$ *is concave, homogenous of degree one, and continuous for* $(k, r, L) \geq 0$; *it is differentiable for* $(k, r, L) \gg 0$.

(A.2) G *is non-decreasing in* k, r, *and* L, *for* $(k, r, L) \geq 0$; *also,* $(G_k, G_r, G_L) \gg 0$ *for* $(k, r, L) \gg 0$.

For $(k, r, L) \gg 0$, we define the capital share (α), the resource share (β) and the labor share (γ), in current output $Q = G(k, r, L)$ by

$$\alpha = (kG_k)/G(k, r, L); \quad \beta = (rG_r)/G(k, r, L),$$

$$\gamma = (LG_L)/G(k, r, L). \tag{8.2.2}$$

By (A.1) and (A.2), it is clear that for $(k, r, L) \gg 0, \alpha, \beta$ and γ are positive, and $(\alpha + \beta + \gamma) = 1$. We denote $\inf_{(k,r,L)\gg 0} \beta$ by $\underline{\beta}$.

It is assumed that L_t, the quantity of labor available in period t is exogenously given and constant over time: $L_t = \bar{L} > 0$. In fact, we will normalize and set $\bar{L} = 1$. Also, for simplification of typing, we shall write S_t to denote the stock of the resource in period t (rather than $\underline{S}_{-t}$).

A *feasible program* from $(\underline{k}, \underline{S}) \gg 0$ is a sequence $\langle \mathbf{k}, \mathbf{S} \rangle = (k_t, S_t)$, satisfying

$$(k_0, S_0) = (\underline{k}, \underline{S}),$$

$$k_t \geq 0, \text{ and } 0 \leq S_{t+1} \leq S_t \text{ for } t \geq 0,$$

$$F(k_t, S_t - S_{t-1}, L_t) - k_{t+1} \geq 0 \text{ for } t \geq 0. \tag{8.2.3}$$

In (8.2.3), S_t is to be interpreted as the resource stock at time t. Associated with a program $(\mathbf{k}, \mathbf{S})$ from $\langle \underline{k}, \underline{S} \rangle$ is a sequence $(\mathbf{r}, \mathbf{y}, \mathbf{c})$ defined by

$$r_t = S_t - S_{t+1}, y_{t+1} = F(k_t, r_t, L_t) \quad \text{for } t \geq 0,$$

$$c_{t+1} = y_{t+1} - k_{t+1} \quad \text{for } t \geq 0. \tag{8.2.4}$$

In (8.2.4), y_{t+1} is to be interpreted as the *total output* in period $(t + 1)$; c_{t+1} as the consumption in period $(t + 1)$, and r_t as the resource use at period t. Note that (8.2.3) and (8.2.4) imply that $r_t \geq 0$ for $t \geq 0$ and $\sum_{t=0}^{\infty} r_t \leq \underline{S}$. A program $(\mathbf{k}, \mathbf{S})$ from $(\underline{k}, \underline{S})$ is *interior* if $(k_t, r_t) \gg 0$ for $t \geq 0$. It is said to *sustain a positive consumption level* if $\inf_{t \geq 1} c_t > 0$.

A program $(\mathbf{k}, \mathbf{S})$ from $(\underline{k}, \underline{S})$ *dominates* a program $(\bar{\mathbf{k}}, \bar{\mathbf{S}})$ from $(\underline{k}, \underline{S})$ if $c_t \geq \bar{c}_t$, for $t \geq 1$ and $c_t > \bar{c}_t$ for some t. A program $(\bar{\mathbf{k}}, \bar{\mathbf{S}})$

from $(\underline{k}, \underline{S})$ is *inefficient* if there is a program from $(\underline{k}, \underline{S})$ which dominates it. It is *efficient* if it is not inefficient.

The production possibilities can be viewed in the "stock version" as given by a technology set $\mathscr{T}$ of input–output pairs in the following way:

$$\mathscr{T} = \{[(k, S, L), (y, S', 0)],$$

$$0 \le y \le F(k, r, L); 0 \le r \le (S - S'); (k, r, L, S') \ge 0\}. \quad (8.2.5)$$

It is clear that if $(\mathbf{k}, \mathbf{S})$ is a program from $(\underline{k}, \underline{S})$, then $[(k_t, S_t, L_t), (y_{t+1}, S_{t+1}, 0)] \in \mathscr{T}$ for $t \ge 0$.

A program $(\mathbf{k}, \mathbf{S})$ from $(\underline{k}, \underline{S})$ is called *competitive* if there is a non-null sequence of non-negative prices $\langle \mathbf{p}, \mathbf{q}, \mathbf{w} \rangle = \langle p_t, q_t, w_t \rangle$, such that for $t \ge 0$,

$$p_{t+1} y_{t+1} + q_{t+1} S_{t+1} - p_t k_t - q_t S_t - w_t L_t$$

$$\ge p_{t+1} y + q_{t+1} S' - p_t k - q_t S - w_t L$$

$$\text{for all } [(k, S, L), (y, S', 0)] \in \mathscr{T}. \quad (8.2.6)$$

In other words, the *intertemporal profit maximization condition* (8.2.6) is satisfied at each date. The prices $(\mathbf{p}, \mathbf{q}, \mathbf{w})$ are said to be competitive or Malinvaud prices *supporting* the program $(\mathbf{k}, \mathbf{S})$. A competitive program is said to satisfy the *transversality condition* at the price sequence $(\mathbf{p}, \mathbf{q}, \mathbf{w})$ if

$$\lim_{t \to \infty} (p_t k_t + q_t S_t) = 0. \quad (8.2.7)$$

It is said to have *finite consumption value* if

$$\sum_{t=1}^{\infty} p_t c_t < \infty. \quad (8.2.8)$$

A program $(\bar{\mathbf{k}}, \bar{\mathbf{S}})$ from $(\underline{k}, \underline{S})$ is a *maximin program* if

$$\inf_{t \ge 1} \bar{c}_t \ge \inf_{t \ge 1} c_t \quad (8.2.9)$$

for every program $(\mathbf{k}, \mathbf{S})$ from $(\underline{k}, \underline{S})$. It is a *non-rival maximin program* if it is a maximin program *and* maintains a positive consumption level (i.e., $\inf_{t \geq 1} \bar{c}_t > 0$). It is an *equitable* or *stationary program* if

$$\bar{c}_t = c_{\overline{t+1}} \quad \text{for } t \geq 1. \tag{8.2.10}$$

It is a *non-trivial equitable* program if it is an equitable program *and* maintains a positive consumption level, i.e., $\bar{c}_t = \bar{c}_{t+1} = \bar{c} > 0$.

8.3 Some Preliminary Properties of Efficient and Equitable Programs

In this section, we note some properties of efficient and equitable programs, which will be useful in the analysis of the following two sections. First, we establish that if there is an equitable program, then there is an efficient program which has the same constant consumption as the original program for $t = 2$ onwards (Theorem 8.3.1). Second, we prove that if an efficient program has a positive non-decreasing consumption level, then it is necessarily interior (Theorem 8.3.2).

For our purpose, we will assume that capital is essential in production:

(A.3′) $G(0, r, L) = 0$ for $(r, L) \geq 0$.

Lemma 8.3.1. *Under* (A.1), (A.2) *and* (A.3′), *if* $(\bar{\mathbf{k}}, \bar{\mathbf{S}})$ *is an inefficient program from* $(\underline{k}, \underline{S})$, *then there is a program* $(\mathbf{k}, \mathbf{S})$, *such that* $c_t \geq \bar{c}_t$ *for all* $t \geq 1$, *and* $c_1 > \bar{c}_1$.

Proof. We provide an informal sketch. Note, first, that if $(\mathbf{k}', \mathbf{S}')$ is a program from $(\underline{k}, \underline{S})$ and $c'_s > c$ for some $c \geq 0$ and $s > 1$, then clearly we can find a program (k'', S'') from $(\underline{k}, \underline{S})$ with $c''_t = c'_t$ for $t \neq s - 1, s, c''_s = c$ and $c''_{s-1} > c'_{s-1}$. This is because, by (A.3′), $k'_{s-1} > 0$. So, we can be reducing c'_s to $c''_s = c$, also reduce k'_{s-1} to k''_{s-1}, by just enough (by (A.3′) again), so that $k''_s = k'_s$. This makes

c''_{s-1} larger than c'_{s-1} by $(k'_{s-1} - k''_{s-1}) > 0$. This procedure leaves $k'_t = k''_t$ for $t \neq s - 1$, and $S'_t = S''_t$ for all t, so $c'_t = c''_t$ for $t \neq s - 1, s$.

If $(\bar{\mathbf{k}}, \bar{\mathbf{S}})$ is an inefficient program from $(\underline{k}, \underline{S})$, then there is a program $(\mathbf{k}', \mathbf{S}')$ from $(\underline{k}, \underline{S})$, such that $c'_t \geq \bar{c}_t$ for all t and $c'_s > \bar{c}_s$ for some $s \geq 1$. If $s > 1$, then by using the above argument a finite number of times, the result is established. □

Lemma 8.3.2. *Under* (A.1), *if* $(\mathbf{k^n}, \mathbf{S^n})$ *is a sequence of programs from* $(\underline{k}, \underline{S})$, *then there is a subsequence* $(\mathbf{k}^{n'}, \mathbf{S}^{n'})$ *which converges coordinatewise to a program* $(\bar{\mathbf{k}}, \bar{\mathbf{S}})$ *from* $(\underline{k}, \underline{S})$.

Proof. Consider the sequence $(\mathbf{x_t})$ defined by $x_0 = k$, $x_{t+1} = F(x_t, \underline{S}, 1)$ for $t \geq 0$. Clearly, for each n, $(0, 0, 0) \leq (k^n_t, y^n_t, c^n_t) \leq (x_t, x_t, x_t)$ for $t \geq 1$; $k^n_0 = \underline{k}$, $L^n_t = 1, 0 \leq S^n_t \leq \underline{S}$ for $t \geq 0$. Hence, for each t, $(k^n_t, S^n_t, L^n_t, y^n_{t+1}, c^n_{t+1})$ is a bounded sequence. By the Cantor Diagonalization Argument, there is a subsequence of n, say n', such that $(k^{n'}_t, S^{n'}_t, L^{n'}_t, y^{n'}_{t+1}, c^{n'}_{t+1})$ converges to some $(\bar{k}_t, \bar{S}_t, \bar{L}_t, y_{\bar{t}+1}, c_{\bar{t}+1})$ for each t. Using (A.1), (8.2.3) and (8.2.4), $(\bar{\mathbf{k}}, \bar{\mathbf{S}})$ is a program from $(\underline{k}, \underline{S})$. Note that since $S^{n'}_t \to \bar{S}$ for $t \geq 0$, so $r^{n'}_t \to \bar{r}_t$ for $t \geq 0$. □

Theorem 8.3.1. *Under* (A.1), (A.2) *and* (A.3'), *if there is an equitable program* $(\hat{\mathbf{k}}, \hat{\mathbf{S}})$ *from* $(\underline{k}, \underline{S})$, *with* $\hat{c}_t = d$ *for* $t \geq 1$, *then there is an efficient program* $(\bar{\mathbf{k}}, \bar{\mathbf{S}})$ *from* $(\underline{k}, \underline{S})$ *with* $\bar{c}_t = d$ *for* $t \geq 2$ *and* $\bar{c}_1 \geq d$.

Proof. Let $A = [(\mathbf{k}, \mathbf{S}) : (\mathbf{k}, \mathbf{S})$ is a program from $(\underline{k}, \underline{S})$ and $c_t \geq d$ for $t \geq 2]$. Let $A_1 = [c : c = c_1$ for some $(\mathbf{k}, \mathbf{S})$ in $A]$.

A_1 is non-empty since d belongs to A_1. Also, A_1 is bounded, since for any program $(\mathbf{k}, \mathbf{S})$ from $(\underline{k}, \underline{S})$, we have $0 \leq c_1 \leq F(\underline{k}, \underline{S}, 1)$. Let $\tilde{c}_1$ be the l.u.b. of A_1. Then there is a sequence $(\mathbf{k}', \mathbf{S}')$ in A, such that c^n_1 converges to $\tilde{c}_1$. By Lemma 8.3.2, there is a program $(\bar{\mathbf{k}}, \bar{\mathbf{S}})$ from $(\underline{k}, \underline{S})$ with $\bar{c}_1 = \tilde{c}_1$ and $\bar{c}_t \geq d$ for all $t \geq 2$.

We must have $\bar{c}_t = d$ for $t \geq 2$. Otherwise, using the argument in the proof of Lemma 8.3.1, we can find a program $(\mathbf{k}', \mathbf{S}')$, such that $c'_t \geq d$ for $t \geq 2$ and $c'_1 > \bar{c}_1 = \tilde{c}_1$, a contradiction to $\tilde{c}_1$ being the l.u.b. of A_1. Using exactly the same argument, $(\bar{k}, \bar{S})$ must also be efficient. □

Lemma 8.3.3. *Under* (A.2), *if* $(\mathbf{k}, \mathbf{S})$ *is an efficient program from* $(\bar{k}, \bar{S})$, *with* $c_{t+1} \geq c_t$ *for* $t \geq 1$, *then* $k_{t+1} \geq k_t$ *for* $t \geq 0$.

Proof. Suppose, on the contrary, that there is $\tau \geq 0$, such that $k_{\tau+1} < k_\tau$. Clearly, there is a program $(\mathbf{k}', \mathbf{S}')$ from $(k_{\tau+2}, S_{\tau+2})$ with $(k'_t, S'_t) = (k_{t+\tau+2}, m_{t+\tau+2})$ for $t \geq 0$ and $c'_t = c_{t+\tau+2}$ for $t \geq 1$. Since $S_\tau \geq S_{\tau+1} \geq S_{\tau+2}$, there is a program $(\mathbf{k}'', \mathbf{S}'')$ from (k_τ, S_τ) with $(k''_0, S''_0) = (k_\tau, S_\tau), (k''_t, S''_t) = (k'_{t-1}, S'_{t-1})$ for $t \geq 1$ and $c''_1 = G(k_\tau, r_\tau + r_{\tau+1}, 1) + (k_\tau - k_{\tau+2}) \geq G(k_{\tau+1}, r_{\tau+1}, 1) + (k_\tau - k_{\tau+1}) + (k_{\tau+1} - k_{\tau+2}) > G(k_{\tau+1}, r_{\tau+1}, 1) + (k_{\tau+1} - k_{\tau+2}) = c_{\tau+2} \geq c_{\tau+1}$, and $c''_{t+1} = c'_t = c_{t+\tau+2} \geq c_{t+\tau+1}$ for $t \geq 1$. This proves that $(\mathbf{k}, \mathbf{S})$ is inefficient. This contradiction proves that $k_{t+1} \geq k_t$ for $t \geq 0$. □

For our next result, we strengthen (A.3′) and assume that both capital and resource are essential in production.

(A.3) $G(k, 0, L) = 0 = G(0, r, L), \quad$ for $(k, L) \geq 0$.

Theorem 8.3.2. *Under* (A.1)–(A.3), *if* $(\mathbf{k}, \mathbf{S})$ *is an efficient program from* $(\underline{k}, \underline{S})$, *with* $c_{t+1} \geq c_t$ *for* $t \geq 1$, *and* $c_1 > 0$, *then* $k_{t+1} > k_t$ *for* $t \geq 0$, *and* $S_{t+1} < S_t$ *for* $t \geq 0$; *furthermore, the program is interior.*

Proof. By Lemma 8.3.3, $k_{t+1} \geq k_t$ for $t \geq 0$. If $S_t = S_{t+1}$ for some $t = \tau$, then $r_\tau = 0$ and by (A.3), $G(k_\tau, r_\tau, 1) = 0$. Since $c_{\tau+1} > 0$, $k_{\tau+1} < k_\tau$, a contradiction. Since $S_{t+1} \leq S_t$ for $t \geq 0$, $S_{t+1} < S_t$ for $t \geq 0$.

Clearly, $\underline{k} > 0$; otherwise, if $\underline{k} = 0$, then by (A.3), $c_1 = G(\underline{k}, r_0, 1) + \underline{k} - k_1 = -k_1 < 0$, since $c_1 > 0$, a contradiction. Since $k_{t+1} \geq k_t$ for $t \geq 0$, by Lemma 8.3.3, $k_t > 0$ for $t \geq 0$. Since $S_{t+1} < S_t$ for $t \geq 0$, $r_t > 0$. Hence, $(\mathbf{k}, \mathbf{S})$ is interior.

To prove that $k_{t+1} > k_t$ for $t \geq 0$, suppose, on the contrary, that $k_{t+1} = k_t$ for some $t = T$. Then there is a program $(\mathbf{k}', \mathbf{S}')$ from (k_T, S_T) with $(k'_0, S'_0) = (k_T, S_T)$ and $(k'_t, S'_t) = (k_{t+T+1}, S_{t+T+1})$ for $t \geq 1$ and $c'_1 = G(k_T, r_T + r_{T+1}, \bar{L}) + k_T - k_{T+2} = G(k_{T+1}, r_T + r_{T+1}, \bar{L}) + k_{T+1} - k_{T+2} > G(k_{T+1}, r_{T+1}, 1) + k_{T+1} - k_{T+2}$ (since $(\mathbf{k}, \mathbf{S})$ is interior and (A.2) holds) $= c_{T+2} \geq c_{T+1}$; and $c'_{t+1} = c_{t+T+2} \geq c_{t+T+1}$ for $t \geq 1$. Hence, $(\mathbf{k}, \mathbf{S})$ is inefficient, a contradiction. □

As a final result of this section, we note that for an interior competitive program, the present value price of the exhaustible resource is a constant. This is a fundamental duality result that repeatedly appears in other contexts (recall Theorem 7.2.3).

Theorem 8.3.3. *Under* (A.1) *and* (A.2), *if an interior program* $(\mathbf{k}, \mathbf{S})$ *from* $(\underline{k}, \underline{S})$ *is competitive at the price sequence* $(\mathbf{p}, \mathbf{q}, \mathbf{w})$, *then*

$$q_t = q_{t+1} \quad for\ t \geq 0. \tag{8.3.1}$$

Proof. Let $(\mathbf{k}, \mathbf{S})$ be a competitive program with supporting prices $(\mathbf{p}, \mathbf{q}, \mathbf{w})$. Since G is homogeneous of degree one, $\mathcal{T}$ is a cone, and the left side of the intertemporal profit maximization condition (8.2.6) is zero for $t \geq 0$. Since $[(0, S, 0), (0, S, 0)] \in \mathcal{T}$, for $S > 0$, we have $q_{t+1}S - q_t S \leq 0$, implying $q_{t+1} \leq q_t$ for all $t \geq 0$. Now, suppose that $q_{t+1} < q_t$ for some t. Since $(\mathbf{k}, \mathbf{S})$ is interior, $S_t > r_t > 0$. Choose $\bar{S}$ such that $S_t > \bar{S} > r_t > 0$. Then surely $[(K_t, \bar{S}, L_t), (Y_{t+1}, \bar{S} - r_t, 0)] \in \mathcal{T}$, and using this in (8.2.6), we get

$$q_{t+1}(S_t - r_t) - q_t S_t \geq q_{t+1}(\bar{S} - r_t) - q_t \bar{S},$$

$$\text{or} \quad (q_{t+1} - q_t)(S_t - \bar{S}) \geq 0.$$

This contradicts $q_{t+1} < q_t$ and $S_t > \bar{S}$. Hence, $q_{t+1} = q_t$, establishing (8.3.1). □

The next result is a characterization of competitive programs in terms of marginal productivities.

Theorem 8.3.4. *Under* (A.1) *and* (A.2), *an interior program* $(\mathbf{k}, \mathbf{S}))$ *is competitive if and only if*

$$F_{r_{t+1}} = F_{r_t} \cdot F_{k_{t+1}} \quad for\ t \geq 0. \tag{8.3.2}$$

Proof. Suppose that the marginal productivity condition (8.3.2) holds. By concavity of F, we have, $(k, r, L) \geq 0$,

$$F(k, r, L) - F(k_t, r_t, L_t) \leq F_{k_t}(k - k_t) + F_{r_t}(r - r_t) + F_{L_t}(L - L_t). \tag{8.3.3}$$

Using this, one gets, for all $t \geq 0$,

$$F(k_t, r_t, L_t) - F_{k_t} k_t - F_{r_t} r_t - F_{L_t} L_t$$
$$\geq F((k, r, L) - F_{k_t} k - F_{r_t} r - F_{L_t} L \quad \text{for } (k, r, L) \geq 0. \quad (8.3.4)$$

Define the prices as

$$p_0 = (F_{k_0}/F_{r_0}), \; p_{t+1} = (p_t/F_{k_t}) \quad \text{for } t \geq 0,$$
$$w_t = F_{L_t} \cdot p_{t+1} \quad \text{for } t \geq 0, \; q_t = 1 \quad \text{for } t \geq 0. \quad (8.3.5)$$

Now, using (8.3.3) and (8.3.4), we get, for all $t \geq 0$,

$$p_{t+1} F(k_t, r_t, L_t) - p_t k_t - q_t r_t - w_t L_t \geq p_{t+1} F(k, r, L)$$
$$- p_t k - q_t r - w_t L \quad \text{for } (K, R, L) \geq 0. \quad (8.3.6)$$

Rewriting, we obtain (observe that $q_t = q_{t+1} = 1$):

$$p_{t+1} y_{t+1} + q_{t+1} S_{t+1} - p_t k_t - q_t S_t - w_t L_t \geq p_{t+1} y + q_{t+1} S'$$
$$- p_t k - q_t S - w_t L \quad \text{for } [(k, S, L), (y, S', 0] \in \mathscr{T}, \quad (8.3.7)$$

and this completes the verification of the intertemporal profit maximization condition (8.2.6).

To go in the other direction, let $(\mathbf{k}, \mathbf{S})$ be an interior competitive program with supporting prices $(\mathbf{p}, \mathbf{q}, \mathbf{w})$. Then, $q_t = q_{t+1} = \bar{q}$ (say) for all $t \geq 0$. Hence, for $t \geq 0$,

$$p_{t+1} F(k_t, r_t, L_t) - p_t k_t - \bar{q} r_t - w_t L_t$$
$$\geq p_{t+1} F(k, r, L) - p_t k - \bar{q} r - w_t L \quad \text{for } (k, r, L) \geq 0. \quad (8.3.8)$$

Since the intertemporal profit,

$$[p_{t+1} F(k, r, L) - p_t k - \bar{q} r - w_t L],$$

is maximized at an interior point, the first-order condition leads to, for all $t \geq 0$,

$$p_{t+1} F_{k_t} = p_t; \quad p_{t+1} F_{r_t} = \bar{q}; \quad p_{t+1} F_{L_t} = w_t. \quad (8.3.9)$$

Since $(\mathbf{p}, \mathbf{q}, \mathbf{w})$ is non-null, we have $p_t > 0$, $w_t > 0$ for all $t \geq 0$, and $\bar{q} > 0$. Hence, from (8.3.9),

$$
\begin{aligned}
(F_{r_{t+1}}/F_{r_t}) &= (p_{t+1}F_{r_{t+1}})/(p_{t+1}F_{r_t}) \\
&= (p_{t+1}/p_{t+2}) \cdot [(p_{t+2}F_{r_{t+1}})/(p_{t+1}F_{r_t})] \\
&= (p_{t+1}/p_{t+2}) \cdot (\bar{q}/\bar{q}) = (p_{t+1}/p_{t+2}) = F_{k_{t+1}}.
\end{aligned}
$$

This completes the verification of the marginal productivity condition (8.3.2). □

Remark 8.3.1. For an interior competitive program, the supporting price sequence $(\mathbf{p}, \mathbf{q}, \mathbf{w})$ is determined uniquely (up to positive scalar multiplication) and can be expressed in terms of marginal productivities as in (8.3.5). In what follows, for an interior competitive program, prices will be defined as in (8.3.5), and $(\mathbf{p}, \mathbf{q}, \mathbf{w}) \gg \mathbf{0}$.

8.4 Existence of an Efficient Equitable Program

Our objective in this section is to establish the existence of an efficient equitable program, given arbitrary positive initial capital and resource stocks. In the process, we will note that such a program is precisely the maximin program from these initial stocks.

First, we consider the set of all constant consumption programs, and choose the one with maximum constant consumption. Such a program is shown to exist. Second, we show that this program is efficient.

The difficult step in this procedure is, obviously, the second. It requires that if the maximum constant consumption path is inefficient (so we can improve the lot of *one* generation without worsening the lot of any other), then we can increase the consumption of *every* generation by a *constant* positive amount.

For our existence proof, we need, in addition to (A.1)–(A.3), an assumption which says that the resource is "important" in production (in a sense made precise in (A.4)). We also need to assume that there exists a non-trivial equitable program from positive initial stocks (see Condition E).

We start the analysis by proving the existence of a maximin program.

Theorem 8.4.1. *Under* (A.1) *and* (A.2), *there exists an equitable maximin program from* $(\underline{k}, \underline{S})$.

Proof. Let $B = \{(\mathbf{k}, \mathbf{S}) : (\mathbf{k}, \mathbf{S})$ is a program from $(\underline{k}, \underline{S})$ and $c_t = c_{t+1}$ for $t \geq 1\}$. B is non-empty, since the equitable program with zero consumption is in B. Let $B_1 = \{c : (\mathbf{k}, \mathbf{S})$ is in B and $c_1 = c\}$. B_1 is bounded since for any program $(\mathbf{k}, \mathbf{S})$ from $(\underline{k}, \underline{S}), 0 \leq c_1 \leq F(\underline{k}, \underline{S}, 1)$. Let c_1^* be the l.u.b. of B_1. Then there exists a sequence $(\mathbf{k}^n, \mathbf{S}^n)$ in B, such that c_1^n converges to c_1^*. By Lemma 8.3.2, there is a program $(\bar{\mathbf{k}}, \bar{\mathbf{S}})$ from $(\underline{k}, \underline{S})$ with $\bar{c}_t = \bar{c}_{t+1}$ for $t \geq 1$ and $\bar{c}_1 = c_1^*$. We claim that this is a maximin program from $(\underline{k}, \underline{S})$. If not, then there is a program $(\tilde{\mathbf{k}}, \tilde{\mathbf{S}})$ from $(\underline{k}, \underline{S})$, such that $\inf_{t \geq 1} \tilde{c}_t > \bar{c}_1 = c_1^*$. But then clearly, there exists a program $(\mathbf{k}', \mathbf{S}')$ from $(\underline{k}, \underline{S})$ with $c_t' = \inf_{t \geq 1} \tilde{c}_t$ for $t \geq 1$, which contradicts the definition of c_1^*.

We now turn to the basic question of the existence of an efficient program. We proceed by introducing the following condition:

Condition E. *There exists a non-trivial equitable program from any* $(\underline{k}, \underline{S}) \gg 0$. □

It should be stressed that the assumptions (A.1)–(A.3) (or (A.4) stated below) are not sufficient for this condition to hold. When the resource is essential or important, one needs enough substitution possibilities among inputs that can compensate for a diminishing availability of the resource. A broad class of Cobb–Douglas economies that satisfy Condition E is identified in Chapter 9. For a complete characterization of production functions, G, for which Condition E is satisfied, the interested reader is referred to Cass and Mitra (1992).

Theorem 8.4.2. *Under* (A.1)–(A.3) *and Condition E, there exists an efficient equitable program* $(\bar{\mathbf{k}}, \bar{\mathbf{S}})$ *from some* $(\bar{k}, \bar{S})$ *with* $\bar{c}_t > 0$ *for all* $t > 0$.

Proof. By Condition E, there is an equitable program $(\hat{\mathbf{k}}, \hat{\mathbf{S}})$ from $(\underline{k}, \underline{S})$ with $\hat{c}_t = d > 0$ for $t \geq 1$. By Theorem 8.3.1, there is an efficient program $(\mathbf{k}', \mathbf{S}')$ from $(\underline{k}, \underline{S})$ with $c_t' = d$ for $t \geq 2$ and $c_1' \geq d$. Let $(k_1, S_1) = (\bar{k}, \bar{S})$, then the program $\langle \bar{k}, \bar{S} \rangle$ from (k_1', S_1') defined by $(\bar{k}_t, \bar{S}_t) = (k_{t+1}', S_{t+1}')$ for $t \geq 0$ is an efficient and equitable program with $\bar{c}_t = d > 0$ for $t \geq 1$. □

It should be stressed that Theorem 8.4.2 asserts the existence of an efficient, equitable program from *some initial stocks* $(\bar{k}, \bar{S})$.

For our next result, we need an additional assumption, which conveys (along with (A.3)) that the exhaustible resource is "important" in production (recall $\bar{L} = 1$).

(A.4) *Given any* $(\tilde{k}, \tilde{r}) \gg 0$, *there is* $\tilde{\eta} > 0$, *such that for all* (k, r), *satisfying* $k \geq \tilde{k}, 0 \leq r \leq \tilde{r}$, *we have* $\{[rG_r(k, r, 1)]/ G_z(k, r, 1)\} \geq \tilde{\eta}$.

This assumption says, roughly speaking, that the ratio of the share of resource in output to the share of labor (recall $\bar{L} = 1$) is bounded away from zero.

Lemma 8.4.1. *Under* (A.1)–(A.4), *if there exists an efficient equitable program* $(\mathbf{k}, \mathbf{S})$ *from* $(\underline{k}, \underline{S})$, *with* $c_t > 0$ *for all* t, *then given any* $\delta > 1$, *there exist* $\lambda > 1$ *and a program* $(\mathbf{k}', \mathbf{S}')$ *from* $(\delta\underline{k}, \delta\underline{S})$ *with* $c_t' = \lambda c_t$ *for* $t \geq 1$.

Proof. Since $c_t = d > 0$ for all t, $k > 0$ by (A.3). By Theorem 8.3.2, $k_{t+1} > k_t$, $0 \leq S_{t+1} \leq S_t$ for $t \geq 0$ and the program is interior. Hence, $k_t \geq \underline{k}$ for $t \geq 0$ and $0 < r_t \leq \underline{S}$ for $t \geq 0$. By (A.4), given $\delta > 1$, there is $\eta > 0$ such that for $k \geq \underline{k}$, $0 < r \leq \delta\underline{S}$,

$$\{[rG_r(k, r, 1)]/G_z(k, r, 1)\} \geq \eta. \tag{8.4.1}$$

Choose $\delta > \lambda > 1$, such that

$$[(\lambda - 1)/(\delta - \lambda)] < [\eta/\delta]. \tag{8.4.2}$$

Clearly, this can be done. Then, for $t \geq 0$,

$$G(\lambda k_t, \delta r_t, 1) - G(\lambda k_t, \lambda r_t, \lambda) \geq 0. \tag{8.4.3}$$

To see this, we write, for $t \geq 0$,

$$G(\lambda k_t, \delta r_t, 1) - G(\lambda k_t, \lambda r_t, \lambda)$$
$$\geq G_r(\lambda k_t, \delta r_t, 1)(\delta - \lambda)r_t - G_L(\lambda k_t, \delta r_t, 1)(\lambda - 1)$$
$$= (\lambda - 1)G_L(\lambda k_t, \delta r_t, 1)\left[\left\{\frac{(\delta - \lambda)G_r(\lambda k_t, \delta r_t, 1)\delta r_t}{\delta(\lambda - 1)G_L(\lambda k_t, \delta r_t, 1)}\right\} - 1\right]$$
$$\geq (\lambda - 1)G_L(\lambda k_t, \delta r_t, 1)\left[\left\{\frac{(\delta - \lambda)\eta}{\delta(\lambda - 1)}\right\} - 1\right] > 0,$$

since $\lambda k_t \geq \underline{k}, 0 < \delta r_t \leq \delta \underline{S}$ and by using (8.4.2). This verifies (8.4.3).

Construct a sequence $(\mathbf{k}', \mathbf{S}')$ in the following way: $k_0' = \delta \underline{k}, k_t' = \lambda k_t$ for $t \geq 1$ and $S_t' = \delta S_t$ for $t \geq 0$. Then

$$\sum_{t=0}^{\infty} r_t' = \delta \sum_{t=0}^{\infty} r_t \leq \delta \underline{S}.$$

Also, for $t \geq 0, c_{t+1}' = F(k_t', r_t', 1) - k_{t+1}' \geq G(\lambda k_t, \delta r_t, 1) + \lambda k_t - \lambda k_{t+1} \geq G(\lambda k_t, \lambda r_t, \lambda) + \lambda k_t - \lambda k_{t+1}$ (by using (8.4.3)) $= \lambda[G(k_t, r_t, 1) + k_t - k_{t+1}] = \lambda c_{t+1}$. Hence, $(\mathbf{k}', \mathbf{m}')$ is a program from $(\delta \underline{k}, \delta \underline{S})$ and $c_{t+1}' \geq \lambda c_{t+1}$ for $t \geq 0$. $\square$

We now state and prove the main result of this section.

Theorem 8.4.3. *Under* (A.1)–(A.4), *there exists an efficient equitable program from* $(\underline{k}, \underline{S}) \gg 0$ *if and only if Condition E holds.*

Proof (Necessity).   Suppose there exists an efficient equitable program $(\bar{\mathbf{k}}, \bar{\mathbf{S}})$ from $(\underline{k}, \underline{S}) \gg 0$. If Condition E does not hold, then $\bar{c}_t = 0$ for $t \geq 1$. Since $(\underline{k}, \underline{S}) \gg 0$, this implies that $(\bar{\mathbf{k}}, \bar{\mathbf{S}})$ is inefficient, a contradiction.

(Sufficiency) By Theorem 8.4.1 and Condition E, there exists a non-trivial equitable maximin program $(\mathbf{k}, \mathbf{S})$ from $(\underline{k}, \underline{S})$. Suppose this is not efficient, then by Theorem 8.3.1, there is an efficient program $(\bar{\mathbf{k}}, \bar{\mathbf{S}})$ from $(\underline{k}, \underline{S})$ with $\bar{c}_1 > c_1 = d$ and $\bar{c}_t = d$ for $t \geq 2$. Then, clearly, there are programs $(\mathbf{k}', \mathbf{S}')$ from $(\bar{\mathbf{k}}_1, \bar{\mathbf{S}}_1)$ with $(k_t', r_t', z_t', y_{t+1}', c_{t+1}') = (\bar{k}_{t+1}, \bar{r}_{t+1}, \bar{z}_{t+1}, \bar{y}_{t+2}, \bar{c}_{t+2})$ for $t \geq 0$

and $(\hat{\mathbf{k}}, \hat{\mathbf{S}})$ from $(\bar{\mathbf{k}}_2, \bar{\mathbf{S}}_2)$ with $(\hat{k}_t, \hat{r}_t, \hat{z}_t, \hat{y}_{t+1}, \hat{c}_{t+1}) = (\bar{k}_{t+2}, \bar{r}_{t+2}, \bar{z}_{t+2},$ $\bar{y}_{t+3}, \bar{c}_{t+3})$ for $t \geq 0$. Clearly, both are efficient and equitable with $\hat{c}_t = c'_t = d > 0$ for all $t \geq 0$. By Theorem 8.3.2, therefore, both are interior and $k'_{t+1} > k'_t$ for $t \geq 0$, $\hat{k}_{t+1} > \hat{k}_t$ for $t \geq 0$.

Let $\bar{c}_1 - d = \epsilon_1 > 0$. Then clearly there is a program $(\mathbf{k}'', \mathbf{S}'')$ from $(\underline{k}, \underline{S})$, such that $c''_1 = d + \epsilon/2$, $k''_1 = \bar{k} + \epsilon_1/2$, $S''_1 = \bar{S}_1$ and $(k''_t, S''_t) = (\bar{k}_t, \bar{S}_t)$ for $t \geq 2$. Since $(\mathbf{k}', \mathbf{S}')$ is interior, then $0 < r'_0 = \bar{r}_1 = \bar{S}_1 - \bar{S}_2 = S''_1 - S''_2 = r''_1$. Since $k''_1 > \bar{k}_1$, then by (A.2), $c''_2 = F(k''_1, r''_1, 1) - k''_2 > F(\bar{k}_1, \bar{r}_1, 1) - \bar{k}_2 = \bar{c}_2$ (since $k''_2 = \bar{k}_2$)$= d$. Let $\epsilon_2 = c''_2 - d > 0$. By (A.1) and (A.2), we can find $0 < \theta < 1$, such that $F(k''_1, \theta r''_1, 1) - k''_2 = d + \epsilon_2/2$. Hence, there is a program $(\tilde{\mathbf{k}}, \tilde{\mathbf{S}})$ from $(\underline{k}, \underline{S})$, such that $\tilde{c}_1 = c''_1 = d + \epsilon_1/2$, $\tilde{c}_2 = d + \epsilon_2/4$, $\tilde{k}_1 = k''_1$, $\tilde{k}_2 = k''_2 + \epsilon_2/4 = \bar{k}_2 + \epsilon_2/4$, $\tilde{S}_1 = S''_1 = \bar{S}_1$, $\tilde{S}_2 = S''_2 + (1 - \theta)r''_1 = \bar{S}_2 + (1 - \theta)r''_1$, $(\tilde{k}_t, \tilde{S}_t) = (k''_t, S''_t) = (\bar{k}_t, \bar{S}_t)$ for $t \geq 3$.

It is clear then that there is a $\delta > 1$ such that $\tilde{k}_2 \geq \delta\bar{k}_2$ and $\tilde{S}_2 \geq \delta\bar{S}_2$. Since $(\hat{k}, \hat{m})$ is an efficient and equitable program from $(\bar{k}_2, \bar{S}_2)$ with $\hat{c}_t = d > 0$ for all t, then, by Lemma 8.4.1, there is a $\lambda > 1$ and a program $(\hat{\hat{\mathbf{k}}}, \hat{\hat{\mathbf{S}}})$ from $(\delta\bar{k}_2, \delta\bar{S}_2)$ such that $\hat{\hat{c}} \geq \lambda\hat{c}_t = \lambda d$ for all $t \geq 1$. Hence there is a program $(\mathbf{k}^*, \mathbf{S}^*)$ from $(\underline{k}, \underline{S})$, such that $k^*_1 = \tilde{k}_1$, $k^*_2 = \delta\bar{k}_2$, $S^*_1 = \tilde{S}_1$, $S^*_2 = \delta\bar{S}_2$, $(k^*_t, S^*_t) = (\hat{\hat{k}}_{t-2}, \hat{\hat{S}}_{t-2})$ for all $t \geq 3$, $c^*_1 \geq \tilde{c}_1 \geq d + \epsilon_1/2$, $c^*_2 \geq \tilde{c}_2 \geq d + \epsilon_2/4$ and $c^*_t \geq \lambda d$ for $t \geq 3$. But then $(\mathbf{k}^*, \mathbf{S}^*)$ is a program from $(\underline{k}, \underline{S})$ with $\inf_t c^*_t > d = \inf_t c_t$ which contradicts (k, S) as a maximin program. Hence, our supposition that (k, S) is inefficient is false. $\qquad\square$

We now note a simple corollary which follows directly from Theorem 8.4.1.

Corollary 8.4.1. *Under* (A.1)–(A.4), *a program* $(\mathbf{k}, \mathbf{S})$ *from* $(\underline{k}, \underline{S}) \gg 0$ *is an efficient equitable program if and only if it is a non-trivial maximin program.*

8.5 Price Characterization of Efficient Programs

We conclude the chapter with a result on price characterization of efficient and equitable programs. Similar results were presented earlier. Duality is a prominent theme in the literature on efficient allocation, and it offers the valuable insights into the role of prices and intertemporal decentralization.

Theorem 8.5.1. *Under* (A.1)–(A.4), *a program* $(\mathbf{k}, \mathbf{S})$ *from* $(\underline{k}, \underline{S}) \gg 0$ *is efficient and equitable if and only if there is a price sequence* $(\mathbf{p}, \mathbf{q}, \mathbf{w})$ *with* $(p_t, q_t, w_t) \gg 0$ *for* $t \geq 0$, *such that* (8.5.1)–(8.5.3) *hold:*

$$0 = p_{t+1} y_{t+1} + q_{t+1} S_{t+1} - p_t k_t - q_t S_t - w_t \bar{L}$$

$$\geq p_{t+1} y + q_{t+1} S' - p_t k - q_t S - w_t L$$

$$\text{for all } [(k, S, L), (y, S', 0)] \in \mathcal{J}, t \geq 0. \tag{8.5.1}$$

$$\lim_{t \to \infty} (p_t k_t + q_t S_t) = 0, \tag{8.5.2}$$

$$c_t = c_{t+1} \quad \text{for } t \geq 1. \tag{8.5.3}$$

Furthermore, the following relations hold along an efficient, equitable program:

$$\sum_{t=0}^{\infty} w_t \equiv \mathcal{W} < \infty, \tag{8.5.4}$$

$$\sum_{t=1}^{\infty} p_t c_t \equiv \mathcal{PVC} < \infty. \tag{8.5.5}$$

Proof. Suppose that $(\mathbf{k}, \mathbf{S})$ is a program from $(\underline{k}, \underline{S})$, such that there is a price sequence $(\mathbf{p}, \mathbf{q}, \mathbf{w}) \gg 0$ that satisfies (8.5.1) and (8.5.2). Consider another program, say, $(\tilde{\mathbf{k}}, \tilde{\mathbf{S}})$ from the same initial $(\underline{k}, \underline{S})$. Then, using (8.5.1) with $t = 0$, we get (for all programs we set $L_t = \bar{L}$ for all t)

$$p_1 y_1 + q_1 S_1 - p_0 \underline{k} - q_0 \underline{S} - w_0 \cdot 1$$

$$\geq p_1 \tilde{y}_1 + q_1 \tilde{S}_1 - p_0 \underline{k} - q_0 \underline{S} - w_0 \cdot 1.$$

Note that $y_1 = k_1 + c_1$ and $\bar{y}_1 = \bar{k}_1 + \bar{c}_1$. Simplifying, we get

$$p_1 k_1 + q_1 S_1 \geq p_1(\bar{c}_1 - c_1) + p_1 \bar{k}_1 + q_1 \tilde{S}_1 \geq p_1(\bar{c}_1 - c_1). \qquad \text{(A)}$$

Now, with $t = 1$, we get

$$p_2(k_2 + c_2) + q_2 S_2 - p_1 k_1 - q_1 S_1 - w_1 \cdot 1$$
$$\geq p_2(\bar{k}_2 + \bar{c}_2) + q_2 \tilde{S}_2 - p_1 \bar{k}_1 - q_1 \tilde{S}_1 - w_1 \cdot 1,$$

or

$$p_2 k_2 + q_2 S_2 - p_1 k_1 - q_1 S_1 - w_1 \cdot 1$$
$$\geq p_2(\bar{c}_2 - c_2) + p_2 \bar{k}_2 + q_2 \tilde{S}_2 - p_1 \bar{k}_1 - q_1 \tilde{S}_1 - w_1 \cdot 1. \qquad \text{(B)}$$

Now, adding (A) and (B) and simplifying,

$$p_2 k_2 + q_2 S_2 \geq p_1(\bar{c}_1 - c_1) + p_2(\bar{c}_2 - c_2) + p_2 \bar{k}_2 + q_2 \tilde{S}_2.$$

More generally, we get for all T, we have

$$p_T k_T + q_T S_T \geq \sum_{t=1}^{T} p_t(\bar{c}_t - c_t) + p_2 \bar{k}_2 + q_2 \tilde{S}_2 \geq \sum_{t=1}^{T} p_t(\bar{c}_t - c_t).$$
$$\text{(8.5.6)}$$

Suppose that $(\tilde{\mathbf{k}}, \tilde{\mathbf{S}})$ dominates $(\mathbf{k}, \mathbf{S})$. Then for all t, $(\bar{c}_t - c_t) \geq 0$. Moreover, there is some $\tau \geq 1$, such that $(\bar{c}_\tau - c_\tau) = \delta > 0$. Hence, for all $T > \tau$, we have (since $p_t > 0$ for all t)

$$p_T k_T + q_T S_T \geq \sum_{t=1}^{T} p_t(\bar{c}_t - c_t) \geq p_\tau \delta > 0.$$

But this contradicts the transversality condition (8.5.2). Hence, $(\mathbf{k}, \mathbf{S})$ is an efficient program from $(\underline{k}, \underline{S})$, and, by (8.5.3) it is equitable.

To go in the other direction, if a program $(\mathbf{k}, \mathbf{S})$ is efficient and equitable, since $(\underline{k}, \underline{S}) \gg 0$, for some $d > 0$, $c_t = d$ for all $t \geq 1$.

Hence, by Theorem 8.3.2, $(\mathbf{k}, \mathbf{S})$ is interior, and

$$k_{t+1} \geq k_t > \underline{k} \quad \text{for } t \geq 0.$$

We now outline of the proof of establishing the existence of prices$(\mathbf{p}, \mathbf{q}, \mathbf{w})$, satisfying (8.5.1). A program $(\mathbf{k}, \mathbf{S})$ from $(\underline{k}, \underline{S})$ is called *short-run inefficient* if there is another program $(\tilde{\mathbf{k}}, \tilde{\mathbf{S}})$ from $(\underline{k}, \underline{S})$ and some finite T, such that

$$(\bar{c}_1, \ldots, \bar{c}_{T+1}, \bar{k}_{T+1}, \bar{S}_{T+1}) > (c_1, \ldots, c_{T+1}, k_{T+1}, S_{T+1}).$$

A program is short-run efficient if it is not short-run inefficient. An efficient program is necessarily short-run efficient, and given the differentiability assumptions, we can use the standard Lagrangian method to show that it satisfies the marginal productivity conditions (recall (8.3.2))

$$F_{r_{t+1}} = F_{r_t} \cdot F_{k_{t+1}}.$$

We can now invoke Theorem 8.3.4 and verify (8.5.1) by defining prices $(\mathbf{p}, \mathbf{q}, \mathbf{w}) \gg 0$ as (recall (8.3.4))

$$\begin{aligned} p_0 &= (F_{k_0}/F_{r_0}), \quad p_{t+1} = (p_t/F_{k_t}) \quad \text{for } t \geq 0, \\ w_t &= F_{L_t} \cdot p_{t+1} \quad \text{for } t \geq 0, \quad q_t = 1 \quad \text{for } t \geq 0. \end{aligned} \tag{8.5.7}$$

The interested reader is referred to Exercise 8.6.1 for details.

Our next step is the verification of (8.5.4). By (A.4), there is $\eta > 0$, such that for $k > \underline{k}$, $0 < r_t \leq \underline{S}$,

$$[rG_r(k, r, 1)]/G_L(k, r, 1) \geq \eta. \tag{8.5.8}$$

Then for $t \geq 0$, since $k_t \geq \underline{k}$ and $0 \leq r_t \leq \underline{S}$, by (8.5.7),

$$w_t = [G_L(k, r, 1)]/G_r(k, r, 1)] \leq (1/\eta) r_t. \tag{8.5.9}$$

Since $\sum_{t=0}^{\infty} r_t \leq \underline{S}$, we get $\sum_{t=0}^{\infty} w_t = \mathcal{W} < \infty$, establishing (8.5.4).

Next, we come to the verification of (8.5.5). Using (8.5.1), verify that for any positive integer $T \geq 1$,

$$\sum_{t=0}^{T} p_{t+1} c_{t+1} = p_0 \underline{k} + \sum_{t=0}^{T} q r_t + \left(\sum_{t=0}^{T} w_t \cdot 1 \right) - p_{T+1} k_{T+1}, \quad (8.5.10)$$

since G is homogeneous of degree one (implying that the maximum competitive profit is zero). Hence, the non-negative sequence $(\sum_{t=0}^{T} p_{t+1} c_{t+1})$ is monotonically non-decreasing and bounded above:

$$\sum_{t=0}^{T} p_{t+1} c_{t+1} \leq p_0 \underline{k} + \underline{S} + \mathcal{W}.$$

Hence, the present value of consumptions (c_t) at prices (p_t), defined as

$$\mathcal{PVC} \equiv \sum_{t=0}^{\infty} p_{t+1} c_{t+1},$$

is finite, confirming (8.5.5).

The transversality condition (8.5.2) remains to be verified. Note that (recall (8.3.1)) $q_t = q > 0$, and $S_t \to 0$ for any efficient program. Rewriting (8.5.9) as

$$-\sum_{t=0}^{T} p_{t+1} c_{t+1} + p_0 \underline{k} + \sum_{t=0}^{T} q r_t + \left(\sum_{t=0}^{T} w_t \cdot 1 \right) = p_{T+1} k_{T+1},$$

$$(8.5.11)$$

we see that the left side has a limit as $T \to \infty$. Hence, $\lim_{t \to \infty} (p_t k_t)$ exists, and it remains to be proved that

$$\lim_{t \to \infty} (p_t k_t) = 0. \qquad (8.5.12)$$

Suppose, to the contrary, that the limit in (8.5.11) is not zero. Then there is some $\beta > 0$, such that $p_t k_t \geq \beta$ for $t \geq 0$. Since the series $\sum_{t=0}^{\infty} p_{t+1} c_{t+1}$ is convergent, there is some finite T_1

$$J_{T_1} \equiv \sum_{t=T_1}^{\infty} p_{t+1} c_{t+1} \leq \beta/2. \qquad (8.5.13)$$

If $J_{T_1} = 0$, the program is clearly inefficient. So, we need to consider the case where $J_{T_1} > 0$. Define $x_{t+1} = (p_{t+1}c_{t+1})/J_{T_1}$ for $t \geq T_1$. Clearly, $\sum_{t=T_1}^{\infty} x_{t+1} = 1$.

Now, construct a sequence $(\bar{k}_t, \bar{r}_t, \bar{L}_t, \bar{y}_t, \bar{c}_t)$ as follows:

$$\bar{k}_0 = \underline{k}, \bar{r}_0 = r_0,$$

for $1 \leq t < T_1$, $(\bar{k}_t, \bar{r}_t, \bar{y}_t, \bar{c}_t) = (k_t, r_t, y_t, c_t)$,

for $t = T_1, \bar{k}_t = (1/2)k_t, \ \bar{r}_t = (1/2)r_t, \ \bar{c}_t = c_t + (1/2)k_t, \ \bar{y}_t = y_t$,

for $t > T_1, \bar{k}_t = k_t\left[1/2 - (1/2)\left[\displaystyle\sum_{s=T_1}^{t-1} x_{s+1}\right]\right]$

$$\equiv \lambda_t k_t, \bar{r}_t = \lambda_t r_t, \ \bar{y}_t = F(\bar{k}_{t-1}, \bar{r}_{t-1}, \bar{L}), \ \bar{c}_t = \bar{y}_t - \bar{k}_t,$$

for $t \geq 0, \quad L_t = 1$.

Clearly, $(\bar{k}_t, \bar{r}_t) \gg 0$ for all $t \geq 0$; $\bar{c}_t = c_t$ for $1 \leq t < T_1$ and $\bar{c}_T > c_T$. It will be shown that $\bar{c}_t \geq c_t$ for $t > T_1$. This will contradict the efficiency of the program $(\mathbf{k}, \mathbf{S})$ and establish that

$$\lim_{t \to \infty} p_t k_t = 0.$$

For $t > T_1$, we have

$p_t\bar{c}_t = p_t F(\bar{k}_{t-1}, \bar{r}_{t-1}, \bar{L}) - p_t\bar{k}_t$

$\qquad \geq p_t F(\lambda_{t-1}k_{t-1}, \lambda_{t-1}r_{t-1}, \lambda_{t-1}) - p_t\lambda_t k_t$

$\qquad = \lambda_{t-1}p_t F(k_{t-1}, r_{t-1}, 1) - \lambda_t p_t k_t$

$\qquad = \lambda_{t-1}p_t(c_t + k_t) - \lambda_t p_t k_t$

$\qquad = \lambda_{t-1}p_t c_t + (1 - \lambda_{t-1})p_t c_t + \lambda_{t-1}p_t k_t - \lambda_t p_t k_t - (1 - \lambda_{t-1})p_t c_t$

$\qquad \geq p_t c_t + p_t k_t[\lambda_{t-1} - \lambda_t - (1/2)(1 - \lambda_{t-1})x_t]$

$\qquad \geq p_t c_t + p_t k_t[(1/2)x_t - (1/2)x_t]$

$\qquad = p_t c_t. \hfill (8.5.14)$

Since $p_t > 0$ for $t \geq 0$, we have $\bar{c}_t \geq c_t$ for $t > T_1$, thus completing the proof. $\qquad\qquad\qquad\square$

8.6   Complements and Details

The exposition in this chapter is based partly on the early definitive paper by Mitra (1978) and a subsequent paper by Dasgupta and
Mitra (1983). A comprehensive review of a number of issues related
to efficiency, (alternative notions of) optimality, consumption value
maximization and the transversality condition in a multisector framework appeared in Cass and Majumdar (1979). Appendix B of Cass
and Majumdar studied the role of an exhaustible resource. The treatment is unavoidably technical. The interested reader may compare
the definitions in Stavins, Wagner and Wagner (2003). The standard
treatment of optimal growth in a continuous-time model in which
an exhaustible resource and a producible "capital" good appear as
inputs is Dasgupta and Heal (1974). See also Dasgupta and Heal
(1979).

Exercise 8.6.1. Complete the proof of the existence of competitive
prices $(\mathbf{p}, \mathbf{q}, \mathbf{w})$ satisfying (8.5.1).
   [▲ *Hints*: See Mitra (1978, pp. 120–121) for details. ▼]

Exercise 8.6.2. Show that under (A.1), (A.2) and (A.4(i)), an inte-
rior program $(\mathbf{k}, \mathbf{S})$ from $(\underline{k}, \underline{S})$ is efficient if and only if there is a
sequence of prices $\mathbf{p} = (p_t) \gg 0$, such that

$$\infty > \sum_{t=0}^{\infty} p_{t+1} c_{t+1} \geq \sum_{t=0}^{\infty} p_{t+1} \bar{c}_{t+1} \tag{8.6.1}$$

for every program $(\bar{\mathbf{k}}, \bar{\mathbf{S}})$ from $(\underline{k}, \underline{S})$.
   [▲ *Hints*: If $(\mathbf{k}, \mathbf{S})$ from $(\underline{k}, \underline{S})$ is an interior efficient program, use
the prices $\mathbf{p} = (p_t)$ defined by (8.3.4), satisfying the condition (8.3.5).
Recall

$$\sum_{t=0}^{T} p_{t+1} c_{t+1} = p_0 \underline{k} + \sum_{t=0}^{T} q r_t + \left(\sum_{t=0}^{T} w_t \cdot \bar{L} \right) - p_{T+1} k_{T+1}$$

$$\leq p_0 \underline{k} + \sum_{t=0}^{T} q r_t + \left(\sum_{t=0}^{T} w_t \cdot \bar{L} \right).$$

Using (A.4(i)), show that $\sum_{t=0}^{\infty} w_t \cdot \bar{L}$ is well defined. Hence,

$$\sum_{t=0}^{\infty} p_{t+1} c_{t+1} < \infty.$$

Take any other program $(\bar{\mathbf{k}}, \bar{\mathbf{S}})$ from $(\underline{k}, \underline{S})$. for any finite T, consider

$$\sum_{t=0}^{T} p_{t+1} (c_{t+1} - \bar{c}_{t+1}).$$

Use the competitive conditions to obtain $\sum_{t=0}^{T} p_{t+1} (c_{t+1} - \bar{c}_{t+1}) \geq -p_{T+1} k_{T+1}$.

$$\sum_{t=0}^{T} p_{t+1} (c_{t+1} - \bar{c}_{t+1}) \geq -p_{T+1} k_{T+1}.$$

Use the transversality condition to complete the proof.

To go in the other direction, suppose that an interior program $(\mathbf{k}, \mathbf{S})$ from $(\underline{k}, \underline{S})$ satisfies (8.6.1) relative to a strictly positive price sequence $\mathbf{p} = (p_t) \gg 0$.

If it is inefficient, there is some other program $(\bar{\mathbf{k}}, \bar{\mathbf{S}})$ from $(\underline{k}, \underline{S})$, such that

$\bar{c}_t \geq c_t$ for all $t \geq 0, \bar{c}_\tau = c_\tau + \varepsilon$, for some period τ, and some $\varepsilon > 0$.

Note that, for any finite $T > \tau$,

$$\sum_{t=0}^{T} p_{t+1} \bar{c}_{t+1} \geq \sum_{t=0}^{T} p_{t+1} c_{t+1} + p_\tau \varepsilon,$$

which contradicts (8.6.1) as $T \to \infty$.▼]

Exercise 8.6.3. Let $G(k, r, L) = k^{(1/2)} r^{(1/4)} L^{(1/4)} \cdot L_t = \bar{L} = 1$ for all $t \geq 0; \underline{k} = 1, \underline{S} = 4$.

Consider the sequence defined by

$k_t = 1$ for all $t \geq 0$; $L_t = 1$ for all $t \geq 0$;
$r_0 = 1, (1/2) r_{t+1} + (r_{t+1})^{(3/4)} = (r_t)^{(3/4)}$ for all $t \geq 0$;
$c_{t+1} = G(k_t, r_t, 1)$ for all $t \geq 0$.

To show that the sequence is a program from the prescribed initial stocks, one only needs to show that $\sum_{t=0}^{\infty} r_t \leq 4$. Note that

$$(1/2)r_{t+1} = (r_t)^{(3/4)} - (r_{t+1})^{(3/4)}, \text{ leading to}$$

$$(1/2) \sum_{t=0}^{T} r_{t+1} = (r_0)^{(3/4)} - (r_{T+1})^{(3/4)} \leq (r_0)^{(3/4)}.$$

Hence, $\left(\sum_{t=0}^{T} r_{t+1} \right)$ converges as $T \to \infty$ and $r_t \to 0$. Upon taking limits,

(8.6.2)

$$\sum_{t=0}^{\infty} r_{t+1} = 2(r_0)^{(3/4)} = 2,$$

$$\sum_{t=0}^{\infty} r_t = r_0 + \sum_{t=0}^{\infty} r_{t+1} = 3 < 4.$$

Verify that the marginal productivity conditions (8.3.2) hold (▲ *Exercise* ▼):

$$F_{r_{t+1}} = F_{r_t} \cdot F_{k_{t+1}} = (1/4)r_{t+1}^{(-3/4)}.$$

Hence, the sequence specified is an *interior competitive program* from $(1, 4)$. Finally,

$$p_{t+1}k_{t+1} = 1/F_{r_t} = 4(r_t)^{(3/4)} \to 0, \quad \text{as } t \to \infty.$$

Thus, the capital value transversality holds, but not the resource exhaustion condition. □

Chapter 9

A Cobb–Douglas Economy

As I recovered, I picked up the subject once more, and discovered that the old opponents were now relatively silent, that the Cobb–Douglas function was being widely used, and that a host of younger scholars led by my former student, Paul Samuelson, his colleague Solow and Marc Nerlove... were all pushing forward into new and sophisticated fields.

—Paul H. Douglas

9.1   Introduction

We continue to explore some basic issues related to sustainability in the framework developed in Chapter 8. Two primary differences in the formal structure of the model that we introduce in this chapter are as follows: (a) the net output function $G(K, R, L)$ is assumed to have the Cobb–Douglas form $(G(K, R, L) = K^\alpha R^\beta L^\gamma)$ and (b) the exogenous supply of labor is not treated as a constant.

Indispensable in vast areas of empirical economics, the Cobb–Douglas form has been a powerful pedagogical tool in economic theory. It is hardly necessary to mention that a carefully chosen functional form opens up possibilities of explicit computations that allow us to have *some* insight into questions that can perhaps be attacked, if at all, with considerable analytical difficulties in a more general context. This will be clear from our experience in the present chapter. We need to keep in mind though — and this point was raised in Chapter 2 — that a conclusion derived from direct calculations becomes more persuasive if it continues to hold for a broader class of models.

It is important to recognize that continued growth of population (either a high rate of growth as in some African countries, or a lower rate on a staggering base, like India and China) and population density are making and will continue to make it difficult to sustain an improvement in the quality of life for the billions in the world. Thus, a model based on the assumption of a constant level of population may be far from a useful approximation to capture the challenges involved in achieving sustainability. Indeed, it is worthwhile in general, and imperative for policy makers in many countries, in particular, to explore the nature of *limits on population growth* when an exhaustible resource is important in production. The Cobb–Douglas framework allows us to derive some exact bounds.

One of the central results (Theorem 8.4.3) in the last chapter was to assert that the existence of an efficient and equitable program was equivalent to (Condition E) the existence of a non-trivial equitable program. The first major result in this chapter identifies the conditions under which (with a Cobb–Douglas net output function) Condition E holds. The result is particularly ingenious as it captures the relative importance of inputs as well as the limits on the population growth rate. When the labor supply is constant (and, with a proper choice of units, taken as equal to one), the condition is expressed as a simple inequality involving the exponents of capital and resource in the Cobb–Douglas form ($\alpha > \beta$).

The second aim of this chapter is revisiting Ramsey optimality in the Cobb–Douglas economy. We choose the one period felicity function of the parametric form $u(c) = -(1/c^\sigma)$. A complete set of conditions for the existence of a Ramsey optimal program is given in Theorem 9.4.1.

9.2   The Model

We reconsider the model of the earlier chapter and indicate the changes introduced. First, we assume that the net output function

$G : R_+^3 \to R_+$ has the Cobb–Douglas form:

$$Q = G(K, R, L) = K^\alpha R^\beta L^\gamma, (\alpha, \beta, \gamma) \gg 0, \alpha + \beta + \gamma = 1.$$

$$(9.2.1)$$

Here, (K, R, L) are the quantities of capital, resource and labor used as input, and Q is the net output.

Assuming no depreciation of capital, the total or gross output function (see (8.2.1)) is defined as $F(K, R, L) = G(K, R, L) + K$.

The initial stocks of capital and resource are given by $(\underline{K}, \underline{S}) \gg 0$. In contrast with parts of the earlier chapter in which the population or labor force was assumed to be constant over time, the exogenously given labor force $\mathbf{L}_t$ (in period t) is assumed to satisfy

$$\mathbf{L}_0 = \mathbf{L} > 1; \mathbf{L}_{t+1} \geq \mathbf{L}_t \text{ for } t \geq 0 : \sup_{t \geq 0}[\mathbf{L}_{t+1}/\mathbf{L}_t] \leq \mathbf{B} \text{ (finite)}.$$

$$(9.2.2)$$

A *program* from $(\underline{K}, \underline{S})$ is a sequence $(\mathbf{K}, \mathbf{R}, \mathbf{L}, \mathbf{Y}, \mathbf{C}) = (K_t, R_t, L_t, Y_{t+1}, C_{t+1})$

$$K_0 = \underline{K}, \sum_{l=0}^{\infty} R_t \leq \underline{S}, L_t = \mathbf{L}_t \qquad \text{for} \quad t \geq 0,$$

$$Y_{t+1} = F(K_t, R_t, L_t), C_{t+1} = Y_{t+1} - K_{t+1} \text{ for} \quad t \geq 0,$$

$$(K_t, R_t, L_t, Y_{t+1}, C_{t+1}) \geq 0 \qquad \text{for} \quad t \geq 0.$$

$$(9.2.3)$$

Associated with a program is the corresponding sequence of resource stocks $\mathbf{S} = (S_t)$ defined by

$$S_0 = \underline{S}, S_{t+1} = S_t - R_t \quad \text{for } t \geq 0. \qquad (9.2.4)$$

A program $(\mathbf{K}, \mathbf{R}, \mathbf{L}, \mathbf{Y}, \mathbf{C})$ is *interior* if $(K_t, S_t) \gg 0$ for all $t \geq 0$. It is *regular interior* if $(K_t, S_t, C_{t+1}) \gg 0$ for all $t \geq 0$.

Given a program $(\mathbf{K}, \mathbf{R}, \mathbf{L}, \mathbf{Y}, \mathbf{C})$, we write (its per capita version):

$$k_t = K_t/L_t, r_t = R_t/L_t, g_{t+1} = L_{t+1}/L_t \quad \text{for } t \geq 0,$$

$$y_{t+1} = Y_{t+1}/L_{t+1}, c_{t+1} = C_{t+1}/L_{t+1} \quad \text{for } t \geq 0. \tag{9.2.5}$$

A program $(\hat{\mathbf{K}}, \hat{\mathbf{R}}, \hat{\mathbf{L}}, \hat{\mathbf{Y}}, \hat{\mathbf{C}})$ from $(\underline{K}, \underline{S})$ can sustain a positive per-capita consumption level if

$$\inf_{t \geq 1}(\hat{c}_t) > 0. \tag{9.2.6}$$

It is (non-trivial) *equitable* if

$$\hat{c}_t = \hat{c}_{t+1} > 0 \text{ for all } t \geq 1. \tag{9.2.7}$$

It is a maximin program if

$$\inf_{t \geq 1}(\hat{c}_t) \geq \inf_{t \geq 1}(c_t) \tag{9.2.8}$$

for every program $(\mathbf{K}, \mathbf{R}, \mathbf{L}, \mathbf{Y}, \mathbf{C})$ from $(\underline{K}, \underline{S})$. It is a non-trivial maximin program if it is a maximin program that maintains a positive consumption.

9.3   Sustaining Positive Consumption

It is clear that if there is an efficient, equitable program, it is necessarily a non-trivial maximin program. In view of the results on the existence of an efficient, equitable program in the earlier chapter, we are led to the following basic question: find necessary and sufficient condition on the sequence (L_t) and the function G such that there is a program that can maintain a positive per capita consumption level. Informally, these conditions must capture two aspects of the problem: the substitution possibilities specified in G and the bound on the growth rate of population.

We will denote

$$\frac{\gamma}{(1-\beta)} \quad \text{by } \theta,$$

$$\frac{\alpha}{(1-\beta)} \quad \text{by } \delta,$$

$$\frac{(1-\beta)}{(1-\alpha)} \quad \text{by } \mu,$$

$$\frac{(\alpha-\beta)}{(1-\alpha)} \quad \text{by } \nu,$$

$$\sup_{t\geq 0}\frac{\mathbf{L}_{t+1}}{\mathbf{L}_t} \quad \text{by } M,$$

$$\sum_{s=0}^{t}\mathbf{L}_s^{\theta} \quad \text{by } A_t.$$

For $0 < e < \alpha$, denote

$$(\alpha - e) \quad \text{by } a,$$

$$(\beta + e) \quad \text{by } b,$$

$$\frac{(1-b)}{(1-a)} \quad \text{by } m,$$

$$\frac{(a-b)}{(1-a)} \quad \text{by } n.$$

Lemma 9.3.1. *Suppose $\alpha > \beta$; if e is a number, such that $0 < e < \alpha$, and $a > b$, then there is a feasible program $\langle \mathbf{K}, \mathbf{R}, \mathbf{L}, \mathbf{Y}, \mathbf{C} \rangle$ and a scalar $E > 0$, such that*

$$C_{t+1} \geq E(A_t^n / L_t^\delta) \quad \text{for } t \geq 0. \tag{9.3.1}$$

Proof. Choose $\phi > 0$ with ϕ sufficiently close to zero to ensure that

$$e(1-a) + e(1-b) \geq \phi(1-a)b. \tag{9.3.2}$$

Clearly, this can be done. Then, it can be verified that

$$m\alpha - (1 + \phi)\beta \geq n. \tag{9.3.3}$$

[▲ Hint: To verify (9.3.3), we may proceed as follows.
By substituting (9.3.3) with our notations above, we have, equivalently,

$$\frac{1-b}{1-a}(a+e) - (1+\phi)(b-e) \geq \frac{a-b}{1-a},$$

or

$$\frac{1-b}{1-a}(a+e) - (b-e) - \phi(b-e) \geq \frac{a-b}{1-a}.$$

Multiplying $(1 - a)$, we have

$$(1-b)(a+e) + (e-b)(1-a) + \phi(e-b)(1-a) \geq a - b.$$

Expanding the left-hand side, we have

$$-eb + a + e - ab - b + ab + e - ae + \phi(e - ae - b + ab) \geq a - b.$$

Canceling some terms, we have

$$-eb + e + e - ae + \phi(e - ae - b + ab) \geq 0.$$

Arranging terms to make a similar form with (9.3.2), we have

$$e(1-b) + e(1-a) + \phi(e-ae) \geq \phi(1-a)b. \ \blacktriangledown]$$

Denote $[\mathbf{L}_t^\theta / A_t^{(1+\phi)}]$ by H_t for $t \geq 0$. Then by the Abel–Dini Theorem (see Theorem 13.1.5), $\sum_{t=0}^{\infty} H_t < \infty$. Define $B^b = 2m(1 + M)^n M^\theta$. Choose $\bar{h} > 0$, such that

$$\sum_{t=0}^{\infty} \bar{h} B H_t = \underline{S}. \tag{9.3.4}$$

Define $h = \min[\frac{1}{2}, \bar{h}, \underline{K}]$.

Define a sequence $\langle \mathbf{K}, \mathbf{R}, \mathbf{L}, \mathbf{Y}, \mathbf{C} \rangle$ as follows: $L_t = \mathbf{L}_t$ for $t \geq 0$: $K_0 = \underline{K}$, $K_t = hA_t^m$ for $t \geq 1$; $R_t = hBH_t$ for $t \geq 0$; $Y_{t+1} = G(K_t, R_t, L_t) + K_t$, $C_{t+1} = Y_{t+1} - K_{t+1}$ for $t \geq 0$. The sequence $\langle \mathbf{K}, \mathbf{R}, \mathbf{L}, \mathbf{Y}, \mathbf{C} \rangle$ will be a feasible program if we show that $C_{t+1} \geq 0$ for all $t \geq 0$.

For $t \geq 0$, $G(K_t, R_t,\ L_t) = h^\alpha A_t^{m\alpha} h^\beta B^\beta H_t^\beta L_t^\gamma = h^{\alpha+\beta} B^\beta A_t^{[m\alpha-(1+\phi)\beta]} L_t^\theta \geq h^{(\alpha+\beta)} \cdot B^\beta A_t^n L_t^\theta$ [by (9.3.3)] $\geq hB^\beta A_t^n L_t^\theta$ (since $0 < \alpha + \beta < 1$ and $0 < h < 1$). Now, for $t \geq 0$, $(K_{t+1} - K_t \leq hA_{t+1}^m - hA_t^m \leq hm A_{t+1}^n L_{t+1}^\theta$ (use binomial expansion and $m-1 = n$). For $t \geq 0$, we have

$$[A_{t+1}/A_t] = 1 + [L_{t+1}^\theta/A_t] \leq [1 + M].$$

So,

$$(K_{t+1} - K_t) \leq hm \left(\frac{A_{t+1}}{A_t} \right)^n A_t^n \left(\frac{L_{t+1}}{L_t} \right)^\theta L_t^\theta \leq hm(1+M)^n$$

$$\times A_t^n M^\theta L_t^\theta = \frac{1}{2} hB^\beta A_t^n L_t^\theta.$$

Therefore, $C_{t+1} \geq \frac{1}{2} hB^b A_t^n L_t^\theta$ for $t \geq 0$, and $\langle \mathbf{K}, \mathbf{R}, \mathbf{L}, \mathbf{Y}, \mathbf{C} \rangle$ is a feasible program. Also, $(C_{t+1}/L_{t+1}) \geq hB^b(A_t^n/L_t^\delta)(1/2M)$ for $t \geq 0$, so choosing $E = (hB^b/2M)$, (9.3.1) is satisfied. □

Theorem 9.3.1. *There exists a program which can maintain a positive per capita consumption level if*

(i) $\alpha > \beta$,
(ii) $\sup_{t \geq 0}[\mathbf{L}_t^\delta/A_t^{\nu-\varepsilon}] < \infty$ *for some $\varepsilon > 0$.*

Proof. Choose $0 < e < \varepsilon$, with e sufficiently close to zero, to ensure that $a > b$, and $n \geq \nu - \varepsilon$. By Lemma 9.3.1, there is a feasible program $\langle \mathbf{K}, \mathbf{R}, \mathbf{L}, \mathbf{Y}, \mathbf{C} \rangle$ and a scalar $E > 0$, such that (9.3.1) holds. Since $n \geq \nu - \varepsilon$, then by (ii) $\inf_{t \geq 0} A_t^n/L_t^\delta > 0$. Hence, $\inf_{t \geq 0}(c_{t+1}) > 0$. □

One can go in the other direction to show that

Theorem 9.3.2. *If there is a feasible program which can maintain a positive percapita consumption level, then*

(i) $\alpha > \beta$,
(ii) $\sup\limits_{t \geq 0}[L_t^\delta / A_t^\nu] < \infty$.

Proof.   We need several steps that are spelled out.   □

First, we have the following lemma.

Lemma 9.3.2. *If* $(\mathbf{K}, \mathbf{R}, \mathbf{L}, \mathbf{Y}, \mathbf{C})$ *from* $(\underline{K}, \underline{S})$ *is a program, then there is a positive scalar V (finite), such that*

$$K_t \leq V \cdot A_t^\mu \quad for\ all\ t \geq 0. \tag{9.3.5}$$

Proof.   Consider the pure accumulation program $(\hat{\mathbf{K}}, \hat{\mathbf{R}}, \hat{\mathbf{L}}, \hat{\mathbf{Y}}, \hat{\mathbf{C}})$ from $(\underline{K}, \underline{S})$ defined by

$$\hat{K}_0 = \underline{K}, \ \hat{K}_{t+1} = \hat{K}_t + G(\hat{K}_t, \hat{R}_t, \hat{L}_t) \quad \text{for } t \geq 0, \ \hat{L}_t = \mathbf{L}_t,$$

$$\hat{R}_t = R_t \quad \text{for } t \geq 0,$$

$$\hat{Y}_{t+1} = \hat{K}_{t+1}, \hat{C}_{t+1} = 0 \quad \text{for } t \geq 0.$$

Then, for all $t \geq 0$,

$$\hat{K}_{t+1} - \hat{K}_t = \hat{K}_t^\alpha \hat{R}_t^\beta \hat{L}_t^\gamma.$$

It follows that

$$(\hat{K}_{t+1}^{1-\alpha} - \hat{K}_t^{1-\alpha}) \leq [\hat{K}_{t+1} - \hat{K}_t]/\hat{K}_t^\alpha = \hat{R}_t^\beta \hat{L}_t^\gamma. \tag{9.3.6}$$

Writing L_t^γ as $[L_t^\theta]^{(1-\beta)}$ and using Holder's inequality in (9.3.6) (see Section 9.5.2) we get, for $T \geq 0$,

$$(\hat{K}_{T+1}^{1-\alpha} - \hat{K}_0^{1-\alpha}) \leq \left[\sum_{t=0}^{T} \hat{R}_t\right]^\beta \left[\sum_{t=0}^{T} \hat{L}_t^\theta\right]^{(1-\beta)} \leq (\underline{S}^\beta)A_T^{(1-\beta)},$$

$$\text{or} \quad \hat{K}_{T+1}^{1-\alpha} \leq \hat{K}_0^{1-\alpha} + (\underline{S}^\beta)A_T^{(1-\beta)}.$$

Hence, we can find some positive, finite V, such that for $t \geq 0$,

$$\hat{K}_t \leq V A_t^{[(1-\beta)/[(1-\alpha)]} = V A_t^\mu.$$

Clearly, $K_t \leq \hat{K}_t$ for $t \geq 0$, so $K_t \leq V A_t^\mu$ for $t \geq 0$. $\square$

Next, we note the following lemma.

Lemma 9.3.3. *If there is an efficient program* $(\mathbf{K}, \mathbf{R}, \mathbf{L}, \mathbf{Y}, \mathbf{C})$, *such that* $c_1 > 0$, $c_{t+1} \geq c_t$ *for* $t \geq 1$, *then*

(i) $c_{t+1} \leq k_t^\alpha r_t^\beta$ *for* $t \geq 0$,
(ii) $\alpha > \beta$.

Proof. We use two results from the previous chapter. First, the efficient program $(\mathbf{K}, \mathbf{R}, \mathbf{L}, \mathbf{Y}, \mathbf{C})$ is interior. Then, it is competitive and satisfies

$$G_{R_{t+1}} = G_{R_t} F_{K_{t+1}}, \quad \text{for } t \geq 0. \tag{9.3.7}$$

To prove claim (i), we suppose to the contrary and arrive at a contradiction.

Suppose, then, that there is some $t = \tau$ for which (i) does not hold, i.e.,

$$c_{\tau+1} = k_\tau^\alpha r_\tau^\beta + \varepsilon \quad \text{where } \varepsilon > 0. \tag{9.3.8}$$

For any program, and any t, we have

$$k_{t+1} - k_t \leq g_{t+1} k_{t+1} - k_t = k_t^\alpha r_t^\beta - g_{t+1} c_{t+1}. \tag{9.3.9}$$

Now,

$$g_{\tau+1} c_{\tau+1} \geq c_{\tau+1} = k_\tau^\alpha r_\tau^\beta + \varepsilon. \tag{9.3.10}$$

Hence, by (9.3.9) and (9.3.10),

$$k_{\tau+1} \leq k_\tau - \varepsilon < k_\tau. \tag{9.3.11}$$

Using (9.3.7) and noting that $F_{K_{t+1}}(= 1 + G_{K_{t+1}}) > 1$, we get $G_{R_{t+1}} > G_{R_t}$, so that

$$(k_{t+1}^\alpha / k_t^\alpha) \geq (r_{t+1}^{1-\beta} / r_t^{1-\beta}) \text{ for } t \geq 0. \tag{9.3.12}$$

Since $k_{\tau+1} < k_\tau$, then $r_{\tau+1} < r_\tau$ (by (9.3.12)). Hence, $k_{t+1}^\alpha r_{\tau+1}^\beta < k_\tau^\alpha r_\tau^\beta$. It follows that

$$g_{\tau+2} c_{\tau+2} \geq c_{\tau+2} \geq c_{\tau+1} = k_\tau^\alpha r_\tau^\beta + \varepsilon > k_{t+1}^\alpha r_{\tau+1}^\beta + \varepsilon. \qquad (9.3.13)$$

From (9.3.9) and (9.3.13),

$$k_{\tau+2} \leq k_{\tau+1} - \varepsilon. \qquad (9.3.14)$$

Continuing this procedure for each successive period,

$$k_{t+1} \leq k_t - \varepsilon \quad \text{for } t \geq \tau. \qquad (9.3.15)$$

But this implies that $k_t < 0$ for large t, i.e., a contradiction, establishing claim (i).

We now turn to claim (ii). We have for $t \geq 0$,

$$K_t^\alpha R_t^\beta L_t^\gamma \geq c_1 L_t. \qquad (9.3.16)$$

Using the bound (9.3.5) in (9.3.16), we get

$$R_t \geq [c_1^{(1/\beta)} L_t^{(1-\gamma)/\beta}] / [V^{(\alpha/\beta)} A_t^{(\alpha\mu/\beta)}]. \qquad (9.3.17)$$

Now,

$$(1-\gamma)/\beta = (\alpha+\beta)/\beta > 1 > \gamma/(\alpha+\gamma) = \gamma/(1-\beta) = \theta,$$

$$[\alpha\mu/\beta] = [\alpha(1-\beta)/\beta(1-\alpha)].$$

Hence, rewriting (9.3.17), we get

$$R_t \geq [c_1^{(1/\beta)}/V^{(\alpha/\beta)}][L_t^\theta/A_t^{[\alpha(1-\beta)/\beta(1-\alpha)]}]. \qquad (9.3.18)$$

Since $\sum_{t=0}^\infty R_t$ is finite, by the Abel–Dini theorem (see Theorem 13.1.5)

$$[\alpha(1-\beta)/\beta(1-\alpha)] > 1,$$

$$\text{i.e.,} \quad \alpha > \beta.$$

Thus, claim (ii) is established.   □

We need one more step.

Lemma 9.3.4. *If $\alpha > \beta$, and $(\mathbf{K}, \mathbf{R}, \mathbf{L}, \mathbf{Y}, \mathbf{C})$ from $(\underline{K}, \underline{S})$ is an interior efficient program, there is a scalar $\hat{E} > 0$, such that*

$$k_t^\alpha r_t^\beta \leq \hat{E}[A_t^\nu / L_t^\delta] \quad for \ t \geq 0. \tag{9.3.19}$$

Proof. Recall from Theorem 8.5.1 that if $(\mathbf{K}, \mathbf{R}, \mathbf{L}, \mathbf{Y}, \mathbf{C})$ from $(\underline{K}, \underline{S})$ is an interior efficient program, the transversality condition holds:

$$K_t / G_{R_t} \to 0 \quad \text{as } t \to \infty.$$

Hence, there is a (finite) positive integer T, such that

$$K_t / G_{R_t} \leq 1 \quad \text{for all } t \geq T,$$

$$\text{or} \quad [K_t^{1-\alpha} R_t^{1-\beta} / L_t^\gamma] \leq 1 \quad \text{for all } t \geq T. \tag{9.3.20}$$

This leads to the bound,

$$R_t \leq [L_t^\theta / K_t^{(1/\mu)}] \quad \text{for all } \ t \geq T. \tag{9.3.21}$$

Using (9.3.5) and (9.3.21), we see that for all $t \geq T$,

$$G(K_t, R_t, L_t) = K_t^\alpha R_t^\beta L_t^\gamma \leq [K_t^\alpha L_t^{(\beta\theta+\gamma)}] / K_t^{(\beta/\mu)}$$

$$= K_t^{[(\alpha-\beta)/(1-\beta)]} L_t^\theta$$

$$\leq V^{[(\alpha-\beta)/(1-\beta)]} A_t^\nu L_t^\theta \ (\text{using } \alpha > \beta).$$

Thus, for all $t \geq T$,

$$k_t^\alpha r_t^\beta \leq V^{[(\alpha-\beta)/(1-\beta)]} [A_t^\nu / L_t^\delta]. \tag{9.3.22}$$

Since T is finite, it is possible to choose $\hat{E} > 0$ satisfying (9.3.19).

At last, we come to the proof of Theorem 9.3.2. Observe that if $(\mathbf{K}^*, \mathbf{R}^*, \mathbf{L}^*, \mathbf{Y}^*, \mathbf{C}^*)$ is a program that maintains a positive per capita consumption level, then there is an equitable program

$(\acute{K}, \acute{R}, \acute{L}, \acute{Y}, \acute{C})$ with

$$\acute{C}_t/\acute{L}_t = \inf_{t \geq 1}[C_t^*/L_t^*] > 0 \quad \text{for all } t \geq 1.$$

Using Theorem 8.3.1, there is an efficient equitable program $(\mathbf{K}, \mathbf{R}, \mathbf{L}, \mathbf{Y}, \mathbf{C})$ with $c_t = c_1 > 0$ for $t \geq 1$. By Lemma 9.3.3, claim (i) holds. Moreover, (9.3.19) also holds. Combining Lemma 9.3.3(i) with (9.3.19), we have, for $t \geq 0$,

$$c_1 \leq c_{t+1} \leq k_t^\alpha r_t^\beta \leq \hat{E}[A_t^\nu/L_t^\delta]. \tag{9.3.23}$$

Claim (ii) follows from (9.3.23).   □

9.4   Ramsey Optimality

We turn to Ramsey optimality. One period return or felicity from per capita consumption is generated by a function $u : \mathbb{R}_+ \to \mathbb{R}$, which is assumed to be of the form

$$u(c) = -(1/c^\sigma). \tag{9.4.U}$$

For convenience, we denote $-u(c)$ by $\vartheta(c)$.

A program $(\mathbf{K}^*, \mathbf{R}^*, \mathbf{L}^*, \mathbf{Y}^*, \mathbf{C}^*)$ is optimal if

$$\limsup_{T \to \infty} \sum_{t=1}^{T} [L_t u(c_t) - L_t^* u(c_t^*)] \leq 0.$$

Our objective is to characterize conditions under which the precise limitations on population growth with other specified parameters of the model are consistent with attainment of Ramsey optimality.

Theorem 9.4.1. *There exists an optimal program if*

(i) $\alpha > \beta$,

(ii) $\nu\sigma > 1$,

(iii) $\sum_{t=0}^{\infty} \dfrac{L_t^{(1+\sigma\delta)}}{A_t^{\nu\sigma - \hat{\varepsilon}}} < \infty$ *for some $\hat{\varepsilon} > 0$.*

Proof. By (iii), there is $\hat{\varepsilon}$, satisfying $\nu\sigma > \hat{\varepsilon} > 0$. Let ε denote $(\hat{\varepsilon}/\sigma)$; then $\nu > \varepsilon > 0$. Given (i) choose $0 < e < \alpha$, with e sufficiently close to zero, to ensure that $a > b$, and $n \geq \nu - \varepsilon$.

By Lemma 9.3.1, there is a feasible program $\langle \mathbf{K}, \mathbf{R}, \mathbf{L}, \mathbf{Y}, \mathbf{C} \rangle$, and a scalar $E > 0$, such that (9.3.1) holds. Since $n \geq \nu - \varepsilon$, then

$$\mathbf{C}_{t+1} \geq E(A_t^{\nu-\varepsilon}/L_t^{\delta}) \quad \text{for } t \geq 0. \tag{9.4.1}$$

Then, for $t \geq 0$, $L_{t+1}\vartheta(c_{t+1}) = K \cdot [L_t^{(1+\delta\sigma)}/A_t^{\sigma\nu-\hat{\varepsilon}}]$ where K is a constant.

Then by (iii),

$$\sum_{t=0}^{\infty} L_{t+1}\vartheta(c_{t+1}) < \infty. \tag{9.4.2}$$

Then, by Brock and Gale (1969, Lemma 2), there exists an optimal program. $\qquad\square$

Theorem 9.4.2. *If there exists an optimal program, then*

(i) $\alpha > \beta$,
(ii) $\nu\sigma > 1$,
(iii) $\sum_{t=0}^{\infty} \frac{L_t^{(1+\sigma\delta)}}{A_t^{\nu\sigma}} < \infty.$

Proof. Suppose there is an optimal program, call it $\langle \mathbf{K}, \mathbf{R}, \mathbf{L}, \mathbf{Y}, \mathbf{C} \rangle$. Then, for each $t \geq 1$, the expression $L_t u\{[F(K_{t-1}, R_{t-1}, L_{t-1}) - K]/L_t\} + L_{t+1}u\{[F(K_t, R_t, L_t) - K]/L_{t+1}\}$ must be a maximum at $K = K_t$. The maximum must be at an interior point, i.e., $(C_{t+1}, K_t, R_t) \gg 0$ for $t \geq 0$. So,

$$u'(c_t) = u'(c_{t+1})F_{K_t} \quad \text{for } t \geq 1. \tag{9.4.3}$$

Also, for each $t \geq 1$, and for $0 < R < R_{t-1} + R_t$, the expression $L_t u\{[F(K_{t-1}, R_{t-1}, L_{t-1}) - K]/L_t\} + L_{t+1}u\{[F(K_t, R_t, L_t) - K]/L_{t+1}\}$ must be a maximum at $R = R_{t-1}$. Since the maximum is at the interior point, then

$$u'(c_t)F_{R_{t-1}} = u'(c_{t+1})F_{R_t} \quad \text{for } t \geq 1. \tag{9.4.4}$$

Define a sequence $\langle \mathbf{p}, \mathbf{q} \rangle$ in the following way:

$$p_0 = u'(c_1)F_{k_0}, p_t = u'(c_t) \quad \text{for } t \geq 1,$$

$$q_t = u'(c_1)F_{R_0} \quad \text{for } t \geq 0. \tag{9.4.5}$$

Note that, by (9.4.4), (9.4.5),

$$p_{t+1} = [q_0/F_{R_t}] \quad \text{for } t \geq 0. \tag{9.4.6}$$

By (9.4.3), $u'(c_t) \geq u'(c_{t+1})$, so $c_{t+1} \geq c_t$ for $t \geq 0$. Hence, by Lemma 9.3.3 (using the fact that $\langle \mathbf{K}, \mathbf{R}, \mathbf{L}, \mathbf{Y}, \mathbf{C} \rangle$ is efficient), (i) holds, and

$$c_{t+1} \leq k_t^\alpha d_t^\beta \quad \text{for } t \geq 0. \tag{9.4.7}$$

Using (9.4.6) and (8.5.5) in Theorem 8.5.1, we get

$$\sum_{t=0}^{\infty} p_{t+1}C_{t+1} < \infty. \tag{9.4.8}$$

Using (9.4.5) and (9.4.8),

$$\sum_{t=1}^{\infty} L_t u'(c_t)c_t < \infty. \tag{9.4.9}$$

Note that $u'(c)c = \sigma c^{-\sigma} = \sigma \vartheta(c)$ for $c > 0$.

Hence, by (9.4.9), we have (since $L_{t+1} \geq L_t$ for $t \geq 0$)

$$\sum_{t=1}^{\infty} L_t \vartheta(c_{t+1}) < \infty. \tag{9.4.10}$$

By (9.4.7) and Lemma 9.3.4,

$$c_{t+1} \leq \hat{E}[A_t^\nu / L_t^\delta] \quad \text{for } t \geq 0. \tag{9.4.11}$$

Using (9.4.11) and (9.4.10), we have

$$\sum_{t=1}^{\infty} \frac{\mathbf{L}_t^{(1+\delta\sigma)}}{\hat{E}^\sigma A_t^{\nu\sigma}} < \infty. \tag{9.4.12}$$

Now, (iii) follows directly from (9.4.12). Since $(1 + \sigma\delta) > 1 > \theta$, then (iii) implies by the Abel–Dini theorem, that $\nu\sigma > 1$, which establishes (ii). □

Remark. Note that if an optimal program exists, then the period social welfares $[L_t u(c_t)]$ are summable by (9.4.10).

Solow (1974) and Dasgupta and Heal (1979) have shown (in a continuous-time model) that if (9.2.1) and (9.4.U) hold, and $\mathbf{L}_t = \mathbf{L}n^t$, where $n \geq 1$, there exists an optimal program if and only if (i) $\alpha > \beta$, (ii) $n = 1$, and (iii) $\nu\sigma > 1$. This result also follows directly from Theorems 9.4.1 and 9.4.2.

We now consider an example in which population grows to infinity over time, and still there exists an optimal program.

Example 9.4.1. Consider the functional forms for G and u, and let $\alpha > \beta$, $\nu\sigma > 1$, $\mathbf{L}_t = \mathbf{L}(t+1)$, and $0 < \lambda < [\sigma\alpha/(1 - \alpha + \beta\sigma)] - 1$. (A numerical example would be the following: $G(K, D, L) = K^{0.20} D^{0.05} L^{0.75}$; $u(c) = -(1/c^{24})$; $\mathbf{L}_t = \mathbf{L}(t+1)$. Then $\alpha = 0.20 > 0.05 = \beta$; $\sigma = 24$, $\nu = (3/16)$, so $\nu\sigma = (9/2) > 1$; $[\sigma\alpha/(1-\alpha+\beta\sigma)] = (4.8/2) = 2.4$; $\lambda = 1$. Since $0 < 1 < 2.4 - 1$, then λ satisfies the inequality $0 < \lambda < [\sigma\alpha/(1 - \alpha + \beta\sigma)] - 1$.) Then, it follows from Theorem 9.4.1 that there exists an optimal program. Note also that $\mathbf{L}_t \to \infty$ as $t \to \infty$, since $\lambda > 0$.

9.5 Complements and Details

The material in this chapter is drawn from Mitra (1983). Cobb–Douglas production functions appeared in the early papers by Solow (1974) and Stiglitz (1974 a,b) (all the models treating time as a continuous variable) on the role of an exhaustible resource as an essential input.

9.5.1 Continuous-time model

We review a continuous-time model and derive one of the basic results on sustaining a constant consumption (Solow (1974b)). To simplify, we write Y_t, C_t, K_t, etc. instead of $Y(t), C(t), K(t)$, etc. Consider an economy with a production function given by

$$Y_t = F(K_t, R_t) = K_t^{\alpha_1} \cdot R_t^{\alpha_2} \quad \text{where } \alpha_1 > 0, \alpha_2 > 0, \alpha_1 + \alpha_2 < 1.$$
(9.5.1)

The initial stocks of capital $\underline{K} > 0$ and the resource $\underline{S} > 0$ are given. A program $\langle \mathbf{c}, \mathbf{k}, \mathbf{r} \rangle = (C_t, K_t, R_t)$ from $(\underline{K}, \underline{S})$ satisfies

$$C_t + \dot{K}_t = K_t^{\alpha_1} \cdot R_t^{\alpha_2},$$
(9.5.2)

$$\underline{S} = \int_0^\infty R_\tau d\tau,$$
(9.5.3)

$$C_t \geq 0, K_t \geq 0, R_t \geq 0 \quad \text{for all } t \geq 0.$$

Here, all functions C_t, K_t, R_t are assumed to be twice continuously differentiable with respect to t at $t > 0$. The equality in (9.5.3) is to be interpreted as an efficiency condition. We refer to C_t, K_t, R_t as a consumption, capital and resource utilization path, respectively.

Our objective is to show the following.

Proposition 9.5.1. *If $\alpha_1 > \alpha_2$, there is some $\bar{C} > 0$, such that one can construct a program with $C_t = \bar{C} > 0$ for all $t \geq 0$.*

It is useful to prove the following lemma first.

Lemma 9.5.1. *Let $f : R_+ \to R_+$ be a twice continuously differentiable function that satisfies $f(0) = 0$; $f'(x) > 0$ and $f''(x) < 0$ for all $x > 0$, and $f'(x) \to \infty$ as $x \to 0$. Then, for any constant $A > 0$, there is some $\tilde{x} > 0$, such that*

$$f(x) \cdot A > x \text{ for all } x \text{ satisfying } 0 < x < \tilde{x}.$$

Proof. By the Mean Value theorem, for any $x > 0$,

$$f(x) = f(0) + xf'(z), \quad \text{where } 0 < z < x.$$

Using the fact that $f''(x) < 0$, we get

$$f'(z) > f'(x), \text{ so that for any } x > 0, \ f(x) > xf'(x). \qquad (9.5.4)$$

Now, given any constant $A > 0$, there is some $\tilde{x} > 0$ such that $f'(\tilde{x}) > (1/A)$. This means that

$$\text{for all } 0 < x < \tilde{x}, f'(x) > f'(\tilde{x}) > (1/A). \qquad (9.5.5)$$

From (9.5.4) and (9.5.5), for all $0 < x < \tilde{x}$, $f(x) > xf'(x) > (x/A)$.
$\square$

Proof of Proposition 9.5.1. To complete the proof of Proposition 9.5.1, we experiment with a capital path of the form,

$$K_t = K_0 + mt, \quad m > 0, \qquad (9.5.6)$$

and see whether we can choose *some* m that is consistent with a consumption path of the form $C_t = \bar{C} > 0$ for all $t \geq 0$ and for *some* $\bar{C} > 0$. For a consumption path $C_t = \bar{C} > 0$, from (9.5.1) and (9.5.2), we must have

$$\bar{C} + \dot{K}_t = Y_t, \qquad (9.5.7)$$

and if the capital path is (9.5.6), we get (as $\dot{K}_t = m$)

$$\bar{C} + m = Y_t > 0. \qquad (9.5.8)$$

This leads to (with $R_t > 0$)

$$0 = \dot{Y}_t/Y_t = d\,[\log Y_t]/dt = d\,[\alpha_1 \log K_t + \alpha_2 \log R_t]/dt \qquad (9.5.9)$$

$$= \alpha_1(\dot{K}_t)/K_t + \alpha_2(\dot{R}_t)/R_t = [\alpha_1 m/(K_0 + mt)] + \alpha_2(\dot{R}_t)/R_t. \qquad (9.5.10)$$

Hence,

$$\alpha_1 m/(K_0 + mt) + \alpha_2(\dot{R}_t)/R_t = 0. \qquad (9.5.11)$$

Consequently,

$$(\dot{R}_t)/R_t = -[\alpha_1 m/(K_0 + mt)\alpha_2$$

$$= -\{(\alpha_1 m/\alpha_2)[1/(K_0 + mt)]\}.$$

Using separation of variables,

$$d(\dot{R}_t/R_t) = -\{(\alpha_1 m/\alpha_2)[1/(K_0 + mt)]\}\,dt,$$

$$d(= \log R_t) = d(\log[K_0 + mt]^{-\alpha_1/\alpha_2}).$$

Integrating,

$$R_t = B(K_0 + mt)^{-\alpha_1/\alpha_2}, \tag{9.5.12}$$

where B is a constant of integration. We shall choose $B = (\bar{C} + m)^{1/\alpha_2}$. Verify, by direct substitution, that the solution suggested by (9.5.12) in fact satisfies equation (9.5.11). Thus, a choice of K_t according to (9.5.6) and the requirement that $C_t = \bar{C} > 0$ imply a resource utilization path given by (9.5.12). Now, the question is: "can we choose $\bar{C} > 0$ and $m > 0$ that will satisfy the efficiency condition (9.5.3)?" So, let us examine the equation (given $\underline{K} > 0$, and $\underline{S} > 0$):

$$\underline{S} = \int_0^\infty R_\tau d\tau = (\bar{C} + m)^{1/\alpha_2} \int_0^\infty (\underline{K} + mt)^{-\alpha_1/\alpha_2}\,dt. \tag{9.5.13}$$

It can be shown that when $\alpha_1 > \alpha_2$, the integral on the right side of (9.5.13) is in fact convergent. Use the substitution $z = K_0 + mt$, then

$$\int_0^\infty (K_0 + mt)^{-\alpha_1/\alpha_2}\,dt = (1/m)\int_{K_0}^\infty z^{-\alpha_1/\alpha_2}\,dz = (1/m)\int_{\underline{K}}^\infty z^{-\theta}\,dz,$$

$$\text{say, where } \theta = \alpha_1/\alpha_2 > 1.$$

For any $b > \underline{K}$, we know that $\int_{\underline{K}}^b z^{-\theta}\,dz = [1/(1-\theta)][b^{1-\theta} - K_0^{1-\theta}]$. Noting that $(1 - \theta) < 0$, letting $b \to \infty$, we get

$$\int_{K_0}^\infty z^{-\theta}\,dz = (K_0^{1-\theta})/(\theta - 1) = 1/[K_0^{\theta-1} \cdot (\theta - 1)] = 1/J, \text{ say, where}$$

$$J = (K_0^{([\alpha_1 - \alpha_2]/\alpha_2)}) \cdot [(\alpha_1/\alpha_2) - 1)]. \tag{9.5.14}$$

Using (9.5.14), we go back to equation (9.5.13) and get

$$\underline{S} = (\bar{C} + m)^{1/\alpha_2} \cdot [1/(m \cdot J)],$$
$$\underline{S} \cdot m \cdot J = (\bar{C} + m)^{1/\alpha_2},$$

or

$$m^{\alpha_2} \cdot (\underline{S}J)^{\alpha_2} = (\bar{C} + m) \quad \text{where } 0 < \alpha_2 < 1. \qquad (9.5.15)$$

Now, the second term on the left side of (9.5.15) is simply a constant. By using Lemma 9.5.1, we see that for sufficiently small values of m, the left side is greater than m, and this, in turn, means that we can satisfy (9.5.15) for (sufficiently small values) $m > 0$ and $\bar{C} > 0$. To be sure, there are many combinations of m and $\bar{C}$ that will satisfy (9.5.15).

To summarize, when $1 > \alpha_1 > \alpha_2 > 0$, there is a class of programs from $\underline{S}$ and $\underline{K}$ of the following forms:

$K_t = K_0 + mt, m > 0,$
$R_t = (\bar{C} + m)^{1/\alpha_2}(K_0 + mt)^{-\alpha_1/\alpha_2},$
$C_t = \bar{C} > 0; Y_t = C_t + \dot{K}_t = \bar{C} + m,$

and $m^{\alpha_2} \cdot (S_0 J)^{\alpha_2} = (\bar{C} + m)$, where $J = (K_0^{([\alpha_1 - \alpha_2]/\alpha_2)}) \cdot [(\alpha_1/\alpha_2) - 1)]$. $\qquad \square$

9.5.2 Application of Hölder's inequality

These remarks are intended to keep the exposition self-contained. Recall the Hölder inequality (Bartle (1964, p. 69)): Let $\{a_1, a_2, \ldots, a_n\}$ and $\{b_1, b_2, \ldots, b_n\}$ be non-negative real numbers. If $r > 1$, $s > 1$ and $(1/r) + (1/s) = 1$, then

$$\sum_{j=1}^{n} a_j b_j \leq \left[\sum_{j=1}^{n} a_j^r\right]^{1/r} \left[\sum_{j=1}^{n} b_j^s\right]^{1/s}.$$

From (9.3.6), we see that

$$\hat{K}_1^{1-\alpha} - \hat{K}_0^{1-\alpha} \leq [\hat{K}_1 - \hat{K}_0]/\hat{K}_0^{\alpha} = \hat{R}_0^{\beta}\hat{L}_0^{\gamma},$$
$$\hat{K}_2^{1-\alpha} - \hat{K}_1^{1-\alpha} \leq \hat{R}_1^{\beta}\hat{L}_1^{\gamma}.$$

Adding up,

$$\hat{K}_2^{1-\alpha} - \hat{K}_0^{1-\alpha} \le \hat{R}_0^{\beta}\hat{L}_0^{\gamma} + \hat{R}_1^{\beta}\hat{L}_1^{\gamma}.$$

Hence, for any $T \ge 0$,

$$\hat{K}_{T+1}^{1-\alpha} - \hat{K}_0^{1-\alpha} \le \sum_{t=0}^{T} \hat{R}_t^{\beta}\hat{L}_t^{\gamma} = \sum_{t=0}^{T} \hat{R}_t^{\beta}(\hat{L}_t^{\theta})^{1-\beta}.$$

Set $r = (1/\beta) > 1$ and $(1/s) = 1 - \beta$, and use Hölder's inequality on the right-hand side and follow the text.

Chapter 10

Technological Transition: An Optimistic Approach

However great the scientific importance may be, its practical value will be no less obvious when we reflect that the supply of solar energy is both without limit and without cost and it will continue to pour down upon us for countless ages after all the coal deposits of the earth have been exhausted and forgotten.

— [on the discovery of photovoltaic effect] Werner von Siemens (1885)

You can't use up creativity. The more you use, the more you have.

— Maya Angelou

10.1 Introduction

A response to the constraints imposed by natural resources that are essential inputs in the production process ought to come from three directions: first, the basic notion of efficiency (avoidance of waste) given a particular technology; second, a search for a technology that allows for a high degree of substitution between an input based on an exhaustible resource and an input that relies on a producible resource; third, and most optimistically, a switch to a technology that is independent of natural resources as inputs. In this chapter, we explore a simple model (basically a discrete-time version of the optimistic model of Vernon Smith (1974)). We consider an economy that has an option of using two alternatives: with a constant labor supply in each period, the first technology uses an exhaustible resource and labor as essential input in production; the second "green technology" uses only labor and has constant returns to scale. The economy tries to maximize a discounted sum of outputs (returns). The principal results are as follows: there is an optimal program, given the structure of the technologies (see (F.1)–(F.4)); in such a program, the economy uses the green technology beyond a period T.

10.2   A Model of Optimistic Transition

A transition from a technology that uses an exhaustible resource as an essential input to one in which the exhaustible resource is not used has been studied by a number of writers (see Smith (1974) or Heal (1976) and the references cited therein). In this section, we consider an economy with an exhaustible resource which can be used as an input in a production function $F(\cdot, \cdot)$. Let the total supply of this resource be $\underline{S} > 0$. At the "beginning" of each period t, the economy also gets a quantity $l_t \equiv \bar{l} > 0$ of another factor of production. Let $q_{1t} = F(r_t, l_{1t})$ be the output (appearing at the "end" of period t) when the input pair is $(r_t, l_{1t}) \geq 0$. Assume the following:

(F.1) $F : R_+^2 \to R_+$ *is continuous.*

(F.2) $F(0, l) = F(r, 0) = 0$, *in other words, both inputs are essential in production.*

(F.3) F *is non-decreasing in each argument, and* $(r, l) \gg (r', l')$ *implies that* $F(r, l) > F(r', l')$.

There is an alternative technology producing the same output, given by $q_{2t} = \theta \cdot l_{2t}$, where

(F.4) $\theta > 1$.

Observe that this technology does not need the exhaustible resource as an input. We shall call this a "green technology".

Given $\langle \underline{S}; l_t \equiv \bar{l} \rangle$ a *program* is a *non-negative* sequence $\langle q_{1t}, q_{2t}; r_t, l_{1t}; l_{2t} \rangle$, satisfying

$$q_{1t} = F(r_t, l_{1t}); \quad q_{2t} = \theta \cdot l_{2t} \quad \text{for all } t \geq 0,$$

$$l_{1t} + l_{2t} = \bar{l} \quad \text{for all } t \geq 0,$$

$$\sum_{t=0}^{\infty} r_t \leq \underline{S}.$$

Thus, a program is a specification of decisions on resource allocation that can be informally described as follows: at the "beginning" of period $t = 0$, it starts with the stock $\underline{S} > 0$, and the supply of $l_0 = \bar{l} > 0$. It decides on the (non-negative) quantities $\langle l_{10}, l_{20}; r_0 \rangle$ subject to the constraint $l_{10} + l_{20} = \bar{l}$ and $r_0 \leq \underline{S}$. As a result, the outputs (q_{10}, q_{20}) appear at the end of that period. At the beginning of the next period $t = 1$, it starts with the stock $\underline{S} - r_0 \geq 0$, and the supply of $l_1 = \bar{l} > 0$, and the decision making task is repeated. Note that if in any period, $l_{2t} = 0$, the corresponding $q_{2t} = 0$. Similarly, if $l_{1t} = 0$, or $r_t = 0$ (or both), the corresponding $q_{1t} = 0$.

We now introduce a notion of optimality. As before, we denote the discount factor by δ. Given $(\underline{S}, \bar{l}) \gg 0$, and $0 < \delta < 1$, a program $\mathbf{x} = \langle q_{1t}^*, q_{2t}^*; r_t^*, l_{1t}^*; l_{2t}^* \rangle$ is *optimal* if

$$\sum_{t=0}^{\infty} \delta^t (q_{1t}^* + q_{2t}^*) \geq \sum_{t=0}^{\infty} \delta^t (q_{1t} + q_{2t})$$

for all programs $\langle q_{1t}, q_{2t}; r_t, l_{1t}; l_{2t} \rangle$. Informally, a program is optimal if it generates the maximum discounted sum of outputs among all programs.

10.2.1 Analysis

The first thing we establish is that there exists an optimal program. Let us note some boundedness properties. As always, we are given

$(\underline{S}, l_t = \bar{l})$. Write $B = F(\underline{S}, \bar{l})$ and $\gamma = \theta \cdot \bar{l}$. Then, for any program $\mathbf{x}$,

$$v(\mathbf{x}) = \sum_{t=0}^{\infty} \delta^t \left(q_{1t} + q_{2t} \right) \leq \sum_{t=0}^{\infty} \delta^t \left(B + \gamma \right) \equiv \frac{B + \gamma}{1 - \delta}.$$

Thus, $\alpha \equiv \sup \{ v(\mathbf{x}) : \mathbf{x}$ is a program given $(\underline{S}, l_t = \bar{l}) \}$ is well defined.

Theorem 10.2.1. *There exists an optimal program* $\mathbf{x}^* = \langle q_{1t}^*, q_{2t}^*; r_t^*, l_{1t}^*; l_{2t}^* \rangle$.

Proof.   We can adapt the proof of Theorem 5.4.1. (▲ *Exercise* ▼)

□

Our next proposition is the principal result on the eventual switch to the green technology.

Theorem 10.2.2. *Let* $\mathbf{x}^* = \langle q_{1t}^*, q_{2t}^*; r_t^*, l_{1t}^*; l_{2t}^* \rangle$ *be an optimal program. Then there is some finite* T, *such that* $l_{2t}^* > 0$, *for all* $t \geq T$.

Proof.   If the claim is *not* true, then there is an infinite subsequence of periods t_n, such that $l_{2t_n}^* = 0$ and $F(r_{t_n}^*, \bar{l}) = q_{1t_n}^* > 0$ for all t_n. Since an optimal program is a (feasible) program, $\sum_t r_t^* \leq S$. This means that given any $\varepsilon' > 0$, there is some $T(\varepsilon')$, such that $0 < r_t^* < \varepsilon'$ for all $t \geq T(\varepsilon')$. In particular, for all $t_n \geq T(\varepsilon')$, it is true that $0 < r_{t_n}^* < \varepsilon'$.

By continuity of F and the assumption that $F(0, \bar{l}) = 0$, there is some $\varepsilon > 0$ sufficiently small, such that $F(r, \bar{l})$ is less than some $\nabla > 0$, where $\nabla < (\theta \cdot \bar{l})$, for all $0 \leq r < \varepsilon$. Consider now a period $t_n \geq T(\varepsilon)$. Then, $F(r_{t_n}^*, \bar{l}) < \nabla$. Hence, $q_{1t_n}^* < \nabla$. Consider an alternative program that agrees with our assumed optimal program for all periods except period t_n, but uses only the green technology in period $t_n \geq T$. The total output in period t_n is $(\theta \cdot \bar{l})) > \nabla > q_{1t}^*$. Clearly, the total output from this alternative program exceeds that of the optimal program, and this contradicts the optimality of $\mathbf{x}^*$ and establishes our claim.   □

10.2.2 Examples

We consider two examples in which technology 1 takes parametric functional forms: one with a perfect complements technology and one with a Cobb–Douglas technology.

Example 10.2.1. For simplicity, we write S in place of $\underline{S}$. Assume that the production function which uses the resource is a perfect complements technology $F(r_t, l_{1t}) = \min\{r_t, al_{1t}\}$, $a > 0$ and $S > 0$. If technology 1 is used in some period t, then it cannot be that $r_t^* < al_{1t}^*$ or else some labor used in technology 1 is not productive and it can be productive in technology 2. If in the optimal program $r_t^* > al_{1t}^*$, then there is also an optimal program with $r_t^{**} = al_{1t}^*$, leaving the remaining $r_t^* - al_{1t}^*$ unused. We therefore focus on the case where in every period, $r_t^* = al_{1t}^*$.

We consider two cases depending on the relative productivity of labor in the two available technologies: (i) $\theta > a$ and (ii) $\theta < a$. We show that the resource is not used at all in case (i) where the green technology is very productive. But, in case (ii), the entire stock of the resource is exhausted.

We will show that for a sufficiently large initial stock of the exhaustible resource, there is a number of initial periods in which only the first technology is used. Then, there is a switch to the green technology, with at most one period of overlapping use of the two technologies. We state these findings more formally.

(i) *If $F(r_t, l_{1t}) = \min\{r_t, al_{1t}\}$ and $\theta > a$, then only technology 2 is used.* (ii) *If $F(r_t, l_{1t}) = \min\{r_t, al_{1t}\}$ and $\theta < a$, then in any optimal program, all of the resource is used, $\sum_t r_t^* = S$. Moreover, only technology 1 will be used for T-periods, where $T = \lfloor \frac{S}{al} \rfloor$ (i.e., the integer part of $\frac{S}{al}$). In period $T+1$, the remaining quantity of the resource (if any) is used $r_{T+1} = S - Ta\bar{l}$, $l_{1T+1} = \frac{S - Ta\bar{l}}{a}$ and for all $t > T+1$, only technology 2 is used, $r_{T+1} = l_{1T+1} = 0$.*

Proof.

(i) We assume here that $\theta > a$. Suppose in an optimal program $\langle r_t^*, l_{1t}^*; l_{2t}^* \rangle$ technology 1 is used in some period t. Recall that

$r_t^* = a l_{1t}^*$. Consider an identical program except that we change allocations in period t by moving $\varepsilon \in (0, l_{1t}^*)$ units of l from technology 1 to technology 2. Production does not change in any other period. In period t, the change in production is

$$\left[\min\{r_t^*, a(l_{1t}^* - \varepsilon)\} + \theta(\bar{l} - l_{1t}^* + \varepsilon)\right]$$
$$- \left[\min\{r_t^*, a l_{1t}^*\} + \theta(\bar{l} - l_{1t}^*)\right]$$
$$= \left[a(l_{1t}^* - \varepsilon) + \theta(\bar{l} - l_{1t}^* + \varepsilon)\right] - \left[a l_{1t}^* + \theta(\bar{l} - l_{1t}^*)\right]$$
$$= (\theta - a)\varepsilon > 0.$$

This is a contradiction to the optimality of the initial program.

(ii) Assume now that $\theta < a$. Suppose by way of contradiction that $\sum_t r_t^* < S$. Since $r_t^* \to 0$, there is at least one period t in which $r_t^* < a\bar{l}$. In period t technology 2 must be used, $l_{2t}^* > 0$, or else some labor is not productive in technology 1, but it could be in technology 2. Recall that $r_t^* = a l_{1t}^*$. Consider the effect of adding $\varepsilon \in (0, \min\{S - \sum_t r_t^*, a l_{2t}^*\})$ of the remaining resource to period t production and moving $\frac{\varepsilon}{a}$ of the labor in period t from technology 2 to technology 1, making no other changes to the program. The change in production in this period is

$$\left[\min\left\{r_t^* + \varepsilon, a\left(l_{1t} + \frac{\varepsilon}{a}\right)\right\} + \theta\left(l_{2t}^* - \frac{\varepsilon}{a}\right)\right]$$
$$- \left[\min\{r_1^*, a l_{1t}^*\} + \theta l_{2t}^*\right]$$
$$= \left[a\left(l_{1t}^* + \frac{\varepsilon}{a}\right) + \theta l_{2t}^* - \theta\frac{\varepsilon}{a}\right] - \left[a l_{1t}^* + \theta l_{2t}^*\right]$$
$$= (a - \theta)\frac{\varepsilon}{a} > 0.$$

This is a contradiction to the optimality of the initial program. This establishes that $\sum_t r_t^* = S$.

We next show that if $S > a\bar{l}$, then $l_{10}^* = \bar{l}$, i.e., in period 0, only technology 1 is used. Suppose, by way of contradiction, that in an optimal program $\langle r_t^*, l_{1t}^*; l_{2t}^* \rangle$, in period 0, $l_{10}^* < \bar{l}$ and $l_{20}^* > 0$. In this period, $r_0^* = a l_{10}^* \geq 0$. We know that all of S is used in the optimal program. Because $S > a\bar{l}$, and for all t, $r_t^* = a l_{1t}^* \leq a\bar{l} < S$, it must be that S is used for at least two periods. Hence, there exists at least one more period where technology 1 is used. Let $t > 0$ be a period where the resource is used, so that $r_t^* = a l_{1t}^* > 0$. Now, consider moving $\varepsilon \in (0, \min\{r_t^*, a l_{20}^*\})$ of the resource from period t to period 0, adjusting labor in both periods accordingly. The new program denoted by ** is everywhere the same as the proposed optimal program, only that $r_0^{**} = r_0^* + \varepsilon$ and $r_t^{**} = r_t^* - \varepsilon$. Also, $l_{10}^{**} = l_{10}^* + \frac{\varepsilon}{a}$ and $l_{20}^{**} = l_{20}^* - \frac{\varepsilon}{a}$ and $l_{1t}^{**} = l_{1t}^* - \frac{\varepsilon}{a}$ and $l_{2t}^{**} = l_{2t}^* + \frac{\varepsilon}{a}$. Clearly, the new program is still feasible. Production does not change in any other periods except in periods 0 and t. In these periods, the change in production is

$$
\left[\min\left\{ r_0^* + \varepsilon, a\left(l_{10}^* + \frac{\varepsilon}{a} \right) \right\} + \theta\left(l_{20}^* - \frac{\varepsilon}{a} \right) \right]
$$
$$
+ \delta^t \left[\min\left\{ r_t^* - \varepsilon, a\left(l_{1t}^* - \frac{\varepsilon}{a} \right) \right\} + \theta\left(l_{2t}^* + \frac{\varepsilon}{a} \right) \right]
$$
$$
- \left[\min\left\{ r_0^*, a l_{10}^* \right\} + \theta l_{20}^* \right] - \delta^t \left[\min\left\{ r_t^*, a l_{1t}^* \right\} + \theta l_{2t}^* \right]
$$
$$
= a l_{10}^* + \varepsilon + \theta l_{20}^* - \frac{\theta \varepsilon}{a} - a l_{10}^* - \theta l_{20}^*
$$
$$
+ \delta^t \left[a l_{1t}^* - \varepsilon + \theta l_{2t}^* + \frac{\theta \varepsilon}{a} - a l_{1t}^* - \theta l_{2t}^* \right]
$$
$$
= \left[1 - \delta^t \right] \frac{a - \theta}{a} \varepsilon > 0.
$$

This contradicts the optimality of the initial program and implies that only technology 1 is used as long as the remaining resource is high enough, and then there is a switch to the green technology. $\square$

Example 10.2.2. For simplicity, we write S in place of $\underline{S}$. Consider technology 1 which is a Cobb–Douglas production function $F(r, l) = r^\alpha l^\beta$ where $\alpha, \beta > 0$. In this example, we will distinguish between production functions with $\beta < 1$, or $\beta \geq 1$, which determine whether the marginal productivity of labor is decreasing, constant or increasing with labor. With diminishing marginal productivity of labor in technology 1, we find that the entire quantity of the resource would be exhausted. With increasing marginal product, if the green technology is productive enough, it is possible that only the green technology is used. We prove the following claims:

Assume $F(r, l) = r^\alpha l^\beta$. (i) If $\beta < 1$, then in an optimal program, all of the resource must be used. (ii) If $\beta \geq 1$ and $S^\alpha \bar{l}^{\beta-1} > \theta$, then in an optimal program, all of the resource must be used. (iii) If $\beta \geq 1$ and $S^\alpha \bar{l}^{\beta-1} < \theta$, then in an optimal program, only technology 2 is used.

Proof.

(i) Assume $\beta < 1$. Suppose first that only technology 2 is used. Consider moving $\varepsilon > 0$ of labor in period 0 from technology 2 to technology 1 and using it with the entire stock of the resource S. Keep all other periods unchanged. Then, in period 0, the change in production is

$$S^\alpha \varepsilon^\beta + \theta \cdot (\bar{l} - \varepsilon) - \theta \cdot \bar{l} = (S^\alpha \varepsilon^\beta - \theta\varepsilon) = (S^\alpha - \theta\varepsilon^{1-\beta})\varepsilon^\beta.$$

Taking $\varepsilon \in (0, \min\{(\frac{S^\alpha}{\theta})^{\frac{1}{1-\beta}}, \bar{l}\})$, we get $S^\alpha - \theta\varepsilon^{1-\beta} > 0$ and thus an increase in production. This contradicts optimality of a program that uses only technology 2. Hence, at least some of the resource must be used in any optimal program. Now, because F is strictly increasing in r for all $l > 0$, it must be that all of the resource is used. Else, increase output by adding the remaining r in a period, where $l_{1t} > 0$.

(ii) Assume $\beta \geq 1$ and $S^\alpha \bar{l}^{\beta-1} > \theta$. Suppose first that only technology 2 is used. Consider moving $\bar{l}$ in period 0 from technology 2 to technology 1 and using it with the entire stock of the resource S. Keep all other periods unchanged. Then, in period 0, the change in production is

$$S^\alpha (\bar{l})^\beta - \theta \cdot \bar{l} = (S^\alpha (\bar{l})^{\beta-1} - \theta) \cdot \bar{l} > 0.$$

This contradicts the optimality of a program that uses only technology 2. Hence, at least some of the resource must be used in any optimal program. Again, since F is strictly increasing in r

for all $l > 0$, it must be that all of the resource is used. Else, increase output by adding the remaining r in a period where $l_{1t} > 0$.

(iii) The third case we need to consider is $\beta \geq 1$ and $S^\alpha(\bar{l})^{\beta-1} < \theta$. Assume by way of contradiction that there is a period t in which technology 1 is used. Consider moving the labor used in this period to technology 2. The change in production is

$$\theta \cdot \bar{l} - [r_t^\alpha l_{1t}^\beta + \theta \cdot (\bar{l} - l_{1t})] = (\theta \cdot l_{1t} - r_t^\alpha l_{1t}^\beta)$$
$$> (S^\alpha \bar{l}^{\beta-1} - r_t^\alpha l_{1t}^{\beta-1})l_{1t} \geq 0.$$

This contradicts the optimality of the optimal program. $\square$

10.3   Complements and Details

This chapter is based entirely on Majumdar and Bar (2013). The literature on research and adoption of new technology is fascinating and quite substantial. See the reviews by Hoppe (2002), Nyrako (2010). Of particular interest are the papers on backstop technology by Heal (1976) and Nordhaus (1973). The question of allocation of resources to research may not be adequately captured by a deterministic model, as the "success" of a research activity may be an essentially random event. See the example in Dasgupta, Heal and Majumdar (1977) based on an elegant theorem proved in Breiman (1961).

Chapter 11

Evolution and Extinction
under Uncertainty

One way which I believe is particularly fruitful and promising is
to study what would become of the solution of a deterministic
dynamic system if it were exposed to a stream of erratic shocks
that constantly upsets its evolution.

—Ragnar Frisch

11.1 Introduction

In Chapter 2, we considered examples of dynamic processes described
by a first-order difference equation:

$$x_{t+1} = \alpha(x_t), \tag{11.1.1}$$

where α (the law of motion) is a function from an appropriate set
S, the set of all "states", into itself. Given an initial x_0, the entire
trajectory,

$$(x_0, \alpha(x_0), \alpha^{(2)}(x_0), \ldots, \alpha^{(j)}(x), \ldots),$$

is determined by (11.1.1). In Chapters 2 and 4, we explored the qualitative properties of trajectories emanating from alternative initial
conditions. Of course, the properties of α are crucial in identifying
the asymptotic behavior of the process (11.1.1).

A simple way to introduce uncertainty has been to consider the
law of motion as a random function. Informally, suppose that instead

of a single α in (11.1.1), we start with an exogenously specified set,

$$\Gamma = \{\gamma_1, \gamma_2, \ldots, \gamma_N\}, \tag{11.1.2}$$

of all possible laws of motion (each element of Γ is a map from S into S). Given the initial state x_0, Tyche ("nature") chooses some element γ_j from Γ according to a probability distribution Q (say, γ_j is chosen with probability $Q_j > 0$, $\sum_{j=1}^{N} Q_j = 1$). We then have the following transition:

$$X_1 = \alpha_1(x_0), \text{ where } \alpha_1(x_0) = \gamma_j(x_0) \text{ with probability } Q_j. \tag{11.1.3}$$

Thus, X_1 is a random variable (with N possible values $\gamma_j(x_0)$, with probabilities Q_j, $j = 1, 2, \ldots, N$). In the next period, independent of α_1, Tyche chooses again some element from Γ according to the same Q and we have

$$X_2 = \alpha_2(X_1), \text{ where } \alpha_2(\cdot) = \gamma_j(\cdot) \text{ with probability } Q_j,$$

and the story is repeated. We are now studying the evolution of a stochastic process:

$$X_{n+1} = \alpha_{n+1}(X_n) \equiv \alpha_{n+1}(\alpha_n \cdot (\ldots \alpha_1(x_0)) \ldots) \ (n \geq 0), \tag{11.1.4}$$

where $\{\alpha_n : n \geq 1\}$ is a sequence of independent maps with the common distribution Q on Γ. It can be verified that $\{X_n : n = 0, 1, \ldots\}$ is a Markov process having the *transition probability* ($x \in S$, and B a (Borel) subset of S):

$$p(x, B) = \text{Prob}(\alpha_1 x \in B) = Q(\{\gamma \in \Gamma : \gamma x \in B\}).$$

One may take X_0 to be random, *independent of* $\{\alpha_n : n \geq 1\}$ and having a given (initial) distribution μ. In order to simplify, we drop the symbols "(" and ")" to denote the composition of maps in (11.1.4) and simply write:

$$X_{n+1} = \alpha_{n+1}(X_t) \equiv \alpha_{n+1}\alpha_n \ldots \alpha_1(x_0) \ (n \geq 0). \tag{11.1.5}$$

In Section 11.2, we provide a formal framework for exploring the evolution of such stochastic processes. The primary theme of Section 11.3 is the existence and uniqueness of a steady state (an invariant distribution) with interesting stability properties. The examples in Section 11.4 deal with the possibilities of growth and extinction. The reader is encouraged to review the relevant parts of Chapter 13 on the results from probability theory and convergence of random variables.

11.2 Random Dynamical Systems

We begin with some formal definitions. Let S be a metric space and $\mathcal{S}$ be the Borel sigma-field of S (see Section 13.4 of Chapter 13 for the definitions and alternative notations). The set of all probability measures on $(S, \mathcal{S})$ is denoted by $\mathcal{P}(S)$. A sequence P_n of probability measures is said to converge weakly to a probability measure P if for every bounded, real-valued continuous function f,

$$\int f dP_n \to \int f dP.$$

Endow Γ, a family of maps from S into S, with a sigma-field Σ, such that the map $(\gamma, x) \to (\gamma(x))$ on $(\Gamma \times S, \ \Sigma \times \mathcal{S})$ into $(S, \mathcal{S})$ is measurable. Let Q be a probability measure on (Γ, Σ).

On some probability space $(\Omega, \mathcal{F}, P)$, let $(\alpha_n)_{n=1}^{\infty}$ be a sequence of *independent* random functions from Γ with *a common distribution* Q. For a given random variable X_0 (with values in S), independent of the sequence $(\alpha_n)_{n=1}^{\infty}$, define

$$X_1 \equiv \alpha_1(X_0) \equiv \alpha_1 X_0, \tag{11.2.1}$$

$$X_{n+1} = \alpha_{n+1}(X_n) \equiv \alpha_{n+1}\alpha_n \dots \alpha_1 X_0 \ (n \geq 0). \tag{11.2.2}$$

We write $X_n(\tilde{x})$ for the case $X_0 = \tilde{x}$. Then X_n is a Markov process with a stationary transition probability $p(x, dy)$ given as follows:

for $x \in S$, $C \in \mathcal{S}$,

$$p(x, C) = Q(\{\gamma \in \Gamma : \gamma(x) \in C\}). \tag{11.2.3}$$

The transition probability $p(x, dy)$ is said to be *weakly continuous* or to have the *Feller property* if for any sequence x_n converging to x, the sequence of probability measures $p(x_n, \cdot)$ converges weakly to $p(x, \cdot)$. One can show that if Γ consists of a family of continuous maps, $p(x, dy)$ has the Feller property.

To study the evolution of the process (11.2.2), it is convenient to recall the definitions of a transition operator and its adjoint. Let $\mathbf{B}(S)$ be the linear space of all bounded real-valued measurable functions on S. The transition operator T on $\mathbf{B}(S)$ is given by

$$(Tf)(x) = \int_S f(y)p(x, dy), \quad f \in \mathbf{B}(S), \tag{11.2.4}$$

and T^* is defined on the space $\mathcal{M}(S)$ of all finite signed measures on $(S, \mathcal{S})$ by

$$T^*\mu(C) = \int_S p(x, C)\mu(dx) = \int_\Gamma \mu(\gamma^{-1}C)Q(d\gamma), \quad \mu \in \mathcal{M}(S). \tag{11.2.5}$$

An element π^* of $\mathcal{P}(S)$ is invariant for $p(x, dy)$ (or for the Markov process X) if it is a fixed point of T^*, i.e.,

$$\pi^* \ is \ invariant \iff T^*\pi^* = \pi^*. \tag{11.2.6}$$

Now, write $p^{(n)}(x, dy)$ for the n-step transition probability with $p^{(1)} \equiv p(x, dy)$. Then $p^{(n)}(x, dy)$ is the distribution of $\alpha_n \ldots \alpha_1 x$. Recall that T^{*n} is the nth iterate of T^*:

$$T^{*n}\mu = T^{*(n-1)}(T^*\mu) \ (n \geq 2), \ T^{*1} = T^*, \ T^{*0} = \text{Identity}, \tag{11.2.7}$$

$$(T^{*n}\mu)(C) = \int_S p^{(n)}(x, C)\mu dx, \tag{11.2.8}$$

so that $T^{*n}\mu$ is the distribution of X_n when X_0 has distribution μ. To express T^{*n} in terms of the common distribution Q of the i.i.d. maps, let Γ^n denote the usual Cartesian product $\Gamma \times \Gamma \times \cdots \times \Gamma$ (n terms), and let Q^n be the product probability $Q \times Q \times \cdots \times Q$ on $(\Gamma^n, \sum^{\otimes n})$, where $\sum^{\otimes n}$ is the product sigma-field on Γ^n. Thus, Q^n is the (joint) distribution of $\alpha = (\alpha_1, \alpha_2, \ldots, \alpha_n)$. For $\gamma = (\gamma_1, \gamma_2, \ldots, \gamma_n) \in \Gamma^n$, let $\tilde{\gamma}$ denote the composition,

$$\tilde{\gamma} := \gamma_n \gamma_{n-1} \cdots \gamma_1. \tag{11.2.9}$$

Then, since $T^{*n}\mu$ is the distribution of $X_n = \alpha_n \ldots \alpha_1 X_0$, one has $(T^{*n}\mu)(A) = \mathrm{Prob}(X_n \in A) = \mathrm{Prob}(X_0 \in \tilde{\alpha}^{-1}A)$, where $\tilde{\alpha} = \alpha_n \alpha_{n-1} \ldots \alpha_1$ and, by the independence of $\tilde{\alpha}$ and X_0,

$$(T^{*n}\mu)(A) = \int_{\Gamma^n} \mu(\tilde{\gamma}^{-1}A)Q^n(d\gamma) \ (A \in \mathcal{S}, \ \mu \in \mathcal{P}(\mathcal{S})). \tag{11.2.10}$$

Finally, we come to the definition of stability. A Markov process X_n is *stable in distribution* if there is a unique invariant probability measure π, such that $X_n(x)$ converges in distribution to π irrespective of the initial state x, i.e., if $p^{(n)}(x, \ dy)$ converges weakly to the same probability measure π for all $x \in S$. In the case one has $(1/n) \sum_{m=1}^{n} p^{(m)}(x, dy)$ converging weakly to the same invariant π for all x, we may define the Markov process to be *stable in distribution on the average*.

11.3   Invariant Distributions

An invariant distribution captures the notion of a steady state of a random dynamical system. Naturally, the existence, uniqueness and stability of an invariant distribution are themes that have been deeply explored by probability theorists. One can write out (11.2.6) explicitly as

$$\pi^* \in \mathcal{P}(S) \text{ is } \textit{invariant if } \pi^*(C) = \int_S p(x, C)\pi^*(dx) \text{ for all } C \in \mathcal{S}.$$
$$\tag{11.3.1}$$

In our applications, S is typically a closed subset of R. In such cases, it is often convenient to deal with the distribution function F_π of a probability measure π on S. Let π^* be an invariant distribution and F^* be its distribution function (dropping the subscript π^* at the moment for the sake of simplicity). By the well-known decomposition theorem (Feller, 1971, pp. 138–142), F^* can be written as a mixture of three probability distribution functions,

$$F^* = \theta_1 F_1^* + \theta_2 F_{ac}^* + \theta_3 F_s^*, \tag{11.3.2}$$

where $\theta_i \geq 0$, $\sum_{i=1}^3 \theta_i = 1$, and F_1^* is atomic (a step function), F_{ac}^* is absolutely continuous (with respect to the Lebesgue measure), and F_s^* is continuous but singular (with respect to Lebesgue measure); we can also write

$$F_{ac}^*(x) = \int_{-\infty}^{x} g(y) \, dy, \tag{11.3.3}$$

where g is an integrable function or the density of F_{ac}^*.

One would like to compute F^* analytically (see the examples in Chapter 12), or, by using some simulation techniques, numerically estimate F^*. More ambitious is the question of sensitivity or comparative statics: how do F^* or its three components in (11.3.2) respond to changes in the parameter of a model? This question leads to deeper mathematical issues that are far beyond the scope of this book.

First, let us summarize some general results on the existence, uniqueness and stability of invariant distributions and indicate a few applications in the following section.

Theorem 11.3.1. *Let S be a (non-empty) compact metric space, and Γ consists of a family of continuous functions from S into S. Then there is an invariant distribution $\pi^* \in \mathcal{P}(S)$.*

Proof (*Sketch*). $\mathcal{P}(S)$ is a compact set in the weak topology. It is convex. The map $T^* : P(S) \to P(S)$ is continuous. Hence, there is a fixed point of T^*.   □

An awkward problem involving the existence question is worth noting. Consider $S = [0, \mathcal{K}]$, where $\mathcal{K}$ is the maximum sustainable stock of a renewable resource. All the regeneration functions $\gamma \in \Gamma$ satisfy $\gamma(0) = 0$. Then the point mass at 0 (the measure δ_0) is obviously an invariant distribution. The challenge here is to figure out whether the structure of the model allows for an invariant distribution with support in $(0, \mathcal{K}]$. Note that such a problem comes up even when $S = R_+$ (where Theorem 11.1 does not apply directly) and when for all the regeneration functions $\gamma(0) = 0$.

We now come to the Dubins–Freedman type splitting conditions applicable to monotone maps. Let S be a non-degenerate interval (finite or infinite, closed, semi-closed, or open) and Γ a set of *monotone* maps from S into S, i.e., each element of Γ is either a non-decreasing function on S or a non-increasing function.

We will assume the following *splitting condition*:
> **(H)** *There exist $z_0 \in S$, $\tilde{\chi}_i > 0$ $(i = 1, 2)$ and a positive N, such that*

(1) $P(\alpha_N \alpha_{N-1} \cdots \alpha_1 x \leqslant z_0 \ \forall x \in S) \geqslant \tilde{\chi}_1$,

(2) $P(\alpha_N \alpha_{N-1} \cdots \alpha_1 x \geqslant z_0 \ \forall x \in S) \geqslant \tilde{\chi}_2$.

Remark 11.3.1. We say that a Borel subset A of S is *closed* under p if $p(x, A) = 1$ for all x in A. If the state space S has two disjoint closed subintervals that are closed under p or under $p^{(n)}$ for some n, then the splitting condition does not hold. This is the case, e.g., when these intervals are invariant under α almost surely.

Denote by $d_K(\mu, \nu)$ the *Kolmogorov distance* on $\mathcal{P}(S)$, i.e., if F_μ, F_ν denote the distribution functions (d.f.) of μ and ν, then

$$d_K(\mu, \nu) := \sup_{x \in R} |\mu((-\infty, x] \cap S) - \nu(-\infty, x] \cap S)|$$

$$\equiv \sup_{x \in R} |F_\mu(x) - F_\nu(x)|, \ \mu, \nu \in \mathcal{P}((S)). \qquad (11.3.4)$$

Theorem 11.3.2. *Assume that the splitting condition* **(H)** *holds on a non-degenerate interval* S.

Then,

(a) *the distribution* $T^{*n}\mu$ *of* $X_n := \alpha_n \cdots \alpha_1 X_0$ *converges to a probability measure* π^* *on* S *exponentially fast in the Kolmogorov distance* d_K *irrespective of* X_0. *Indeed,*

$$d_K(T^{*n}\mu, \pi^*) \leqslant (1 - \tilde{\chi})^{[n/N]} \quad \forall \mu \in \mathcal{P}(S), \qquad (11.3.5)$$

where $\tilde{\chi} := \min\{\tilde{\chi}_1, \tilde{\chi}_2\}$ *and* $[y]$ *denotes the integer part of* y.

(b) π^* *in (a) is the unique invariant probability of the Markov process* X_n.

Proof.   See Bhattacharya and Majumdar (2007, Chapter 3). $\square$

Remark 11.3.2. Suppose that α_n are strictly increasing a.s. Then, one can show that π is *non-atomic*, i.e., its distribution function is continuous.

Remark 11.3.3. Let $S = [a, b]$ and $\alpha_n(n \geqslant 1)$ be a sequence of i.i.d. continuous non-decreasing maps on S into S. Suppose that π is the unique invariant distribution of the Markov process (11.2.1) and (11.2.2). *If* π *is not degenerate, then the splitting condition holds.*

An application of this theorem is given in Example 11.4.4.

We now consider random iteration of Lipschitz maps. Let (S, d) be a compact metric space. A map f from S into S is *Lipschitz* if, for some finite $\mathcal{L} > 0$,

$$d(f(x),\ f(y)) \leq \mathcal{L} d(x, y) \quad \text{for all } x,\ y \in S. \qquad (11.3.6)$$

A family of Lipschitz maps from S into S is denoted by Γ.

Let $\{\alpha_n : n \geq 1\}$ be an i.i.d. sequence of random Lipschitz maps from Γ. Denote by $\mathcal{L}_1^r$ the *random Lipschitz coefficient* of $\alpha_1 \alpha_2 \ldots \alpha_r$:

$$\mathcal{L}_1^r(\cdot) := \sup\{d(\alpha_r \ldots \alpha_1 x, \alpha_r \ldots \alpha_1 y)/d(x, y) : x \neq y\}. \qquad (11.3.7)$$

Theorem 11.3.3. *Let (S, d) be a compact metric space and assume*

$$-\infty \le E \, \log \, \mathcal{L}_1^1 < 0. \tag{11.3.8}$$

Then the Markov process X_n has a unique invariant probability and is stable in distribution.

Proof. See Bhattacharya and Majumdar (2007, Chapter 3). $\square$

Observe that for all $x, y \in S$, one has (writing $\mathcal{L}_1$ for $\mathcal{L}_1^1$)

$$
\begin{aligned}
d(X_n(x), \, X_n(y)) &\equiv \ d(\alpha_n \ldots \alpha_1 x, \alpha_n \ldots \alpha_1 y) \\
&\le \ \mathcal{L}_n d(\alpha_{n-1} \ldots \alpha_1 x, \ \alpha_{n-1} \ldots \alpha_1 y) \\
&\ \ \vdots \\
&\le \ \mathcal{L}_n \mathcal{L}_{n-1} \ldots \mathcal{L}_1 d(x, y) \le \mathcal{L}_n \mathcal{L}_{n-1} \ldots \mathcal{L}_1 M,
\end{aligned}
\tag{11.3.9}
$$

where $M = \operatorname{diam} S = \sup\{d(x, y) : x, y \in S\}$.

By the strong law of large numbers, the logarithm of the last term in (11.3.9) goes to $-\infty$ a.s. since, in view of (11.3.8), $\frac{1}{n} \sum_{j=1}^n \log \mathcal{L}_j \to E \, \log \mathcal{L}_1 < 0$. Hence,

$$\sup_{x, y \varepsilon S} \, d(X_n(x), X_n(y)) \to 0 \text{ a.s. as } n \to \infty. \tag{11.3.10}$$

11.4 Examples

Our first example contrasts the steady state of a random dynamical system (S, Γ, Q) with the steady states of deterministic laws of Γ.

Example 11.4.1. Let $S = [0, 1]$ and consider the maps $\bar{f}$ and $\bar{\bar{f}}$ on S into S defined by

$$\bar{f}(x) = x/2,$$

$$\bar{\bar{f}}(x) = x/2 + \frac{1}{2}.$$

Now, if we consider the deterministic dynamical systems $(S, \bar{f})$ and $(S, \bar{\bar{f}})$, then for each system, all the trajectories converge to a unique fixed point (0 and 1, respectively) independent of initial condition.

Think of the random dynamical system (S, Γ, Q), where $\Gamma = \{\bar{f}, \bar{\bar{f}}\}$ and $Q(\{\bar{f}\}) = p > 0$, $Q(\{\bar{\bar{f}}\}) = 1 - p > 0$. It follows from Theorem 11.3.2 that, irrespective of the initial x, the distributions of $X_n(x)$ converges in the Kolmogorov metric to a unique invariant distribution π which is non-atomic (i.e., the distribution function of π is continuous). If $p = \frac{1}{2}$, then the uniform distribution over $[0, 1]$ is the unique invariant distribution.

Example 11.4.2. This example shows the inevitability of extinction when the loss of population due to a "bad" regeneration function cannot be overcome, even though the "good" regeneration function allows the population just to maintain its size.

Consider $S = R_+$. There are two possible laws of motion f_1, f_2 defined as follows:

(i) $f_1(x) = x$,
(ii) $f_2(0) = 0$, f_2 is increasing and is a uniformly strict contraction, i.e., there is some positive $M < 1$, such that $|f_2(x) - f_2(y)| \leq M \cdot |x - y|$, where $x \neq y$.

We denote the initial population by $x_0 > 0$.

Let $\Gamma = \{f_1, f_2\}$ and $Q = (p, 1 - p)$ be a probability distribution on Γ, where $0 < p < 1$.

Consider a sequence (α_n) of independent identically distributed (with the common distribution Q) random variables with values in Γ.

Denote the interval $[0, x_0]$ by A. Observe that $f_1(A) = A$, and $f_2(A) = [0, y]$ where $y = f_2(x_0) \leq M \cdot x_0$. In fact, for any closed subinterval $[0, x']$ of A, we have $y' = f_2(x') \leq M \cdot x' \leq M \cdot x_0$.

Keeping this in mind, for any "large" n,

$$\alpha_n \alpha_{n-1} \ldots \alpha_j \ldots \alpha_1(x_0) \leq M^\eta x_0, \text{ where } \eta \text{ is the number of periods}$$
in which $[\alpha_j = f_2]$, $0 \leq \eta \leq n$.

However, the events $E_n = [\alpha_n = f_2]$ are independent, occurring with probability $1 - p$. By the Borel–Cantelli lemma, $\text{Prob}\{E_n$ occurs infinitely often$\} = 1$.

Hence, as $n \to \infty$,

$$\alpha_n \alpha_{n-1} \ldots \alpha_j \ldots \alpha_1(x_0) \text{ converges to 0 almost surely.} \qquad (11.4.1)$$

Note that this result does *not* depend on the exact value of p, and x_0, the size of the initial population. $\qquad\square$

Exercise 11.4.1. Prove that one can dispense with the assumption that f_2 is increasing.

[▼ Hints: $|f_2^k(x) - f_2^k(0)| = |f_2(f_2^{k-1}(x)) - f_2(f_2^{k-1}(0)| \leq M |f_2^{k-1}(x) - f_2^{k-1}(0)| \ldots M^k f_2(x) \to 0$ uniformly over $x \leq x_0$. ▲]

Example 11.4.3. Let us go back to Theorem 11.3.3 and let $S = [0, \mathcal{K}]$. Consider a family of Lipschitz maps Γ with the following property: for each $\gamma \in \Gamma$, $\gamma(0) = 0$.

Let $\{\alpha_n : n \geq 1\}$ be an i.i.d. sequence of random Lipschitz maps from Γ. The random Lipschitz coefficients L_1^r of $\alpha_1 \alpha_2 \ldots \alpha_r$ satisfy $-\infty \leq E \log L_1^1 < 0$.

Clearly, the point mass at $0(\delta_0)$ is the unique invariant measure, and $X_n(x)$ converges in probability to 0 for any $x \in [0, \mathcal{K}]$. Using (3.10) and choosing $y = 0$, we see that the law of large numbers ensures extinction: $X_n(x)$ converges to 0 almost surely.

Example 11.4.4 (Stability of Invariant Distributions in Models of Economic Growth). Models of descriptive as well as optimal growth under uncertainty have led to random dynamical systems that are stable in distribution. We look at a "canonical" example and show how Theorem 11.3.2 can be applied.

Recall that, for any function f on S into S, we write f^n for the nth iterate of f.

Consider the case where $S = R_+$, and $\Gamma = \{F_1, F_2, \ldots, F_i, \ldots, F_N\}$ where the distinct laws of motion F_i satisfy the following:

[F.1] F_i *is strictly increasing, continuous, and there is some* $x_i^* > 0$,
       such that $F_i(x) > x$ *on* $(0, x_i^*)$ *and* $F_i(x) < x$ *for* $x > x_i^*$.
       Note that $F_i(x_i^*) = x_i^*$ for all $i = 1, \ldots, N$. Next, assume
[F.2] $x_i^* \neq x_j^*$ *for* $i \neq j$.

In other words, the unique positive fixed points x_i^* of distinct laws of motion are all distinct. We choose the indices $i = 1, 2, \ldots, N$, so that

$$x_1^* < x_2^* < \cdots < x_N^*. \qquad (11.4.2)$$

Let $\text{Prob}(\alpha_n = F_i) = p_i > 0 \quad (1 \leq i \leq N)$.

Consider the Markov process $\{X_n(x)\}$ with the state space $(0, \infty)$. If $y \geq x_1^*$, then $F_i(y) \geq F_i(x_1^*) > x_1^*$ for $i = 2, \ldots, N$, and $F_1(x_1^*) = x_1^*$, so that $X_n(x) \geq x_1^*$ for all $n \geq 0$ if $x \geq x_1^*$. Similarly, if $y \leq x_N^*$, then $F_i(y) \leq F_i(x_N^*) < x_N^*$ for $i = 1, \ldots, N - 1$ and $F_N(x_N^*) = x_N^*$, so that $X_n(x) \leq x_N^*$ for all $n \geq 0$ if $x \leq x_N^*$. Hence, if the initial state x is in $[x_1^*, x_N^*]$, then the process $\{X_n(x) : n \geq 0\}$ remains in $[x_1^*, x_N^*]$ forever. We shall presently see that for a long-run analysis, we can consider $[x_1^*, x_N^*]$ as the effective state space.

We shall first indicate that on the state space $[x_1^*, x_N^*]$, the splitting condition (**H**) is satisfied. If $x \geq x_1^*$, $F_1(x) \leq x$, $F_1^2(x) \equiv F_1(F_1(x)) \leq F_1(x)$, etc. The limit of this decreasing sequence $F_1^n(x)$ must be a fixed point of F_1 and therefore must be x_1^*. Similarly, if $x \leq x_N^*$, then $F_N^n(x)$ increases to x_N^*. In particular,

$$\lim_{n \to \infty} F_1^n(x_N^*) = x_1^*, \qquad \lim_{n \to \infty} F_N^n(x_1^*) = x_N^*.$$

Thus, there must exist a positive integer n_0, such that

$$F_1^{n_0}(x_N^*) < F_N^{n_0}(x_1^*).$$

This means that if $z_0 \in [F_1^{n_0}(x_N^*), F_N^{n_0}(x_1^*)]$, then

$$\text{Prob}(X_{n_0}(x) \leq z_0 \; \forall x \in [x_1^*, x_N^*])$$

$$\geq \text{Prob}(\alpha_n = F_1 \text{ for } 1 \leq n \leq n_0) = p_1^{n_0} > 0,$$

$$\text{Prob}(X_{n_0}(x) \geq z_0 \ \forall x \in [x_1^*, x_N^*])$$

$$\geq \text{Prob}(\alpha_n = F_N \text{ for } 1 \leq n \leq n_0) = p_N^{n_0} > 0.$$

Hence, considering $[x_1^*, x_N^*]$ as the state space, and using Theorem 11.3.2, there is a unique invariant probability π with the stability property holding for all initial $x \in [x_1^*, x_N^*]$.

Now, fix the initial state $x \in (0, x_1^*)$, and define $m(x) = \min_{i=1,\ldots,N} F_i(x)$.

One can verify that (i) m is continuous, (ii) m is strictly increasing, (iii) $m(x_1^*) = x_1^*$ and $m(x) > x$ for $x \in (0, x_1^*)$, and $m(x) < x$ for $x > x_1^*$. Let $x \in (0, x_1^*)$. Clearly, $m^n(x)$ increases with n, and $m^n(x) \leq x_1^*$. The limit of the sequence $m^n(x)$ must be a fixed point of m and is, therefore, x_1^*. Since $F_i(x_1^*) > x_1^*$ for $i = 2, \ldots, N$, there exists some $\varepsilon > 0$, such that $F_i(y) > x_1^*$ ($2 \leq i \leq N$) for all $y \in [x_1^* - \varepsilon, x_1^*]$. Clearly, there is some n_ε, such that $m^{n_\varepsilon}(x) \geq x_1^* - \varepsilon$. If $\tau_1 = \inf\{n \geq 1 : X_n(x) > x_1^*\}$, then it follows that, for all $k \geq 1$,

$$\text{Prob}(\tau_1 > n_\varepsilon + k) \leq p_1^k.$$

Since p_1^k goes to zero as $k \to \infty$, it follows that τ_1 is finite almost surely. Also, $X_{\tau_1}(x) \leq x_N^*$, since for $y \leq x_1^*$, (i) $F_i(y) < F_i(x_N^*)$ for all i and (ii) $F_i(x_N^*) < x_N^*$ for $i = 1, 2, \ldots, N-1$ and $F_N(x_N^*) = x_N^*$ (in a single period, it is not possible to go from a state less than or equal to x_1^* to one larger than x_N^*). By the strong Markov property, and our earlier result, $X_{\tau+m}(x)$ converges in distribution to π as $m \to \infty$ for all $x \in (0, x_1^*)$. Similarly, one can check that as $n \to \infty$, $X_n(x)$ converges in distribution to π for all $x > x_N^*$.

▲**Exercise 11.4.2.** Consider the case where $S = R_+$ and $\Gamma = \{F, G\}$, where

(i) F is strictly increasing, continuous, $F(0) > 0$ and there is $\bar{x}^*$, such that $F(x) > x$ on $(0, \bar{x}^*)$ and $F(x) < x$ for $x > \bar{x}^*$.

(ii) G is strictly increasing, continuous, $G(x) < x$ for all $x \geq 0$.

Let $Q(\alpha_n = F) = p$, $Q(\alpha_n = G) = 1 - p$. Analyze the long-run behavior of the process (11.1.2). ▼

Example 11.4.5 (Growth Models: Multiplicative shocks).

Here, we consider an example of an uncountable Γ.

Let $F : R_+ \to R_+$ satisfy

[F.1] *F is strictly increasing and continuous.*

We shall keep F fixed.

Consider $\Theta = [\theta_1, \theta_2]$, where $0 < \theta_1 < \theta_2$, and assume the following concavity and "end-point" conditions:

[F.2] $F(x)/x$ *is strictly decreasing in* $x > 0$, $\frac{\theta_2 F(x'')}{x''} < 1$ *for some* $x'' > 0$, $\frac{\theta_1 F(x')}{x'} > 1$ *for some* $x' > 0$.

Since $\frac{\theta F(x)}{x}$ is also strictly decreasing in x, [F.1] and [F.2] imply that for each $\theta \in \Theta$, there is a unique $x_\theta > 0$, such that $\frac{\theta F(x_\theta)}{x_\theta} = 1$, i.e., $\theta F(x_\theta) = x_\theta$. Observe that $\frac{\theta F(x)}{x} > 1$ for $0 < x < x_\theta$, $\frac{\theta F(x)}{x} < 1$ for $x > x_\theta$. Now, $\theta' > \theta''$ implies $x_{\theta'} > x_{\theta''}$:

$$\frac{\theta' F(x_{\theta''})}{x_{\theta''}} > \frac{\theta'' F(x_{\theta''})}{x_{\theta''}} = 1 = \frac{\theta' F(x_{\theta'})}{x_{\theta'}}, \text{ implying "} x_{\theta'} > x_{\theta''} \text{".}$$

Write $\Gamma = \{f : f = \theta F, \ \theta \in \Theta\}$ and $f_1 \equiv \theta_1 F$, $f_2 \equiv \theta_2 F$.

Assume that θ_n is chosen i.i.d. according to a density function $g(\theta)$ on Θ which is positive and continuous on Θ.

In our notation, $f_1(x_{\theta_1}) = x_{\theta_1}$; $f_2(x_{\theta_2}) = x_{\theta_2}$. If $x \geq x_{\theta_1}$, and $\tilde{\theta}$ has a density $\dot{g}$, then $f(x) \equiv \tilde{\theta} F(x) \geq f(x_{\theta_1}) \geq f_1(x_{\theta_1}) = x_{\theta_1}$. Hence, $X_n(x) \geq x_{\theta_1}$ for all $n \geq 0$ if $x \geq x_{\theta_1}$. If $x \leq x_{\theta_2}$,

$$f(x) \leq f(x_{\theta_2}) \leq f_2(x_{\theta_2}) = x_{\theta_2}.$$

Hence, if $x \in [x_{\theta_1}, x_{\theta_2}]$, then the process $X_n(x)$ remains in $[x_{\theta_1}, x_{\theta_2}]$ forever. Now, $\lim_{n \to \infty} f_1^n(x_{\theta_2}) = x_{\theta_1}$ and $\lim_{n \to \infty} f_2^n(x_{\theta_1}) = x_{\theta_2}$, and there must exist a positive integer n_0, such that $f_1^{n_0}(x_{\theta_2}) < f_2^{n_0}(x_{\theta_1})$. Choose some $z_0 \in (f_1^{n_0}(x_{\theta_2}), f_2^{n_0}(x_{\theta_1}))$. There exist intervals $[\theta_1, \theta_1 + \delta]$, $[\theta_2 - \delta, \theta_2]$, such that for all $\theta \in [\theta_1, \theta_1 + \delta]$ and $\hat{\theta} \in [\theta_2 - \delta, \theta_2]$,

$$(\theta F)^{n_0}(x_{\theta_2}) < z_0 < (\hat{\theta} F)^{n_0}(x_{\theta_1}).$$

Then the splitting condition (**H**) holds on the state space $[x_{\theta_1}, x_{\theta_2}]$. Now, fix x, such that $0 < x < x_{\theta_1}$, then

$$\tilde{\theta} F(x) \geq \theta_1 F(x) > x.$$

Let m be any given positive integer. Since $(\theta_1 F)^n(x) \to x_{\theta_1}$ as $n \to \infty$, there exists $n' \equiv n'(x)$, such that, $(\theta_1 F)^n(x) > x_{\theta_1} - \frac{1}{m}$ for all $n \geq n'$. This implies that $X_n(x) > x_{\theta_1} - \frac{1}{m}$ for all $n \geq n'$. Therefore, $\liminf_{n\to\infty} X_n(x) \geq x_{\theta_1}$. We now argue that with probability one, $\liminf_{n\to\infty} X_n(x) > x_{\theta_1}$. For this, note that if we choose $\eta = \frac{\theta_2 - \theta_1}{2}$ and $\varepsilon > 0$, such that $x_{\theta_1} - \varepsilon > 0$, then $\min\{\theta F(y) - \theta_1 F(y) : \theta_2 \geq \theta \geq \theta_1 + \eta, \ y \geq x_{\theta_1} - \varepsilon\} = \eta F(x_{\theta_1} - \varepsilon) > 0$. Write $\eta' \equiv \eta F(x_{\theta_1} - \varepsilon) > 0$. Since with probability one, the i.i.d. sequence $\{\theta^{(n)} : n = 1, 2, \ldots\}$ takes values in $[\theta_1 + \eta, \ \theta_2]$ infinitely often, one has $\liminf_{n\to\infty} X_n(x) > x_{\theta_1} - \frac{1}{m} + \eta'$. Choose m, so that $\frac{1}{m} < \eta'$. Then, with probability one, the sequence $X_n(x)$ exceeds x_{θ_1}. Since $x_{\theta_2} = f_2^{(n)}(x_{\theta_2}) \geq X_n(x_{\theta_2}) \geq X_n(x)$ for all n, it follows that with probability one, $X_n(x)$ reaches the interval $[x_{\theta_1}, x_{\theta_2}]$ and remains in it thereafter. Similarly, one can prove that if $x > x_{\theta_2}$, then with probability one, the Markov process $X_n(x)$ will reach $[x_{\theta_1}, x_{\theta_2}]$ in finite time and stay in the interval thereafter.

Example 11.4.6 (Concavity and Extinction). We will now consider an example in which the regeneration functions are concave as well as monotone non-decreasing and throw light on extinction.

Let $\Gamma = \{\gamma : R_+ \to R_+ : \gamma$ is non-decreasing, concave and $\gamma(0)\}$.

On some probability space $(\Omega, \mathcal{F}, P)$, let $(\alpha_n)_{n=1}^\infty$ be a sequence of *independent* random functions from Γ with *a common distribution* Q. Define

$$g(x) = E\alpha_1(x). \tag{11.4.3}$$

Assume that

(G.1) *g is continuous at $x = 0$ and satisfies $g(x) < x$ for all $x > 0$.*

Proposition 11.4.1. *The Markov process $X_n(x)$ converges in L^1 to 0 as $n \to \infty$, for $x > 0$. It also follows that $X_n(x)$ converges to 0 in probability.*

Proof. Note first that

$$g \text{ is non-decreasing and concave.} \tag{11.4.4}$$

To prove the concavity property asserted in (11.4.4), let $0 < \theta < 1$, $x_1 < x_2$. Then,

$$\begin{aligned}
g(\theta x_1 + (1-\theta)x_2) &= E\alpha_1(\theta x_1 + (1-\theta)x_2) \\
&\geq E(\theta \alpha_1(x_1) + (1-\theta)\alpha_1(x_2) \\
&= \theta g(x_1) + (1-\theta)g(x_2).
\end{aligned}$$

Also, to complete the proof of (11.4.4), observe that

$$g(x_1) = E\alpha_1(x_1) \leq E\alpha_1(x_2) = g(x_1).$$

Moreover,

$$\alpha_1(\alpha_2(\ldots, \alpha_n(\cdot))\ldots) \text{ is non-decreasing and concave.} \tag{11.4.5}$$

We shall verify (11.4.5) for $\alpha_1(\alpha_2(\cdot))$. The general case follows by induction. Again, let $0 < \theta < 1$, $x_1 < x_2$. Since α_1 is non-decreasing and α_2 is concave,

$$\alpha_1(\alpha_2(\theta x_1 + (1-\theta)x_2)) \geq \alpha_1(\theta \alpha_2(x_1) + (1-\theta)\alpha_2(x_2)).$$

Since α_1 is concave,

$$\alpha_1(\theta \alpha_2(x_1) + (1-\theta)\alpha_2(x_2)) \geq \theta \alpha_1(\alpha_2(x_1)) + (1-\theta)\alpha_1(\alpha_2(x_2)).$$

Next, consider the backward process $Y_n(x) = \alpha_1 \alpha_2 \ldots \alpha_n(x)$ $[n \geq 1]$ and $\mathcal{F}_n = \sigma\{\alpha_1, \alpha_2, \ldots, \alpha_n\}$. One has

$$\begin{aligned}
E[Y_{n+1}(x)|\mathcal{F}_n] &= E[Y_n(\alpha_{n+1}(x))|\mathcal{F}_n] \leq Y_n E\alpha_{n+1}(x) \\
&= Y_n(g(x) \leq Y_n(x). \tag{11.4.6}
\end{aligned}$$

Thus, the process $Y_n(x)$ $(n \geq 1)$ is a non-negative supermartingale, so that $Y_n(x) \to Y$ (a.s. as well as in L^1). Also, (4.9) implies

$$E[Y_{n+1}(x)] \leq E(Y_n(g(x)))E(Y_{n-1}(g \circ g(x))) \cdots$$
$$\leq E_1(g^n(x)) = g(g^n(x)) = g^{n+1}(x). \quad (11.4.7)$$

Observe that $g^n(x) = g^{n-1}(g(x)) \leq g^{n-1}(x)$ (strict inequality for all $x > 0$). The sequence $g^n(x)$ decreases to some x^* as $n \to \infty$. Using the continuity of g and the relation,

$$g^{n=1}(x) = g(g^n(x)).$$

We take the limit as $n \to \infty$ to get $g(x^*) = x^*$, but this means that $x^* = 0$. Using (11.4.7), it follows that $E[Y_n(x)] \to 0$ as $n \to \infty$. Hence, $Y_n(x) \to 0$ in L^1 as $n \to \infty$. But $X_n(x)$ has the same distribution as $Y_n(x)$. Hence, $X_n(x) \to 0$ in L^1 as $n \to \infty$.

Example 11.4.7 (Iterates of Quadratic Maps). We return to the family of quadratic maps introduced in Chapter 2 and explore the problem of stability in distribution on the average of random iterates of the form,

$$X_n(x) = F_{\varepsilon_n} F_{\varepsilon_{n-1}} \cdots F_{\varepsilon_1}(x), \quad (11.4.8)$$

where $\varepsilon_n, n \geq 1$ is an i.i.d. sequence with values in the parameter space $[0, 4]$ of the quadratic map:

$$F_\theta = \theta x(1 - x), \quad (11.4.9)$$

and x is the initial state. To exclude the trivial invariant probability $\delta_{(0)}$ (that assigns the unit mass at 0), we restrict the state space to $S = (0, 1)$.

The problem turns out to be analytically challenging. The interested reader is referred to Bhattacharya and Majumdar (2004) for an in-depth analysis and related references. Following is one of the important results.

Proposition 11.4.2. *If ε_1 has a non-zero density component which is bounded away from zero on some non-degenerate interval contained in $(1, 4)$ and if, in addition,*

$$E \log \varepsilon_1 > 0 \quad and \quad E \, |\log(4 - \varepsilon_1)| < \infty, \tag{11.4.10}$$

then $\{X_n, n \geq 0\}$ has a unique invariant distribution π and is stable on the average, i.e., for all $B \in \mathcal{S}$,

$$\sup_B \left| (1/n) \sum_{n=1}^{N} p^{(n)}(x, B) - \pi(B) \right| \to 0 \quad as \ N \to \infty.$$

11.5   Complements and Details

For an extended discussion of random dynamical systems, see Bhattacharya and Majumdar (1999, 2001, 2004, 2007). Athreya (2003), Ellner (1984), Yahav (1974) have more complete treatments and definitive results that shed light on the examples.

Many examples of the process (11.1.4) are descriptive stochastic models. Constant yield harvesting also generates (11.1.4) (recall the introductory remarks following (1.5.14) in Chapter 1 and see the more elaborate discussion in Chapter 12). Another important route that leads to (11.1.4) begins with a dynamic programming model of optimization under uncertainty. One first derives an optimal policy function (Maitra (1968)), which together with the stochastic law of transition describes the optimal evolution of the process of states in the form (11.1.4). An example of this approach is spelled out in detail in Majumdar, Mitra and Nyarko (1989). Of interest in this context are results on the "inverse optimal problem under uncertainty" due to Mitra (1998) and Montrucchio and Privileggi (1999), which assert that a broad class of random dynamical systems (11.1.4) can be so generated. Identifying the possibility of extinction is quite appropriately a major issue: see Olson and Roy (2000), Mitra and Roy (2006, 2007).

Chapter 12

Sustainable Consumption
and Uncertainty

We have always known that the long oil-driven expansion would
come to an end. We have now reached that juncture. The
Norwegian economy has enjoyed an exceptionally long summer.
Winter is coming.

— Governor Oeystein Olsen, Central Bank, Norway (2016).

Water, water, everywhere
Nor any drop to drink.

— Samuel Coleridge, *The Rime of the Ancient Mariner.*

12.1 Introduction

In this chapter, we explore two models that deal with the sustainability of a target consumption level under uncertainty. The first deals
with living off a sovereign "Wealth Fund" (often known as the "oil
fund") and the second addresses the recurrent water crisis in reservoirs that are "stressed" on the average (expected recharge is short
of the target withdrawal).

For a number of countries that rely on the export revenues generated by an exhaustible resource (oil and natural gas are prime
examples), a consensus has been reached to develop a sufficiently
large Wealth Fund, the returns from which can be used to safeguard and build financial wealth in order to meet the *needs of future*

generations. As the Norges Bank that oversees one of the largest such funds announces in its report: "The Fund is saving for future generations in Norway. One day the oil will run out, but the return on the Fund will continue to benefit the Norwegian Population. Its focus areas are children's rights, climate change and water management."

To begin with, we briefly recapitulate the problem of constant harvesting in a deterministic model that was developed in Section 2.4 of Chapter 2 (only case (ii) indicated in the following was developed there). An economy starts with a positive initial stock y (initial reserve of a sovereign "Wealth Fund"). From this, a positive quantity c is subtracted. The parameter c is a datum: it is a target consumption that the economy wishes to *sustain*. If the remainder $x_0 = y - c$ is negative, the economy is "ruined". If the remainder is non-negative, it is interpreted as an *input* into some productive activity (i.e., as an "investment"). The *output* of this activity is then the stock at the beginning of the next period and is given by $y_1 = f(x_0) = f(y - c)$, where $f : R \to R_+[f(i) = 0$ for $i \le 0; f(x) > 0$, for $x > 0]$. This f is (the regeneration function in Section 2.4) the return function. Again, in period one, if $y_1 - c < 0$, the economy is ruined in period 1; otherwise, the investment $x_1 = y_1 - c$ generates the output $y_2 = f(x_1)$ and the story is repeated. Let T be the first period, if any, such that $x_T < 0$. If T is finite, we say that the economy can *sustain* c up to (but not including) the period T (or that the economy *survives* up to period T). If T is infinite (i.e., $x_t > 0$ for all t), we say that the consumption target c is *sustainable* (or the economy *sustains* c "forever").

Now, consider two types of f:

(i) *Linear:* $f(x) = \varepsilon \cdot x$, where $\varepsilon > 0$ (ε is the output–capital ratio, the return factor from investment).

(ii) *Nonlinear:* f is strictly concave, satisfying appropriate "boundary" (*Uzawa*) conditions.

To capture the notion that the returns from investment are "uncertain", as in Chapter 11, we let f to be chosen by Tyche in every period (independently, with an identical distribution).

Then, the period T is the (random) first time, such that $Y_t < 0$; if T is finite, c is sustainable up to (but not including) T; otherwise $T = \infty$. We say that the economy *sustains* c from initial y *with probability* $\rho(y, c)$ if $\mathrm{Prob}(T = \infty) = \rho(y, c)$. Intuitively, such a probability depends on y, c and, of course, the distribution of the random function f.

In Section 12.2.1, we explore case (i). We introduce an i.i.d. sequence ε_t of non-negative random variables. An investment x_t in period t generates output Y_{t+1} according to the rule $Y_{t+1} = (\varepsilon_{t+1}) \cdot x_t$. With an initial stock y, and a target consumption c, it is ruined if $(y - c) \leq 0$. If $y - c > 0$, then the stock in period one is $Y_1 = \varepsilon_1(y - c)$. We use a capital letter to stress that Y_1 is a random variable. Again, if $Y_1 - c \leq 0$, it is ruined. Otherwise, after consumption, $Y_1 - c$ is invested to generate $Y_2 = \varepsilon_2 \cdot (Y_1 - c)$, and the story is repeated. In general, one studies the following process:

$$Y_0 = y, Y_{t+1} = (\varepsilon_{t+1})(Y_t - c)^+, \text{ where } a^+ = \max(a, 0). \quad (12.1.1)$$

If $y > c$, the probability of sustaining c is defined as

$$\rho(y, c) = P(Y_t > c \text{ for all } t \geq 0 | Y_0 = y). \quad (12.1.2)$$

It is shown that

$$\rho(y, c) = P \left\{ \sum_{t=1}^{\infty} (\varepsilon_1 \varepsilon_2 \dots \varepsilon_t)^{-1} < (y/c) - 1 \right\}. \quad (12.1.3)$$

The basic formula (12.1.3) can be used to identify conditions on the common distribution of ε_t, under which the value of $\rho(y, c)$ can be derived. For example, if $E\varepsilon_1 \leq 1$, then $\rho(y, c) = 0$ for all y and c. The case $E \log[\varepsilon_1] > 0$ is perhaps the most interesting and turns out to be challenging. We identify a condition (see (12.2.11)) on the distribution of ε_1 that ensures that $\rho(y, c) < 1$, *no matter how large y is*. From the policy point of view, it suggests a need for prudent management that avoids undue exposure to risk.

Note that if we define the random variable ζ as

$$\zeta = \sum_{t=1}^{\infty}(\varepsilon_1\varepsilon_2\ldots\varepsilon_t)^{-1}, \qquad (12.1.4)$$

we realize that the distribution of ζ is crucial (though intractable analytically) in determining $\rho(y,c)$. To this effect, we derive a recursive relation that facilitates computing the moments of ζ.

In Section 12.2.3, we turn to case (ii): the reader is referred to Section 2.4 of Chapter 2 for details of the deterministic counterpart. Let $\Gamma = \{f_1, f_2, \ldots, f_j, \ldots, f_N\}$ be a given set of return functions, where *each* f_j satisfies (F.1–F.3) spelled out in Section 12.2.3, and $N \geq 2$ (finite!). A probability distribution Q on Γ is given, and we assume that $Q(f_j) > 0$ for all j. In the initial period, if the input $x \ (= y - c)$ is positive, the stock in period 1 is given by

$$Y_1 = \alpha_1(x),$$

where α_1 is an element of Γ chosen by Tyche according to the distribution Q (independent of y and c). Again, the economy is ruined if $Y_1 - c \leq 0$. Otherwise, writing $X_1 = Y_1 - c$, the stock Y_2 in period 2 is given by

$$Y_2 = \alpha_2(X_1),$$

where α_2 is chosen by Tyche independent of α_1 and so on. An ordering assumption (O) plays an essential role in simplifying the exposition. Let f_1 be the least favorable return function, and write $h_1(x) = f_1(x) - x$, to denote the corresponding net return function. Our assumptions guarantee that there is unique $\bar{k}_1 > 0$, such that $f_1(\bar{k}_1) = \bar{k}_1$. Let $H_1 = \max_{[0,\bar{k}_1]} h(x) > 0$. It is seen that if $c > H_1$, then $\rho(y,c) = 0$ for all y. If $c \leq H_1$, two tipping points $\eta_1(c) < \eta_2(c)$ (see (12.2.30) and (12.2.31)) emerge: $\rho(y,c) = 0$ for $y \leq \eta_1(c)$; $\rho(y,c) = 1$ for $y > \eta_2(c)$; $0 < \rho(y,c) < 1$ for $\eta_1(c) < y < \eta_2(c)$.

The model discussed in Section 12.3 can best be interpreted as one of sustaining a target of "water consumption" from a reservoir (or aquifer) when the augmentation (recharge) in each period (net increment due to precipitation) is random. Let $X_t \geq 0$ be the stock ("level") at the end of period t, $\{\mathfrak{R}_{t+1}\}$ a sequence of i.i.d. non-negative real-valued random variables (new supply or augmentation of the stock during the period $t + 1$) and $c > 0$ a given parameter: a target level of consumption. At the end of period $t + 1$, if $X_t + \mathfrak{R}_{t+1} > c$, the planner withdraws (consumes c), leaving $X_{t+1} = X_t + (\mathfrak{R}_{t+1} - c)$ as the stock at the end of period $t + 1$. If, on the other hand, $X_t + \mathfrak{R}_{t+1} \leq c$, the entire available stock is withdrawn, leaving $X_{t+1} = 0$ (the community fails to meet the target c and faces a *crisis*). Writing $Z_{t+1} = \mathfrak{R}_{t+1} - c$, we get

$$X_{t+1} = (X_t + Z_{t+1})^+ \quad \text{where} \quad z^+ = \max(z, 0). \tag{12.1.5}$$

The initial x_0 can be taken as an arbitrary non-negative random variable independent of the sequence ($Z_{t+1:} \geq 0$). A relatively non-technical exposition (which does not involve integration theory at all) of the dynamic behavior of the process is given in Section 12.3.1, Here, the variables X_t and c are assumed to take only non-negative integer values, and $\mathfrak{R}_t$ assumes two values, $c + 1$ and $c - 1$, with probabilities p and $1 - p$. The exposition of the more general model focuses on the case when the reservoir is "stressed" ($E\mathfrak{R}_{t+1} < c$) and proves that the stock level X_t convergence to a unique stochastic equilibrium (an invariant distribution). The analysis provides insights on the probability of stocks remaining in a specific interval of R_+ or the expected time for a crisis to return.

12.2 Living off a Wealth Fund

12.2.1 The deterministic case: A linear model

Let us first look at the deterministic case and state the main result. Here, starting with an initial wealth or "fortune" $y > 0$, the economy is ruined if $y - c < 0$. If $y > c$, the investment $x_0 = y - c$ generates the fortune $y_1 = \varepsilon \cdot i_0 = \varepsilon(x - c)$ at the beginning of period 1, and the story is repeated. If the economy can sustain c up to (but not including) period 2, we know that

$$c + x_0 = y,$$
$$c + x_1 = \varepsilon \cdot x_0. \tag{12.2.1}$$

It follows that $c(1 + 1/\varepsilon) + x_1/\varepsilon = y$, leading to

$$c(1 + 1/\varepsilon) < y.$$

Hence, if the economy can sustain c up to period N, we must have

$$\sum_{n=0}^{N-1} (1/\varepsilon^n) < y/c. \tag{12.2.2}$$

We immediately conclude as follows.

Proposition 12.2.1. *If $\varepsilon \leq 1$, then for any $c > 0$, there is no $y > 0$, such that c can be sustained forever. If $\varepsilon > 1$, the economy sustains $c > 0$ forever if and only if*

$$\varepsilon/[\varepsilon - 1] \leq y/c. \tag{12.2.3}$$

The formula (12.2.3) gives the relationship among the return factor ε, the initial fortune y and the target withdrawal c that enables the economy to sustain c forever.

12.2.2 A stochastic linear model

We now turn to the situation when the returns to investment are random. To avoid undue repetition, the economy starts with an initial wealth fund or fortune $y > 0$ and plans to sustain a consumption level $c > 0$. Focus on the case $y > c$ and we are led to the following process:

$$Y_0 = y, \quad Y_{t+1} = \varepsilon_{t+1}(Y_t - c)^+ \quad (t \geq 0), \quad a^+ = \max(a, 0), \tag{12.2.4}$$

where $\{\varepsilon_t : t \geq 1\}$ is an i.i.d. sequence of non-negative random variables. The *probability* of sustaining c *forever* with an initial fortune

$y > c$ is

$$\rho(y, c) := P(Y_t > c, \quad \text{for all} \quad t \geq 0 | Y_0 = y). \tag{12.2.5}$$

Suppose $P(\varepsilon_1 > 0) = 1$. For otherwise, it is simple to check that that eventual ruin is certain, i.e. $\rho(y, c) = 0$. From (12.2.4), successive iteration yields

$$Y_{t+1} > c \Leftrightarrow Y_t > c + \frac{c}{\varepsilon_{t+1}} \Leftrightarrow Y_{t-1} > c + \frac{c + c/\varepsilon_{t+1}}{\varepsilon_t} + \cdots$$

$$\Leftrightarrow Y_0 \equiv y > c + \frac{c}{\varepsilon_1} + \frac{c}{\varepsilon_1 \varepsilon_2} + \cdots + \frac{c}{\varepsilon_1 \varepsilon_2 \cdots \varepsilon_{t+1}}.$$

$$\tag{12.2.6}$$

Hence, on the set $\{\varepsilon_t > 0 \text{ for all } t\}$,

$$\{Y_t > c \text{ for all } t\} = \left\{ y > c + c \sum_{j=1}^{t} \frac{1}{\varepsilon_1 \varepsilon_2 \cdots \varepsilon_j} \text{ for all } t \right\} \tag{12.2.7}$$

$$= \left\{ y > c + c \sum_{t=1}^{\infty} \frac{1}{\varepsilon_1 \varepsilon_2 \cdots \varepsilon_t} \right\}$$

$$= \left\{ \sum_{t=1}^{\infty} \frac{1}{\varepsilon_1 \varepsilon_2 \cdots \varepsilon_t} < \frac{y}{c} - 1 \right\}. \tag{12.2.8}$$

In other words,

$$\rho(y, c) = P \left\{ \sum_{t=1}^{\infty} \frac{1}{\varepsilon_1 \varepsilon_2 \cdots \varepsilon_t} < \frac{y}{c} - 1 \right\}. \tag{12.2.9}$$

It is clear that for a fixed y, $\rho(y, c)$ is non-increasing in c, for a given c, $\rho(y, c)$ is non-decreasing in y. We review conditions on the common distribution of ε_t, under which one has (a) $\rho(y, c) = 0$, (b) $\rho(y, c) = 1$, or (c) $\rho(y, c) < 1(y > c)$. Our main result will now be stated and proved.

Theorem 12.2.1. *Let*

$$m := \inf\{z \geq 0 : P(\varepsilon_1 \leq z) > 0\}. \tag{12.2.10}$$

(a) *If* $E \log \varepsilon_1 \leq 0$, *then* $\rho(y, c) = 0$ *for all* $y > 0$ *and* $c > 0$.

(b) *If $E \log \varepsilon_1 > 0$, then*

$$\rho(y,c) \begin{cases} < 1 & \text{if } m \leq 1 \text{ for all } y \\ < 1 & \text{if } m > 1, \text{ for } y < c[m/(m-1)] \\ = 1 & \text{if } m > 1, \text{ for } y \geq c[m/(m-1)] \end{cases} . \qquad (12.2.11)$$

Proof. The Strong Law of Large Numbers gives

$$\frac{1}{n} \sum_{r=1}^{n} \log \varepsilon_r \xrightarrow{\text{a.s.}} E \log \varepsilon_1. \qquad (12.2.12)$$

Thus, if $E \log \varepsilon_1 < 0, \varepsilon_1 \varepsilon_2 \cdots \varepsilon_n \to 0$ a.s. This implies that the infinite series in (12.2.9) diverges a.s., i.e.,

$$\rho(y,c) = 0 \quad \text{for all } y \text{ and } c \text{ if } E \log \varepsilon_1 < 0. \qquad (12.2.13)$$

Now, by the celebrated Jensen's inequality, $E[\log \varepsilon_1] \leq \log E[\varepsilon_1]$, with strict inequality unless ε_1 is degenerate. Therefore, if $E[\varepsilon_1] = 1$, and ε_1 is non-degenerate, then $E[\log \varepsilon_1] < 0$ and the reasoning leading to (12.2.13) applies. If ε_1 is degenerate and $E[\varepsilon_1] = 1$, then $P(\varepsilon_1 = 1) = 1$, and the infinite series in (12.2.9) diverges. To sum up,

$$\rho(y,c) = 0 \text{ for all } (y,c) \text{ if } E[\varepsilon_1] \leq 1. \qquad (12.2.13a)$$

Next, let us consider claim (b). Let $E[\log \varepsilon_1] > 0$. We will prove the first claim in (12.2.11), i.e., prove that

$$\rho(y,c) < 1 \quad \text{for all } y, \text{ if } m \leq 1. \qquad (12.2.14)$$

To this effect, fix $A > 0$, however large. One can find n_0, such that $n_0 > A \prod_{r=1}^{\infty}(1 + r^{-2})$ as $\prod(1 + r^{-2}) < \exp\{\sum r^{-2}\} < \infty$. If $m \leq 1$, then $P(\varepsilon_1 \leq 1 + r^{-2}) > 0$ for all $r \geq 1$. Consequently,

$$0 < P(\varepsilon_r \leq 1 + r^{-2} \text{ for } 1 \leq r \leq n_0)$$

$$\leq P\left(\sum_{r=1}^{n_0} \frac{1}{\varepsilon_1 \varepsilon_2 \cdots \varepsilon_r} \geq \sum_{r=1}^{n_0} \frac{1}{\prod_{j=1}^{r}(1 + 1/j^2)}\right)$$

$$\leq P\left(\sum_{r=1}^{n_0} \frac{1}{\varepsilon_1 \varepsilon_2 \cdots \varepsilon_r} \geq \frac{n_0}{\prod_{j=1}^{\infty}(1 + 1/j^2)}\right)$$

$$\leq P\left((\sum_{r=1}^{n_0}(\varepsilon_1 \varepsilon_2 \cdots \varepsilon_n)^{-1} > A\right) \leq P\left(\sum_{r=1}^{\infty}(\varepsilon_1 \varepsilon_2 \cdots \varepsilon_r)^{-1} > A\right).$$

Since A is arbitrary, $\rho(y, c)$ is less than 1 for all y, proving (12.2.14). We now turn to the other two claims in (12.2.11), i.e., we prove that

$$\rho(y, c) \begin{cases} < 1 & \text{if } y < c\left(\frac{m}{m-1}\right), \\ = 1 & \text{if } y \geq c\left(\frac{m}{m-1}\right) \qquad (m > 1). \end{cases} \tag{12.2.15}$$

Observe that $\sum_{n=1}^{\infty}(\varepsilon_1 \varepsilon_2 \cdots \varepsilon_n)^{-1} \leq \sum_{n=1}^{\infty} m^{-n} = 1/(m-1)$, with probability 1 if $m > 1$. The second relation in (12.2.15) follows from (12.2.9). To establish the first relation in (12.2.15), let $y < cm/(m-1) - c\delta$ for some $\delta > 0$. This implies $y/c - 1 < 1/(m-1) - \delta$. One can choose $n(\delta)$, such that $\sum_{r=n(\delta)}^{\infty} m^{-r} < \delta/2$ and then choose $\delta_r > 0$ $(1 \leq r \leq n(\delta) - 1)$, such that

$$\sum_{r=1}^{n(\delta)-1} \frac{1}{(m + \delta_1) \cdots (m + \delta_r)} > \sum_{r=1}^{n(\delta)-1} \frac{1}{m^r} - \frac{\delta}{2}.$$

Then,

$$0 < P(\varepsilon_r < m + \delta_r \text{ for } 1 \leq r \leq n(\delta) - 1)$$

$$\leq P\left\{\sum_{r=1}^{n(\delta)-1} \frac{1}{\varepsilon_1 \cdots \varepsilon_r} > \sum_{r=1}^{n(\delta)-1} \frac{1}{m^r} - \frac{\delta}{2}\right\}$$

$$\leq P\left\{\sum_{r=1}^{\infty} \frac{1}{\varepsilon_1 \cdots \varepsilon_r} > \sum_{r=1}^{\infty} \frac{1}{m^r} - \delta\right\}$$

$$= P\left\{\sum_{r=1}^{\infty} \frac{1}{\varepsilon_1 \cdots \varepsilon_r} > \frac{1}{m-1} - \delta\right\}.$$

For $\delta > 0$ small enough, the last probability is smaller than $P(\sum(\varepsilon_1 \cdots \varepsilon_r)^{-1} > (y/c) - 1)$ if $(y/c) - 1 < 1/(m-1)$, i.e., if $y < cm/(m-1)$. For such y, $1 - \rho(y, c) > 0$, completing the verification of all the claims in (b). □

Remark 12.2.1. Compare (12.2.3) with (12.2.15). A more complete description of $\rho(y, c)$ is now established. Write

$$\zeta = \sum_{t=1}^{\infty} \frac{1}{\varepsilon_1 \varepsilon_2 \cdots \varepsilon_t},$$

$$d_1 = \inf\{z \geq 0 : P(\zeta \leq z) > 0\},$$

$$d_2 = \sup\{z \geq 0 : P(\zeta \geq z) > 0\},$$

$$M = \sup\{z \geq 0 : P(\varepsilon_1 \geq z) > 0\},$$

and it is useful to recall (12.2.10)

$$m := \inf\{z \geq 0 : P(\varepsilon_1 \leq z) > 0\}.$$

Theorem 12.2.2. *Assume $E[\log \varepsilon_1] > 0$. Then,*

(a) ζ *is finite almost surely.*
(b)

$$\rho(y, c) = \begin{cases} 0 & \text{if } y < c(d_1 + 1) \\ \in (0, 1) & \text{if } c(d_1 + 1) < y < c(d_2 + 1) \\ 1 & \text{if } y > c(d_2 + 1) \end{cases}. \quad (12.2.16)$$

(c) $\rho(y, c) = 0$ *if $y < cM/[M-1]$(when $1 < M < \infty$)' and $\rho(y, c) = 0$ (when $M \leq 1$).*

(d) *One can express the (essential) lower bound d_1 and (essential) upper bound d_2 of ζ in terms of those of ε_1, namely, m and M:*

   (i) $d_1 = \sum_{n=1}^{\infty} M^{-n} = 1/(M-1)$ *if $M > 1$, and $d_1 = \infty$ if $M \leq 1$.*

(ii) $d_2 = \sum_{n=1}^{\infty} m^{-n} = 1/(m-1)$ *if* $m > 1$, *and* $d_2 = \infty$ *if* $m \leq 1$.

Proof.   See Section 12.4.   □

The distribution of the random variable ζ seems intractable. The following result characterizes the moments of ζ in terms of the moments of $1/\varepsilon$:

Proposition 12.2.2. *Let* $E[\log \varepsilon_1] > 0$. *Denote the moments of* ζ *and* $1/\varepsilon_1$ *by* $\beta_r = E\zeta^r$, $\gamma_r = E(1/\varepsilon_1)^r$, *respectively*, $(r = 1, 2, \ldots)$. *Then, for all* r, *such that* $\gamma_r < 1$,

$$\beta_r = \gamma_r \sum_{0 \leq j \leq r} \binom{r}{j} \beta_j; \ (1 - \gamma_r)\beta_r = \gamma_r \sum_{0 \leq j \leq r-1} \binom{r}{j} \beta_j,$$

$$\beta_r = [\gamma_r/(1 - \gamma_r)] \sum_{0 \leq j \leq r-1} \binom{r}{j} \beta_j. \tag{12.2.17}$$

If $\gamma_r \geq 1$ *for some* r, *then* $\beta_r = \infty$.

With the moments of ζ recursively obtained in Proposition 12.2.2, one obtains conservative estimates of ruin and survival probabilities using Chebyshev Inequality:

$$1 - \rho(y, c) = P(\zeta \geq (y/c) - 1) < \frac{\beta_r}{((y/c) - 1)^r}(r = 1, 2, \ldots), (y > c). \tag{12.2.18}$$

Note that the smaller the upper estimate of ruin probability, the better the approximation of the true ruin probability is. Equivalently, the larger the lower estimate of survival probability, the better the approximation. Therefore, the estimate (12.2.18) with r over the one with $r + 1$ should be selected if and only if

$$\frac{\beta_r}{((y/c) - 1)^r} \leq \frac{\beta_{r+1}}{((y/c) - 1)^{r+1}}, \ \text{or} \ y \leq c\left(1 + \frac{\beta_{r+1}}{\beta_r}\right).$$

The upper estimate of ruin probability and the lower of survival probability, consequently, are obtained as follows:

$$1 - \rho(y, c) < \frac{\beta_r}{(y/c - 1)^r}, \quad \rho(y, c) > 1 - \frac{\beta_r}{((y/c) - 1)^r}, \quad (y > c),$$

where r is chosen as follows:

$$\begin{pmatrix} c(1 + \beta_r/\beta_{r-1}) < y \leq c(1 + \beta_{r+1}/\beta_r) & \text{if } r \geq 2 \\ c < y \leq c(1 + \beta_2/\beta_1) & \text{if } r = 1 \end{pmatrix},$$

subject to the restriction $\gamma_r < 1$.

We conclude with an example.

Example 12.2.1 (Lognormal). Let $\varepsilon_1 = e^N$ be lognormal, where N is a normal random variable with mean $\mu > 0$ and a positive variance σ^2. One can show that $0 < \rho(y, c) < 1$ for every $y > c$ (see Bhattacharya, Kim and Majumdar (2015, Corollary 0.5, p. 176)). $1/\varepsilon_1 = e^{-N}$ is also lognormal, $-N$ being Normal with mean $-\mu < 0$ and variance σ^2. Hence, the moments of $1/\varepsilon_1$ are given by

$$\gamma_r = E\varepsilon_1^{-r} = e^{-r\mu + \frac{r^2\sigma^2}{2}} \, (r = 1, 2, \ldots).$$

Using Proposition 12.2.2, the moments β_r of ζ may now be computed. One must require $r < 2\mu/\sigma^2$.

12.2.3   A nonlinear return function

The model considered in this section is the stochastic counterpart of our model in Section 2.4 of Chapter 2. Tyche chooses a return function from a class Γ of possible nonlinear return (or "regeneration": in our earlier interpretation) functions.

First, we assume that Γ has a finite number $N \geq 2$ elements:

$$\Gamma = \{f_1, f_2, \ldots, f_j, \ldots, f_N\}.$$

Each f_j satisfies

- (F.1) $f_j : R_+ \to R_+$ is continuous on $\mathbb{R}_+$, $f_j(x) = 0$ for $x \leq 0$.
- (F.2) $f_j(x)$ is twice continuously differentiable at $x > 0$, $f_j'(x) > 0, f_j''(x) < 0$.
- (F.3) $\lim_{x \to 0} f_j'(x) = 1 + \mu_1$, where $\mu_1 > 0$, and $\lim_{x \to \infty} f_j'(x) = 1 - \mu_2$, where $\mu_2 > 0$.

For each f_j that there is some $\bar{k}_j > 0$, such that $f_j(x) > x$, for all x, satisfying $0 < x < \bar{k}_j$; and $f_j(x) < x$ for $x > \bar{k}_j$. By continuity of

f_j, it follows that $f_j(\bar{k}_j) = \bar{k}_j$. This $\bar{k}_j$ is the maximum sustainable stock for the return function f_j.

Moreover, we assume an "ordering" property:

(O) *For every* $x > 0, f_j(x) < f_{j+1}(x), \ j = 1, 2, \ldots, N.$

We refer to $f_1(x)$ as the *least favorable return function* and $f_N(x)$ as the *most favorable return function*: given (O), these play an important role. This assumption has strong implications, perhaps plausible in many situations.

Let Q be a probability distribution on Γ, and we assume that $Q(f_j) > 0$ for all $j = 1, 2, \ldots, N$.

Consider a sequence $(\alpha_t)_{t=1}^{\infty}$ of independent, identically distributed random functions with values in Γ, the common distribution being Q. We are led to the study of the following sequence of random variables:

$$Y_0 = y,$$
$$X_t = Y_t - c, \quad t \geq 0, \qquad (12.2.19)$$
$$Y_{t+1} = \alpha_{t+1}(X_t).$$

As before, let T (random) denote the first period t, if any, such that $X_t < 0$; otherwise, $T = \infty$. Given the initial y and a target c, the economy sustains c with probability $\rho(y, c)$ if $\text{Prob}\{T = \infty\} = \rho(y, c)$.

It is helpful to look at some examples that will pave the way for the qualitative analysis to follow.

Example 12.2.2. This is a somewhat "degenerate" case, but still illuminating. There are only two elements in $\Gamma = \{f_1, f_2\}$. The favorable return function f_2 is defined as

$$f_2(x) = \begin{cases} 0 & \text{for } x \leq 0 \\ 2x & \text{for } x \in [0, 4] \\ (x/2) + 6 & \text{for } x \geq 4. \end{cases} \qquad (12.2.20)$$

The unfavorable return function f_1 is defined as

$$f_1(x) = \begin{cases} 0 & \text{for } x \leq 0 \\ (3/2)x & \text{for } x \in [0, 4] \\ (x/2) + 4 & \text{for } x \geq 4. \end{cases} \qquad (12.2.21)$$

Observe first that while the ordering assumption (O) holds, (F.2) does not. Let Q be any probability distribution with $Q(f_1) > 0$, $Q(f_2) > 0$.

The random variables α_t take values f_1 and f_2 with probabilities $Q(f_1)$ and $Q(f_2)$, respectively. The domain of these random variables has no bearing in our analysis.

Now, let the initial sock $y = 4$ and the target consumption $c = 2$. Note that if the favorable return function f_2 occurs in *every* period (and this event plainly has zero probability), the (stationary) plan $X_t = 2, c = 2$ achieves survival in that evolution (trajectory) of the environment. However, if f_1 occurs *only once*, (i.e., $\alpha_\tau = f_1$, for some period $\tau \geq 1$, and $\alpha_t = f_2$ for all $t \neq \tau$, one has, with $x_{\tau-1} \leq 2$, $y_\tau = f_1(x_{\tau-1}) \leq 3$. Hence, $x_\tau \leq 1$, leading to $y_{\tau+1} = f_2(x_\tau) \leq 2$. Hence, $x_{\tau+1} = 0$, and the economy cannot sustain $c = 2$ anymore. It should perhaps be stressed that this conclusion does not depend on the values of $Q(f_1)$ and $Q(f_2)$ at all. The extreme conclusion that "bad luck" in just *one* period "spells doom" depends on the structure of the two return functions and is not "typical" . However, as we shall see, the *persistence* of bad luck has serious implications for the sustainability problem.

Exercise 12.2.1. ▼ What happens in this case when we choose a "large" initial $y > 0$ (keeping Γ unchanged and the target $c = 2$)? ▲

Exercise 12.2.2. ▼ Draw the graphs of $f_1(x) - 2$ and $f_2(x) - 2$ and complete a diagrammatic analysis of the sustainability problem. ▲

Example 12.2.3. Here, $\Gamma = \{f_1, f_2\}$, where

$$f_2(x) = \begin{cases} 0 & \text{for } x \leq 0, \\ 2x & \text{for } x \in [0, 4], \\ (x/2) + 6 & \text{for } x \geq 4 \end{cases} \qquad (12.2.22)$$

and

$$f_1(x) = \begin{cases} 0 & \text{for } x \le 0, \\ (3x)/2 & \text{for } 0 \le x \le 5, \\ x/2 + 5 & \text{for } x \ge 5. \end{cases} \qquad (12.2.23)$$

Observe that (12.2.20) and (12.2.22) are the same.

Now, define, for each j,

$$h_j(x) = f_j(x) - x, \qquad (12.2.24)$$

and write

$$H_j = \max_{[0, \bar{k}_j]} h_j(x). \qquad (12.2.25)$$

Solving $h_1(x) = 0$, with $x > 0$, we get $\bar{k}_1 = 10$, and $H_1 = \max_{x \in [0,10]} h_1(x) = 7.5$.

Similarly, solving $h_2(x) = 0$, with $x > 0$, we get $\bar{k}_2 = 12$, and $H_2 = \max_{x \in [0,12]} h_2(x) = 8$.

Finally, one considers for each $c > 0$ and for each $k = 1, 2$, the roots of $h_j(x) = c$, or equivalently, $f_j(x) - x = c$. The roots (denoted by $\xi_j' < \xi_j''$) will be distinct in case $c < H_j$. Taking $c = 2$, one easily verifies that

$$\xi_1' = 4, \ \xi_1'' = 6; \ \xi_2' = 2, \ \xi_2'' = 8. \qquad (12.2.26)$$

It is not an accident that

$$\xi_2' < \xi_1' < \xi_1'' < \xi_2''. \qquad (12.2.27)$$

It is useful to draw the graph of $f_j - 2$ [▲ *Exercise* ▼] and carry out a complete diagrammatic analysis. For example, if the initial $X_0 < \xi_2' = 2$, then X_t will reach or drop below 0 in finite time (i.e., $T < \infty$); hence, sustainability has zero probability. On the other hand, if $X_0 \ge \xi_1' = 4$, then whatever be the evolution of the environment, X_t cannot fall below X_0, and the probability of sustaining $c = 2$ is one. Now, suppose that X_0 is in the open interval

$$I = (\xi_2' < \xi_1'). \qquad (12.2.28)$$

Since I is bounded, a sufficiently long *initial* run of *favorable returns* f_2 will carry X_t above ξ_1', from which point sustainability is assured. Since such an initial run has strictly positive probability, the probability of sustaining $c = 2$ is *strictly positive*. On the other hand, a sufficiently long *initial run of unfavorable returns* f_1 will carry X_t below ξ_2', from which point sustainability has 0 probability. Since such an initial run also has strictly positive probability, the probability of sustaining $c = 2$ is *strictly less than* 1. Finally, consider the case when $X_0 = \xi_2' = 2$. As long as $\alpha_{t+1} = f_2$, the inputs $X_{t+1} = 2$. The first time f_1 occurs, the input will fall below $\xi_2' = 2$, and from that point, the probability of survival is 0. Since f_1 will eventually occur with probability 1, it follows that the probability of sustaining $c = 2$ with initial $X_0 = \xi_2' = 2$ is 0.

The analysis of the general case uses the insights from the examples.

If $c \leq H_1$, define $\eta_1(c)$ and $\eta_2(c)$ by

$$\eta_1(c) = \xi_N' + c, \tag{12.2.29}$$

$$\eta_2(c) = \xi_1' + c. \tag{12.2.30}$$

The main result on sustaining a target consumption is as follows.

Theorem 12.2.3.

(i)   *The probability $\rho(y, c)$ is non-decreasing in y and non-increasing in c.*

(ii) *If $c > H_1$, then $\rho(y, c) = 0$ for all y.*

(iii) *If $c \leq H_1$, then*

$$\begin{array}{lll} \rho(y, c) = 0 & for\ y \leq \eta_1(c), \\ 0 < \rho(y, c) < 1 & for\ \eta_1(c) < y < \eta_2(c), & (12.2.31) \\ \rho(y, c) = 1 & for\ y > \eta_2(c). \end{array}$$

Proof.   To begin with (iii), it is easy to verify that our assumptions imply that for any f_j distinct from f_1 and f_N, one has

$$\xi_N' < \xi_j' < \xi_1' \leq \xi_1'' < \xi_j'' < \xi_N''.$$

Therefore, an argument almost identical to that used for Example 12.2.3 leads to the conclusion.

Turning to case (ii), if $c > H_N$, one recalls using (12.2.1) that

$$X_{t+1} = Y_{t+1} - c$$

$$= \alpha_{t+1}(X_t) - c,$$

$$\text{or } X_{t+1} - X_t = \alpha_{t+1}(X_t) - X_t - c$$

$$\leq H_N - c.$$

Hence, for any initial X_0, the failure time T is bounded above uniformly over all possible realizations of (α_t); hence, $\rho(y, c) = 0$ for all y.

Suppose, then that $c \leq H_N$. Choose $\varepsilon > 0$ sufficiently small. For any X_0, there is some τ, such that for all realizations of (α_t),

$$X_t \leq \xi''_N + \varepsilon \text{ for all } t \geq \tau.$$

Now, one can show that there is an integer N, such that if $X_t \leq \xi''_N + \varepsilon$, and there is a run of consecutive events $\alpha_{t+1} = \alpha_{t+2} = \cdots = \alpha_{t+n} = f_1$, then

$$X_{t+N} < \xi'_N. \tag{12.2.32}$$

As in the example, conditional on (12.2.32), the probability of survival is 0. Hence, to show that the unconditional probability of survival is 0, it suffices to show that, almost surely, there will be a run of N consecutive events some time after period τ. This is a well-known result (see, e.g., Santosh Venkatesh (2013, p. 263)).

Finally, claim (i) follows from the fact that each f_j is increasing. □

12.3 Water Crisis and Sustainable Consumption

We consider a problem of sustaining a constant consumption (or harvesting) of a resource (water from a reservoir, aquifer) that can be stored without costs and is augmented by a random increment (recharge). Let $X_t \geq 0$ be the stock (water level) at the end of period t, $(\mathfrak{R}_t)$ a sequence of i.i.d. non-negative real-valued random variables, and $c > 0$ a given parameter (target consumption, withdrawal from the reservoir). Interpret $\mathfrak{R}_{t+1}$ as the total inflow or augmentation of the stock during period $t + 1$ (rainfall minus evaporation). At the end of period $t + 1$, if $X_t + \mathfrak{R}_{t+1} > c$, the planner withdraws c, leaving $X_{t+1} = X_t + (\mathfrak{R}_{t+1} - c)$ as the stock (the community is able to meet its "minimal" target consumption c). If, on the contrary, $X_t + \mathfrak{R}_{t+1} \leq c$, the entire available stock is withdrawn, leaving $X_{t+1} = 0$ (the community fails to sustain c and faces a *crisis*). Setting $Z_t \equiv \mathfrak{R}_t - c$, we get

$$X_{t+1} = (X_t + Z_{t+1})^+ \text{ where } z^+ = \max(z, 0). \qquad (12.3.1)$$

The initial $x_0 > 0$ can also be taken as an arbitrary non-negative random variable independent of the sequence $(Z_{t+1} : t \geq 0)$. We refer to (12.3.1) as a Lindley–Spitzer process.

Qualitative properties of the process (12.3.1) can be understood without integration theory in the special case when stocks X_t assume only (non-negative) integer values and the net increments Z_t are $+1$ and -1 with probabilities $p > 0$ and $q \equiv (1 - p) > 0$, respectively. This is treated in detail in Section 12.3.1. We return to the general case in Section 12.3.2. We focus on the case in which $E\mathfrak{R}_{t+1} < c$: the reservoir is under "stress". Here, the process converges in distribution ("weakly") to an invariant probability (a stochastic equilibrium) π^*, which provides valuable insights into the expected time for a crisis to return. In Examples 12.3.1 and 12.3.2, corresponding to the specified distribution of Z_t, we are able to calculate π^*, which has an atom at 0 and a density over $(0, \infty)$. Finally, when Z_t has a moment-generating function in a neighborhood of zero or an appropriate number of moments, a stronger convergence is obtained and the estimates of the speed of convergence specified.

12.3.1   Random walk

Consider the case when the stock level X_t assumes values in $S = \{0, 1, 2, \ldots\}$. Let c be a positive integer and assume that $\mathfrak{R}_t = c + 1$ with probability $p > 0$ and $\mathfrak{R}_t = c - 1$ with probability $q = (1 - p) > 0$. Thus, if $X_t = i \geq 1$, $X_{t+1} = i + 1$ with probability p, and $X_{t+1} = i - 1$ with probability q. However, when $X_t = 0$, it can move only to $X_{t+1} = 1$ again with probability p or stay at 0 with probability q. The one-step transition probabilities are summarized as follows:

$$\text{for any } i \geq 0, \ p_{i,i+1} = p,$$
$$\text{for any } i \geq 1, \ p_{i,i-1} = 1 - p, \qquad (12.3.2)$$
$$p_{0,0} = 1 - p.$$

▲ (*Exercise*: Write out the one step transition probability matrix $[p_{i.j}]$ explicitly.) ▼

We shall write $p_{i.j}^n$ to denote the probability of going from i to j in n steps.

A probability distribution $\pi^* = (\pi_i^*)$ on S satisfies $\pi_i^* \geq 0$, $\sum_{i=1}^{\infty} \pi_i^* = 1$. It is *invariant* if

$$\sum_i \pi_i^* p_{i,j} = \pi_j^* \text{ for all } j \in S. \qquad (12.3.3)$$

Writing out (12.3.3) more explicitly, we get a probability distribution $\pi^* = (\pi_i^*)$ on S is invariant if

$$\begin{aligned} \pi_0^* \cdot q + \pi_1^* \cdot q &= \pi_0^* && \text{for } j = 0, \\ \pi_{j-1}^* \cdot p + \pi_{j+1}^* \cdot q &= \pi_j^* && \text{for } j \geq 1. \end{aligned} \qquad (12.3.4)$$

From (12.3.4),

$$\pi_1^* = (p/q) \cdot \pi_0^*. \qquad (12.3.5)$$

For $j = 1$,

$$\pi_0^* \cdot p + \pi_2^* \cdot q = \pi_1^*$$
$$\text{or} \quad \pi_1^* \cdot q + \pi_2^* \cdot q = \pi_1^*$$
$$\text{or} \quad \pi_2^* = (p/q) \cdot \pi_1^*.$$

Hence, we derive

$$\pi_{j+1}^* = (p/q) \cdot \pi_j^*$$
$$\text{or} \quad \pi_j^* = (p/q)^j \cdot \pi_0^*. \qquad (12.3.6)$$

Now, consider the implications of the conditions: $\pi_i^* \geq 0$, $\sum_{i=0}^{\infty} \pi_i^* = 1$. First, it is clear that $\pi_0^* > 0$. Set $\pi_0^* = \kappa > 0$. There are three possibilities to consider:

(i) $p > 1/2$ ($E\Re_t > c$): With $(p/q) > 1, \pi_j^* \to \infty$, so $\sum_{i=1}^{\infty} \pi_i^* = 1$ cannot be satisfied, and there is no invariant distribution.

(ii) $p = 1/2$ ($E\Re_t = c$): With $(p/q) = 1, \pi_j^* = \kappa$, so $\sum_{i=1}^{\infty} \pi_i^* = 1$ cannot be satisfied, and there is no invariant distribution.

These two cases are not relevant for understanding the repeated crises. In case (i), $X_t \to \infty, a.e.$ and long-run sustainability is not a problem, and in case (ii), the expected time for a crisis to return is infinite. Thus, we turn to

(iii) $p < 1/2$ ($E\Re_t < c$): With $(p/q) < 1$, recall that the geometric series $\sum_{k=0}^{\infty} \theta^k = 1/[(1 - \theta)]$ when $0 < \theta < 1$. Taking $\theta = (p/q)$, we see that $\sum_{i=0}^{\infty} \pi_i^* = 1$ is equivalent to $\kappa/(1 - \theta) = 1$, or $\kappa = (1 - \theta) = (q - p)/q$. Hence, (12.3.6) gives us the unique invariant distribution:

$$\pi_j^* = (p/q)^j \cdot [q - p]/q. \qquad (12.3.7)$$

Now, it is easy to see that the process is *irreducible*, for any pair of states (i, j), there is some positive integer $n(i, j)$, such that the probability of going from i to j in $n(i, j)$ steps is positive. Since $p_{0,0} = q > 0$, it is *aperiodic*. Hence, we can invoke a fundamental theorem from the theory of Markov chains.

Theorem 12.3.1. *When $p < 1/2$, $P(X_t = j)$ converges to π_j^* (irrespective of initial x_0).*

Proof.   See Bhattacharya and Majumdar (2007, Theorems 8.1 and 8.2).   □

Let the process start from j, and define $\tau_j = \inf\{t \geq 1 : X_t = 0\}$ as the first return time to j (with the understanding that the infimum is infinite if the process never returns to j). Since the process has an invariant distribution, it follows that $E_j \tau_j$ is finite (and this property called *positive recurrence* is shared by *all* states), and one has the valuable information:

$$\pi_j^* = 1/E_j\tau_j. \tag{12.3.8}$$

In particular, if the process hits the crisis state 0, its expected return time $E_0\tau_0 = 1/\pi_0^*$. A high value of π_0^* calls for remedial measures. On the contrary, attempts can be made to lower c either through recycling, direct rationing or an appropriate use of the price mechanism. Better collection of rain water and investment in infrastructure to develop a source of supply for the community that is independent of the vagaries of $\mathfrak{R}_t$ are parts of a water management policy in this situation.

12.3.2 A Lindley–Spitzer process

To focus on scarcity and crisis, we study the discrete-time Markov process (12.3.1) in detail under the assumption,

$$EZ_t < 0, \quad \text{i.e.,} \quad E[\mathfrak{R}_t] < c. \tag{12.3.9}$$

This means that we are looking at the water source that is "stressed" on an average, i.e., the expected recharge of the reservoir is less than c.

To avoid trivialities, we always assume that

$$P(Z_1 > 0) > 0. \tag{12.3.10}$$

Here, the state space $S = R_+ = [0, \infty)$, and the transition probability function is defined as

$$p(x, B) = P[X_{t+1} \in B | X_t = x] = P[(x + Z_1)^+ \in B]$$
$$\times (x \in R_+, B \in \mathcal{B}(R_+)), \tag{12.3.11}$$

where $\mathcal{B}(S)$ denotes the Borel σ-field of S. It may be shown, as a consequence of a more general result (see Proposition 12.4.1 in Section 12.4), that the transition probability defined according to

(12.3.11) has the required property:

For each fixed B, $p(\cdot, B)$ is $\mathcal{B}(R_+)$-measurable.   (12.3.12)

The initial stock is denoted by x_0 (can be taken as a random variable). One important remark on the scope of our analysis: the results in this section depend only on the common distribution of the i.i.d. random variables Z_t in the model (12.3.11) and not on any specific representation, such as $Z_t = \mathfrak{R}_t - c$. In particular, one can take c_n to be an i.i.d. sequence rather than a constant consumption c.

As in (11.3.11), a steady state of the process (X_t) is an invariant distribution (probability) π^*, which satisfies for each $B \in \mathcal{B}(S)$

$$\pi^*(B) = \int_S p(x, B)\pi^*(dx).$$ (12.3.13)

It has been shown by Spitzer (1956) that the process (12.3.1) has an invariant probability, which is necessarily unique, if and only if

$$\sum_{n=1}^{\infty} \frac{1}{n}P(S_n > 0) < \infty, \quad S_n := Z_1 + \cdots + Z_n.$$ (12.3.14)

If $E[Z_t] > 0$, there is no invariant distribution ($X_t \to \infty$ a.s., so in our economic context, there is no long-term problem of scarcity ("too much" water)). The knife-edge case, $E[Z_t] = 0$ is not discussed here and there is no invariant probability in this case either (see, e.g., Chung (1974, pp. 264–270)).

Theorem 12.3.2. *Assume* (12.3.9) *and* (12.3.10). *The Markov process (X_t) has a unique invariant probability π^*, and X_t converges in distribution to π^*, no matter what x_0 is.*

Proof.   See Section 12.4.   □

It is of interest to study the atomic or absolutely continuous components (recall (11.3.2)) of the distribution function F^* representing the invariant probability π^*, but explicit analytical computation has been a challenging, somewhat elusive issue (warning: F^* does not have any singular component). Numerical approximations of π^* in a number of examples, for which an analytical expression of π^* is unknown, were reported in Iams and Majumdar (2010).

In general, one has $\pi^*(\{0\}) = (E_0\tau_0)^{-1} > 0$ for the process in (12.3.1). In the case where Z_1 has an absolutely continuous distribution (with respect to Lesbegue measure on $[0, \infty)$), the invariant probability π^* has a density on $(0, \infty)$, in addition to the point mass at 0 (see Bhattacharya and Majumdar (2015) for an expanded discussion with proofs).

Here are two examples of invariant distributions.

Example 12.3.1. The distribution of Z_t in (12.3.1) has probability density function $f(x)$ given by

$$f(x) = \begin{cases} \dfrac{ab}{a+b} e^{-bx}, & x > 0, \\ \dfrac{ab}{a+b} e^{ax} & \text{otherwise}, \end{cases} \qquad (12.3.15)$$

where $0 < a < b$ are constant parameters. It can be verified that (see Section 12.4) the invariant distribution π^* associated with X_t has an atom at zero and a continuous density function on $(0, \infty)$:

$$\pi^*(x) = \begin{cases} 1 - \dfrac{a}{b}, & x = 0, \\ \dfrac{b-a}{b} ae^{-(b-a)x}, & x > 0. \end{cases} \qquad (12.3.16)$$

Recall the decomposition in (12.3.15): so, in this case, $F_s^* \equiv 0$.

Example 12.3.2. Consider the process described by (12.3.1), where the i.i.d. sequence $\{\mathfrak{R}_t : t \geq 1\}$ has the exponential distribution with mean $\theta < c$. Then, the distribution of
$Z_t = \mathfrak{R}_t - c$ is described by the pdf

$$f(x) = \begin{cases} \frac{1}{\theta} e^{-\frac{1}{\theta}(x+c)} & \text{for } x \geq -c, \\ 0 & \text{for } x < -c. \end{cases} \qquad (12.3.17)$$

The invariant distribution π^* associated with X_t has an atom at zero and a continuous exponential density function on $(0, \infty)$:

$$\pi^*(x) = \begin{cases} \dfrac{\theta}{\beta}, & x = 0, \\ \beta^{-1}\exp[-(x+c)/\beta], & x > 0, \end{cases} \qquad (12.3.18)$$

where $\beta > \theta$ solves $1 - \frac{\theta}{\beta} = \exp\left(-\frac{c}{\beta}\right)$. The point mass at 0 is, of course, given by $\pi^*(\{0\}) = 1 - \int_0^\infty \pi^*(x)\,dx$.

The verification is spelled out in Section 12.4.

Assume that the *moment-generating function (mgf)* $M(d)$ of Z_t is finite for some $d > 0$. Since $M(0) = 1$ and $M'(0) = EZ_1 < 0$, the quantity $M^* = \inf\{M(d) : d > 0\}$ is less than 1. Let d^* be the point where this minimum is attained. One has the following result on a stronger notion of convergence and its "exponential rate".

Theorem 12.3.3. *Under the above assumptions on Z_t, one has*

$$d_{tv}(p^{(t)}(x, dy), \pi^*) = o(\exp\{-ct\}) \;\; \forall c < \ln\frac{1}{M^*}, \qquad (12.3.19)$$

where $p^{(t)}$ denotes the t-step transition probability of the Markov chain $\{X_t : t \geq 0\}$, and d_{tv} is total variation distance:

$$d_{tv}(\mu_1, \mu_2) = \sup\{|\mu_1(B) - \mu_2(B)| : B \text{ a Borel subset of } S = [0, \infty)\}.$$

We look at the examples again.

Example 12.3.1 (continued). In this example,

$$M(d) = Ee^{dZ_1} = \frac{a}{a+b}\frac{1}{1-\frac{d}{b}} + \frac{b}{a+b}\frac{1}{1+\frac{d}{b}} \quad (d < b),$$

$$d^* = \frac{b-a}{2}, \quad M^* = \frac{4ab}{(a+b)^2}. \qquad (12.3.20)$$

Example 12.3.2 (continued).

$$M(d) = e^{-cd}/(1-\theta d), \quad d < \frac{1}{\theta},$$

$$M^* = M\left(\frac{c-\theta}{c\theta}\right) = (e^{-\frac{c}{\theta}+1})\frac{c}{\theta}. \qquad (12.3.21)$$

Example 12.3.3 (Normal with negative mean). Let Z_1 have the normal distribution $N(\mu, \sigma^2)$ with mean $\mu < 0$ and variance $\sigma^2 > 0$. Then

$$M(d) = e^{d\mu + \sigma^2 d^2/2}, \quad d^* = -\frac{\mu}{\sigma^2}, \quad M^* = e^{-\frac{\mu^2}{2\sigma^2}}. \quad (12.3.22)$$

Real data in economics and finance often seem to be generated by the so-called *heavy-tailed* distributions, namely, those that do not have finite *mgfs* in any neighborhood of zero. In such examples, the path to equilibrium is very slow (slower than the rate captured in (12.3.19)), thus exhibiting *long-range dependence*, as well as large jumps in values of X_t. To obtain a precise estimate of the speed of convergence, we assume

$$\mu_1 \equiv EZ_1 < 0, \quad P(Z_1 > 0) > 0, \quad \rho_s \equiv E\left(\left|\frac{Z_1 - \mu_1}{\sigma}\right|^s\right) < \infty, \quad (12.3.23)$$

where $s \geq 3$ is an integer and $\sigma^2 = E(Z_1 - \mu_1)^2$. We can now state the main result on the speed of convergence in the "heavy-tailed" case.

Theorem 12.3.4. *Assume* (12.3.23), *with* $s \geq 3$, *for the Lindley process* (12.3.1). *Then, for every* $x \in [0, \infty), d_{tv}(p^{(t)}(x, \cdot), \pi^*) = o(t^{-\alpha}) \ \forall \alpha < (s-1)/2$.

12.4 Complements and Details

The quote attributed to Governor Olsen is from *Reuters*, February 18, 2016, article, "Norway to tap rainy fund, seek stimulus as economic boom ends", available on the Internet. I am thankful to Ms. Nivedita Kutty who drew my attention to it. Information on Norway's wealth fund (or other sovereign wealth funds) is also available on the following websites:

www.nbim.no/en/the-fund.

https://en.wikipedia.org/wiki/Sovereign_weath_fund.

The models in Sections 12.2 were first presented in Majumdar and Radner (1992). There were several directions in which the model was extended in this paper. The present exposition, however, relies on Bhattacharya, Kim and Majumdar (2015), which has several numerical examples on the approximation of probability of ruin and expanded discussions on moments of ζ corresponding to alternative specifications of the distribution of ε (Pareto, Gamma). It is the natural supplementary reading.

Proof of Theorem 12.2.2. (a) By the Strong Law of Large Numbers,

$$\left(\sum_{1 \leq j \leq n} \log \varepsilon_j \right) / n \to \mu \text{ (with probability 1)},$$

where $\mu = E \log \varepsilon_1 (>0)$. Therefore, there exists a random variable N which is finite a.s., such that $(\sum_{1 \leq j \leq n} \log \varepsilon_j)/n > \mu/2$ for all $n > N$. In other words, $(\varepsilon_1 \varepsilon_2 \cdots \varepsilon_n)^{-1} < e^{-n\mu/2}$ for $n > N$. This leads to

$$\zeta = \sum_{1 \leq n \leq N} (\varepsilon_1 \varepsilon_2 \cdots \varepsilon_n)^{-1} + \sum_{n > N} (\varepsilon_1 \varepsilon_2 \cdots \varepsilon_n)^{-1}$$

$$< \sum_{1 \leq n \leq N} (\varepsilon_1 \varepsilon_2 \cdots \varepsilon_n)^{-1} + \sum_{n > N} e^{-n\mu/2} < \infty (a.s.).$$

(b) $y < c(d_1 + 1)$ implies $y/c - 1 < d_1$. One can find $\theta > 0$, such that $(y/c) - 1 < d_1 - \theta$, which implies

$$\rho(y, c) = P(\zeta \leq (y/c) - 1) \leq P(\zeta < d_1 - \theta) = 0.$$

Likewise, $y > c(d_2 + 1)$ indicates $(y/c) - 1 > d_2$. Again, one can find $\theta > 0$, such that $(y/c) - 1 > d_2 + \theta$, which indicates

$$1 = P(\zeta < d_2 + \theta) \leq P(\zeta \leq (y/c) - 1) = \rho(y, c).$$

Finally, $c(d_1 + 1) < y < c(d_2 + 1)$ implies $d_1 < (y/c) - 1 < d_2$. Then, $(y/c) - 1 > d_1 + \theta$ for some $\theta > 0$, which means $\rho(y, c) = P(\zeta \leq (y/c) - 1) \geq P(\zeta < d_1 + \theta) > 0$. Similarly, $(y/c) - 1 < d_2 - \theta'$ for some $\theta' > 0$, which turns out to be $\rho(y, c) = P(\zeta \leq (y/c) - 1) \leq P(\zeta < d_2 - \theta') < 1$.

(c) $y < cM/(M-1)$ can be rewritten in the form of $(y/c) - 1 < 1/(M-1)$. Note that $P(\varepsilon_1 > M) = 0$. This is because if $P(\varepsilon_1 > M) > 0$, then $P(\varepsilon_1 \leq M) < 1$. Thus, there exists $\theta > 0$, such that $P(\varepsilon_1 \geq M + \theta) > 0$, a contradiction. Then, $\zeta = \sum_1^\infty (\varepsilon_1 \varepsilon_2 \cdots \varepsilon_n)^{-1} \geq \sum_1^\infty M^{-n} = 1/(M-1) > (y/c) - 1$, so that $\rho(y) = P(\zeta \leq (y/c) - 1) = 0$.

Next, $M \leq 1$ leads to $\varepsilon_1 \varepsilon_2 \cdots \varepsilon_n \leq 1$ for all n. Then, $(\varepsilon_1 \varepsilon_2 \cdots \varepsilon_n)^{-1} \geq 1$ for all n, which implies $\zeta = \infty$ almost surely, and $\rho(y, c) = P(\zeta \leq (y/c) - 1) = 0$, no matter how large y may be.

(d)-(i) For $M \leq 1$, $P(\varepsilon_n \leq M) = 1$ for all n, yielding $\zeta = \sum_1^\infty (\varepsilon_1 \varepsilon_2 \cdots \varepsilon_n)^{-1} \geq \sum_1^\infty M^{-n} = \infty$ almost surely. It follows that $d_1 = \infty$.

For $M > 1$, again, $P(\varepsilon_n \leq M) = 1$ for all n, leading to $\zeta = \sum_1^\infty (\varepsilon_1 \varepsilon_2 \cdots \varepsilon_n)^{-1} \geq \sum_1^\infty M^{-n} = 1/(M-1)$, almost surely. Therefore, $d_1 \geq 1/(M-1)$. To prove $d_1 \leq 1/(M-1)$, note that there exists $\theta > 0$, such that $M - \theta > 1$, and $P(\varepsilon_1 > M - \theta) > 0$ by the definition of M. Since ε_n's are independent, $P(\varepsilon_n > M - \theta$ for all $n = 1, 2, \ldots, N) = \prod_{1 \leq n \leq N} P(\varepsilon_n > M - \theta) > 0$ for every N. This implies $P(\sum_{1 \leq n \leq N} (\varepsilon_1 \varepsilon_2 \cdots \varepsilon_n)^{-1} < \sum_{1 \leq n \leq N} (M - \theta)^{-n}) > 0$ for every N. Besides, $\sum_{1 \leq n \leq N} (\varepsilon_1 \varepsilon_2 \cdots \varepsilon_n)^{-1} \to \zeta$, and $\sum_{1 \leq n \leq N} (M - \theta)^{-n}$ converges to $1/(M - \theta - 1)$ as $N \to \infty$. Hence, $P(\zeta \leq 1/(M - \theta - 1)) > 0$. Therefore, $d_1 \leq 1/(M - \theta - 1)$ for every $\theta > 0$. Letting $\theta \downarrow 0$ gives rise to $d_1 \leq 1/(M-1)$.

(d)-(ii) For $m > 1$, $P(\varepsilon_1 \geq m) = 1$, indicating $\zeta = \sum_1^\infty (\varepsilon_1 \varepsilon_2 \cdots \varepsilon_n)^{-1} \leq \sum_1^\infty m^{-n} = 1/(m-1)$ almost surely. It follows $d_2 \leq 1/(m-1)$. Note that $P(\zeta \geq 1/(m-1) + \theta') = 0$ for all $\theta' > 0$. To prove $d_2 \geq 1/(m-1)$, one obtains $P(\varepsilon_1 < m + \theta) > 0$ for any $\theta > 0$ and by the definition of m. Arguing as in (i), one demonstrates that $P(\sum_{1 \leq n \leq N} (\varepsilon_1 \varepsilon_2 \cdots \varepsilon_n)^{-1} > \sum_{1 \leq n \leq N} (m + \theta)^{-n}) > 0$ for every N, and $P(\zeta \geq 1/(m + \theta - 1)) > 0$ as $N \to \infty$. This proves that $d_2 \geq 1/(m + \theta - 1)$. This is true for every $\theta > 0$, so that $d_2 \geq 1/(m-1)$.

Now, let $m \leq 1$. For every $\theta > 0$, $P(\varepsilon_1 \leq 1 + \theta) > 0$, implying $P(\zeta \geq \sum_{1 \leq n \leq N} (1 + \theta)^{-n}) > 0$. Since $\zeta = \sum_1^\infty (\varepsilon_1 \varepsilon_2 \cdots \varepsilon_n)^{-1} \geq \sum_{1 \leq n \leq N} (1 + \theta)^{-n} \to 1/\theta$ as $N \to \infty$. Hence, $P(\zeta \geq 1/\theta) > 0$ for every $\theta > 0$, implying $d_2 = \infty$. $\qquad\square$

Lindley–Spitzer Process: Proofs. There is a substantial literature on water management. An excellent accessible review of the vital role of water at the current stage of civilization — as an input in production and for direct household consumption — is Black (2016). It is perhaps not always recognized that of the volume of water in the world, roughly "2.5 per cent is fresh. More than two-thirds of fresh water is locked up in polar ice-caps and permanent snow cover. Of the rest, a small proportion is in lakes and streams, and the rest in underground aquifers." For reviews from different perspectives, see Serageldin (1995) and Koundouri (2004). Possible scarcity of water as a consumption good as well as an input in both the agricultural and industrial sectors is now a matter of concern in many parts of the world. Sinking groundwater levels and contamination of groundwater are regularly reported in the news media around the world (e.g., "Across India, High Levels of Toxin in Groundwater" in The *Times of India*, 7/31/18, "The 11 Cities Most Likely to Run Out of Water like Cape Town" in *BBC News*, 2/11/18), and "Is India's Bangalore doomed to be the next Cape Town?" in *BBC News*, 3/06/18).

For a more complete treatment of Example 12.3.1, see Durrett (1999). The general Lindley–Spitzer process described by (12.3.1) is studied in Feller (1971, pp. 194–200), Lindley (1952), Lund and Tweedie (1996) and Spitzer (1956). The present exposition relies on the more technically demanding Bhattacharya, Majumdar and Hashimzade (2010) and Bhattacharya and Majumdar (2015), which contain extensions in several directions and may serve as natural supplementary readings.

A Measurability Issue: We need to establish, strictly speaking, that $p(x, B)$ defined in (12.3.11) is a transition probability. Since $p(x, \cdot)$ is clearly a probability for each $x \geq 0$, one needs to show that $p(\cdot, B)$ is Borel measurable on $S = [0, \infty)$. This follows from the following general proposition.

Proposition 12.4.1. *Suppose that (S, ρ) is a metric space and $\mathcal{B}$ is its Borel sigma-field. Let (Ω, Φ, P) be a probability space and $\alpha(x, \omega)$*

a function on $S \times \Omega$ into S. Assume that for every $x \in S$, the map $\alpha(x, \cdot) : \Omega \to S$ is measurable, and that for every ω, the map $\alpha(\cdot, \omega)$ is continuous on S. Then, for every $B \in \mathcal{B}$, the map $p(., B) \equiv P(\{\omega : \alpha(x, \omega) \in B\})$ is measurable.

Proof. For every real-valued bounded continuous function f on S, the map,

$$x \to Ef(\alpha(x, .)) \equiv \int f(x, \omega) P(d\omega), \qquad (12.4.1)$$

is continuous on S, i.e., $x_n \to x$ implies

$$\int f(x_n, \omega) P(d\omega) \to \int f(x, \omega) P(d\omega) \qquad (12.4.2)$$

by Lebesgue's dominated convergence theorem, establishing continuity and, therefore, Borel measurability of the map (12.4.1) $x \to Ef(\alpha(x, \cdot)$ on $(S, \mathcal{B})$ (into the real line R with its Borel sigma-field).

Given any closed subset C of S, there exists a sequence of continuous functions f_n, $0 \le f_n \le 1$, converging pointwise to the indicator function $\mathbf{1}_C$ (see, e.g., Billingsley (1968, p. 8)). Hence,

$$x \to p(x, C) \equiv P(\{\omega : \alpha(x, \omega) \in C\})$$

$$= E\mathbf{1}_C(\alpha(x, .)) = \int \mathbf{1}_C(\alpha(x, \omega) P(d\omega)$$

is Borel measurable on $(S, \mathcal{B})$, being the limit of a sequence of continuous functions. Let $\mathbf{A}$ be the class of all sets $A \in S$, such that the map $x \to p(x, A)$ is Borel measurable. Then $\mathbf{A}$ is closed under finite intersection [$\mathbf{1}_{A \cap b} = \mathbf{1}_A \mathbf{1}_B$ is measurable]. Hence, $\mathbf{A}$ is a π-system. Also, (i) $\Omega \in \mathbf{A}$, (ii) $\mathbf{A}$ is closed under complementation [$\mathbf{1}_{\Omega - A} = 1 - \mathbf{1}_A$], and countable unions (If A_n is a sequence in $\mathbf{A}$, then $\bigcap A_n^c \in \mathbf{A}$ ($\mathbf{1}_{\bigcap A_n^c} = \Pi \mathbf{1}_{A_n^c}$). Hence, $\mathbf{A}$ is a λ-system. Since $\mathbf{A}$ contains the class of closed sets of S, by Dynkin's π–λ theorem, $\mathbf{A} = \mathcal{B}$ (Billingsley (1986, p. 37)). Hence, the map $p(\cdot, B) \equiv P(\{\omega : \alpha(x, \omega) \in B\})$ is measurable for every $B \in \mathcal{B}$. $\qquad \square$

Proof of Theorem 12.3.2. The random maps $\alpha_n = f_{\varepsilon_n}$ are defined as

$$\alpha_n x = f_{\varepsilon_n}(x), \text{ where } f_\theta(x) = (x + \theta)^+, (\theta \in R). \qquad (12.4.3)$$

Then,

$$X_n(x) = f_{\varepsilon_{n+1}} f_{\varepsilon_n} \cdots f_{\varepsilon_1}(x). \qquad (12.4.4)$$

The maps f_{ε_n} are monotone increasing and continuous, and S has a smallest point, namely, 0. Writing the backward iteration,

$$Y_n(x) = f_{\varepsilon_1} f_{\varepsilon_2} \cdots f_{\varepsilon_n}(x),$$

we see that $Y_n(0)$ increases a.s. to $\underline{Y}$. If $\underline{Y}$ is a.s. finite, then the distribution of $\underline{Y}$ is an invariant probability π of the Markov process X_n. One can show that this is the case if $E\varepsilon_n < 0$, and that if $E\varepsilon_n < 0$, the distribution $p^{(n)}(z, dy)$ of $X_n(z)$ converges weakly to π, whatever be the initial state $z \in [0, \infty)$. To prove this, we first derive the following identity:

$$f_{\theta_1} f_{\theta_2} \cdots f_{\theta_n}(0) = \max \left\{ \theta_1^+, (\theta_1 + \theta_2)^+, \ldots, (\theta_1 + \theta_2 + \cdots + \theta_n)^+ \right\}$$

$$= \max \left\{ (\theta_1 + \cdots + \theta_j)^+ : 1 \leq j \leq n \right\},$$

$$\forall\, \theta_1, \ldots, \theta_n \in R,\ n \geq 1. \qquad (12.4.5)$$

For $n = 1$, (12.4.5) holds since $f_{\theta_1}(0) = \theta_1^+$. As an induction hypothesis, suppose (12.4.5) holds for $n - 1$ for some $n \geq 2$. Then,

$$f_{\theta_1} f_{\theta_2} \cdots f_{\theta_n}(0) = f_{\theta_1}(f_{\theta_2} \cdots f_{\theta_n}(0))$$

$$= f_{\theta_1}(\max\{(\theta_2 + \cdots + \theta_j)^+ : 2 \leq j \leq n\})$$

$$= (\theta_1 + \max\{(\theta_2 + \cdots + \theta_j)^+ : 2 \leq j \leq n\})^+$$

$$= (\max\{\theta_1 + \theta_2^+, \theta_1 + (\theta_2 + \theta_3)^+, \ldots, \theta_1$$

$$+ (\theta_2 + \cdots + \theta_n)^+\})^+$$

$$= \max\{(\theta_1 + \theta_2^+)^+, (\theta_1 + (\theta_2 + \theta_3)^+)^+, \ldots,$$

$$(\theta_1 + (\theta_2 + \cdots + \theta_n)^+)^+\}. \qquad (12.4.6)$$

Now, check that $(y + z^+)^+ = \max\{0, y, y + z\} = \max\{y^+, (y + z)^+\}$. Using this in (12.4.6), we get $f_{\theta_1} f_{\theta_2} \cdots f_{\theta_n}(0) = \max\{\theta_1^+, (\theta_1 + \theta_2)^+ \ (\theta_1 + \theta_2 + \theta_3)^+, \ldots, (\theta_1 + \theta_2 + \cdots + \theta_n)^+\}$, and the proof of (12.4.5) is complete. Hence,

$$Y_n(0) = \max\{(\varepsilon_1 + \cdots + \varepsilon_j)^+, \quad 1 \le j \le n\}. \tag{12.4.7}$$

By the strong law of large numbers, $(\varepsilon_1 + \cdots + \varepsilon_n)/n \to E\varepsilon_1 < 0$ a.s. as $n \to \infty$. This implies $\varepsilon_1 + \cdots + \varepsilon_n \to -\infty$ a.s.

Hence, $\varepsilon_1 + \cdots + \varepsilon_n < 0$ for all sufficiently large n, so that outside a P-null set, $(\varepsilon_1 + \cdots + \varepsilon_n)^+ = 0$ for all sufficiently large n. This means that $Y_n(0) = $ constant for all sufficiently large n, and therefore, $Y_n(0)$ converges to a finite limit $\underline{Y}$ a.s. The distribution π of $\underline{Y}$ is, therefore, an invariant probability.

Finally, note that, by (12.4.5),

$$Y_n(z) = f_{\varepsilon_1} \cdots f_{\varepsilon_{n-1}} f_{\varepsilon_n}(z) = f_{\varepsilon_1} f_{\varepsilon_2} \cdots f_{\varepsilon_{n-1}} f_{\varepsilon_n + z}(0)$$

$$= \max\{\varepsilon_1^+, (\varepsilon_1 + \varepsilon_2)^+, \ldots, (\varepsilon_1 + \varepsilon_2 + \cdots + \varepsilon_{n-1})^+,$$

$$(\varepsilon_1 + \varepsilon_2 + \cdots + \varepsilon_{n-1} + \varepsilon_n + z)^+\}. \tag{12.4.8}$$

Since $\varepsilon_1 + \cdots + \varepsilon_{n-1} \to -\infty$ a.s., $\varepsilon_1 + \cdots + \varepsilon_{n-1} + z < 0$ for all sufficiently large n. Hence, by (12.4.7), $Y_n(z) = Y_{n-1}(0) \equiv \max\{\varepsilon_1^+, (\varepsilon_1 + \varepsilon_2)^+, \ldots, (\varepsilon_1 + \varepsilon_2 + \cdots + \varepsilon_{n-1})^+\}$ for all sufficiently large n, so that $Y_n(z)$ converges to $\underline{Y}$ a.s. as $n \to \infty$. Thus, $X_n(z)$ converges in distribution to π as $n \to \infty$ for every initial state z. In particular, π is the unique invariant probability. $\qquad \square$

Example 12.3.1. Let X be a random variable with distribution π^* given by (12.3.16). Let N be a random variable independent of X with probability density function as in (12.3.15). We wish to compute the distribution of $Y = \max(0, X + N)$ and show that it is π^*, thus verifying that π^* is an invariant distribution of this process.

$$\Pr(Y = 0) = \Pr(X = 0, N \le 0) + \Pr(X > 0, N < -X)$$

$$= \Pr(X = 0) \int_{-\infty}^{0} f(x)dx + \int_{0}^{\infty} \left(\pi^*(x) \int_{-\infty}^{-x} f(y)dy \right) dx$$

$$= \frac{b-a}{b} \int_{-\infty}^{0} \frac{ab}{a+b} e^{ax} dx + \frac{b-a}{b} a$$

$$\times \int_{0}^{\infty} \left(e^{-(b-a)x} \int_{-\infty}^{-x} \frac{ab}{a+b} e^{ay} dy \right) dx$$

$$= \frac{b-a}{b} \left(\frac{b}{a+b} + a \int_{0}^{\infty} \frac{b}{a+b} e^{-(b-a)x} e^{-ax} dx \right)$$

$$= \frac{b-a}{b+a} \left(1 + a \int_{0}^{\infty} e^{-bx} dx \right)$$

$$= \frac{b-a}{b+a} \left(1 + \frac{a}{b} \right) = \frac{b-a}{b} = \pi^*(\{0\}).$$

On $(0, \infty)$, let the distribution of Y have density $f(x)$:

$$f(x) = \pi^*(\{0\}) f(x) + \int_{(0,\infty)} \pi^*(y) f(x-y) dy$$

$$= \frac{b-a}{b} \frac{ab}{a+b} e^{-bx} + \frac{b-a}{b} \frac{ab}{a+b} a$$

$$\times \left(\int_{0}^{x} e^{-(b-a)y} e^{-b(x-y)} dy + \int_{x}^{\infty} e^{-(b-a)y} e^{a(x-y)} dy \right)$$

$$= \frac{b-a}{a+b} a \left(e^{-bx} + ae^{-bx} \int_{0}^{x} e^{ay} dy + ae^{ax} \int_{x}^{\infty} e^{-by} dy \right)$$

$$= \frac{b-a}{a+b} a \left(e^{-bx} + e^{-bx} (e^{ax} - 1) + \frac{a}{b} e^{ax} e^{-bx} \right)$$

$$= \frac{b-a}{a+b} a \left(e^{-(b-a)x} \left(1 + \frac{a}{b} \right) \right) = \frac{b-a}{b} a e^{-(b-a)x} = \pi^*(x).$$

We have verified that $f(x) = \pi(x)$ for $x > 0$ and $\Pr(Y = 0) = \Pr(X = 0)$, so π^* is an invariant distribution.

Example 12.3.2. Let N be a random variable with pdf $f(x) = \theta^{-1} \exp[-(x+c)/\theta]$, for $x \geq -c$, and let X be a non-negative random variable independent of N. Define $Y = \max\{0, N + X\}$, and let X and Y have the same distribution function $G(x)$. We assert that this distribution has an atom at zero, $\Pr[X = 0] = \Pr[Y = 0] = \gamma$, and for $x > 0$, this distribution has a continuous density function $f(x)$.

For this to be true, it has to be the case that

$$\gamma = \Pr[Y = 0]$$

$$= \Pr[X = 0, -c \le N < 0] + \Pr[0 < X \le c, -c < N < -x]$$

$$= \gamma \int_{-c}^{0} f(x)dx + \int_{0}^{c} g(x)dx \int_{-c}^{-x} f(y)\, dy$$

$$= \gamma[1 - \exp(-c/\theta)] + \int_{0}^{c} g(x)[1 - \exp((x-c)/\theta)]\, dx.$$

This gives

$$\gamma = \exp(c/\theta) \int_{0}^{c} g(x)[1 - \exp((x-c)/\theta)]\, dx. \qquad (12.3.9)$$

Moreover,

$$\Pr[0 < Y \le x] = \Pr[X = 0, 0 < N \le x] + \Pr[X > 0, \max\{-c, -X\}$$

$$< N \le x - X]$$

$$= \gamma \int_{0}^{x} f(x)dx + \int_{0}^{c} g(y) \int_{-y}^{-y+x} f(z)\, dzdy$$

$$+ \int_{c}^{c+x} g(y) \int_{-c}^{-y+x} f(z)\, dzdy,$$

and, therefore,

$$g(x) = \gamma f(x) + \int_{0}^{c+x} g(y)f(x-y)\, dy$$

$$= \gamma\theta^{-1}\exp[-(x+c)/\theta] + \int_{0}^{c+x} g(y)\theta^{-1}\exp[-(x+c)/\theta]\, dy.$$

Differentiation with respect to x gives

$$g'(x) = -\gamma\theta^{-2}\exp[-(x+c)/\theta] + \theta^{-1}g(x+c)$$

$$- \theta^{-1} \int_{0}^{c+x} g(y)\theta^{-1}\exp[-(x+c)/\theta]dy$$

$$= -\gamma\theta^{-2}\exp[-(x+c)/\theta] + \theta^{-1}g(x+c)$$
$$- \theta^{-1}[g(x) - \gamma\theta^{-1}\exp[-(x+c)/\theta]]$$
$$= \theta^{-1}[g(x+c) - g(x)].$$

This differential equation has a solution

$$g(x) = \beta^{-1}\exp[-(x+c)/\beta],$$

such that

$$-\beta^{-1} = \theta^{-1}[\exp[-c/\beta] - 1],$$

$$\frac{\theta}{\beta} = [1 - \exp(-c/\beta)].$$

By the properties of the distribution function,

$$1 = \gamma + \int_0^\infty g(x)\,dx = \gamma + \exp(-c/\beta)$$

$$= \gamma + 1 - \frac{\theta}{\beta}, \text{ so that } \gamma = \frac{\theta}{\beta},$$

and it is straightforward to verify that the latter satisfies (12.3.9):

$$\gamma = \exp(c/\theta)\int_0^c \beta^{-1}\exp[-(x+c)/\beta][1 - \exp((x-c)/\theta)]dx$$

$$= \exp(c/\theta - c/\beta)[1 - \exp(-c/\beta)] - \frac{\beta^{-1}}{1/\beta - 1/\theta}\exp(-c/\beta)$$
$$\times [1 - \exp[-c(1/\beta - 1/\theta)]]$$

$$= \exp(c/\theta - c/\beta)[1 - \exp(-c/\beta)] - \frac{\theta/\beta}{\theta/\beta - 1}\exp(-c/\beta)$$
$$\times [1 - \exp[-c(1/\beta - 1/\theta)]]$$

$$= \exp(c/\theta - c/\beta)[1 - \exp(-c/\beta)] + \frac{1 - \exp(-c/\beta)}{\exp(-c/\beta)}\exp(-c/\beta)$$
$$\times [1 - \exp[-c(1/\beta - 1/\theta)]]$$

$$= \exp(c/\theta - c/\beta)[1 - \exp(-c/\beta)][1 + \exp(-c/\theta + c/\beta) - 1]$$

$$= 1 - \exp(-c/\beta) = \frac{\theta}{\beta}. \qquad \square$$

Chapter 13

Mathematical Preliminaries

For a detailed exposition of the material in Sections 13.1 and 13.2, the reader may consult Billingsley (1968), Rudin (1976), Dieudonne (1960) or Royden (1968). Section 13.3 is drawn from Nikaido (1968). For details on probability theory, see Feller (1950), Billingsley (1986), or Bhattacharya and Waymire (2007).

13.1 Metric Spaces

A set S (whose elements are called *points*) is called a *metric space* if with any two points x and y, there is a non-negative number $d(x, y)$ (called the *distance* between x and y), satisfying the following:

(a) $d(x, y) \geq 0$ and $d(x, y) = 0$ if and only if $x = y$;
(b) $d(x, y) = d(y, x)$;
(c) $d(x, y) \leq d(x, z) + d(z, y)$ for any $z \in S$.

Any function with properties (a)–(c) is called a *distance* function or a *metric*. An example of a metric $\ddot{d}$ (may not be useful in many contexts) is

$$\ddot{d}(\mathrm{x}, \mathrm{x}) = 0, \ \ddot{d}(x, y) = 1 \quad \text{for } x \neq y. \qquad (13.1.1)$$

Note that every subset T of a metric space S with the same distance function is a metric space "in its own right". For our purpose, the most important examples of metric spaces are as follows: R (the set of real numbers), R_+ (the set of non-negative real numbers) and a non-degenerate closed interval $[a, b]$, with the standard distance function $d(x, y) = |x - y|$. Many mathematical results have extensions to R^n (the set of all n-vectors), with the distance function (which we use unless mentioned to the contrary):

$$d(x, y) = \left[\sum_{k=1}^{n} (x_k - y_k)^2 \right]^{1/2}. \qquad (13.1.2)$$

Denote by $B(x, r)$ the *open ball* with center x and radius $r > 0$: $B(x, r) = \{y : d(x, y) < r\}$. The *closed* ball is defined by replacing $<$ by $\leq$ in the right-hand expression).

Let A be a subset of S. A point x in A is an *interior point* of A if, for some $r > 0$, there is an open ball $B(x, r)$, such that $B(x, r) \subset A$. A is open if every point of A is an interior point of A. A is closed if the complement of A in S (denoted by $A^c = S \backslash A$) is open.

The *closure* of a subset A of S is denoted by $\bar{A}$, its *interior* by A^0 and its *boundary* by $\partial A (= \bar{A} \backslash A^0)$. $\bar{A}$ is the intersection of all closed sets containing A. A^0 is the union of all open sets contained in A. We define the distance from x to A as

$$d(x, A) = \inf\{d(x, y) : y \in A\}. \qquad (13.1.3)$$

The *diameter* of A is denoted by

$$\mathrm{diam}(A) \equiv \sup_{x,y \in A} d(x, y). \qquad (13.1.4)$$

A is *bounded* if $\mathrm{diam}(A)$ is finite. Note that $\mathrm{diam}(\bar{A}) = \mathrm{diam}(A)$.

If $x = (x_k)$ and $y = (y_k)$ are in R^n, we define the *inner product* of x and y as

$$x \cdot y = \sum_{k=1}^{n} x_k y_k, \qquad (13.1.5)$$

and the Euclidean *norm* by

$$\|x\| = \left(\sum_{k=1}^{n} x_k^2 \right)^{1/2}. \qquad (13.1.6)$$

Note from (13.1.2) that in R^n, $d(x,y) = \|x - y\|$. The following properties of norms and distances in R^n are repeatedly used. For simplification, we shall be using the same symbol to denote a "vector" in R^n, say, $x = (x_k)$ and the real number x. A look at the context should avoid misunderstanding (with the appearance of (x_k)). We write $x + y = (x_k + y_k)$ and $\alpha x = (\alpha x_k)$, where α is a real number.

Lemma 13.1.1. *Suppose $x = (x_k)$, $y = (y_k)$ and $z = (z_k)$ are all in R^n and α is a real number. Then,*

(a) $\|x\| \geq 0$;
(b) $\|x\| = 0$ *if and only if* $x = 0$;
(c) $\|\alpha x\| = |\alpha|\,\|x\|$;
(d) $\|x \cdot y\| \leq \|x\|\,\|y\|$;
(e) $\|x + y\| \leq \|x\| + \|y\|$;
(f) $\|x - z\| \leq \|x - y\| + \|y - z\|$.

13.1.1 Sequences

A sequence is a function f defined either over the set of non-negative or positive integers. Denoting the domain by $\mathcal{D}$, if $f(n) = x_n$ for $n \in \mathcal{D}$, it is customary to write (x_n) or simply $x_0, x_1, \ldots$ (or, x_1, $x_2, \ldots$). If A is a set and $f : \mathcal{D} \to A$, then (x_n) is a *sequence in* A, or a *sequence of elements of* A. The values of f are the *terms of the sequence*. As an example, "a sequence of positive real numbers" is a function $f : \mathcal{D} \to R_+$ where $\mathcal{D}$ is the set of non-negative or positive integers, written as $(x_0, x_1, \ldots)$ or $(x_1, x_2, \ldots)$. Given a sequence $(x_1, x_2, \ldots, x_n, \ldots)$, consider a sequence (n_k) of positive integers, such that $n_1 < n_2 < n_3 < \cdots$, then the sequence (x_{n_k}) is a subsequence of (x_n). The set of all points $\{x_n \ (n = 1, 2, \ldots)\}$ is the *range of the sequence* (x_n). A sequence is *bounded* if its range is bounded.

A sequence (x_n) in a metric space S is said to converge if there is a point $x \in S$ with the following property: for every $\varepsilon > 0$, there is a positive integer N, such that $n \geq N$ implies $d(x_n, x) < \varepsilon$. In this case, we say that "x is the limit of (x_n)", or "(x_n) converges to x" as $n \to \infty$ [written as $x_n \to x$ as $n \to \infty$ or $\lim_{n\to\infty} x_n = x$]. Note that the definition of a convergent sequence involves the distance d as well as the set S. The sequence $(1/n)$ in R with the usual $d(x,y) = |x-y|$ converges to 0, but fails to converge in the set of R_{++}. This sequence $(1/n)$ in R does not converge in the metric $\ddot{d}$.

Following are some useful results that are widely used:

R.1. "(x_n) *converges to x" if and only if "for every $r > 0$, the open ball* $\mathrm{B}(x,r)$ *with center x and radius r contains all but finitely many terms of (x_n)".*

R.2. *If $x \in S$, and $\hat{x} \in S$, and a sequence (x_n) converges to x as well as to $\hat{x}$, then $x = \hat{x}$.*

R.3. *If (x_n) is convergent, then (x_n) is bounded.*

R.4. *If a sequence (x_n) converges to x, then every subsequence (x_{n_k}) converges to x.*

R.5. *If every subsequence of (x_n) has a further subsequence that converges to x, then (x_n) converges to x.*

R.6. *A subset A of S is closed if and only if for every sequence (x_n) in A converging to some x, the limit $x \in A$.*

A sequence (x_n) of real numbers is (a) monotone *non-decreasing* if $x_n \leq x_{n+1}$ for all n and (b) *monotone non-increasing* if $x_n \geq x_{n+1}$ for all n. It is monotone *increasing* (respectively, *decreasing*) if $<$ (respectively, $>$) holds in (a) (respectively, (b)). The class of *monotone sequences* consists of the non-decreasing and non-increasing sequences.

R.7. *Suppose (x_n) is a monotone sequence of real numbers. Then (x_n) converges if and only if it is bounded.*

13.1.2 Separability

A metric space (S, d) is *separable* if it contains a countable, dense subset, i.e., a countable subset A, such that $\bar{A} = S$. A *base* for S is a class of open sets, such that each open subset of S is the union of some of the members of the class. An *open cover* of A is a class of open sets whose union contains A.

Theorem 13.1.1. *The following three conditions are equivalent:*

(i) S *is separable.*

(ii) S *has a countable base.*

(iii) *Each open cover of each subset of S has a countable subcover. Moreover, separability implies*

(iv) S *contains no uncountable set A with*

$$\inf\{d(x,y) : x, y \in A, \ x \neq y\} > 0. \tag{13.1.7}$$

The metric spaces R, R_+ and $[a, b]$ (all with the distance function $d(x,y) = |x - y|$) are separable. R^n (the set of all n-vectors), with the distance function (13.1.2) is also separable.

13.1.3 Completeness

A sequence (x_n) in a metric space (S, d) is a *Cauchy sequence* if, for every $\varepsilon > 0$, there exists a positive integer $\bar{n}(\varepsilon)$, such that $d(x_p, x_q) < \varepsilon$ for all $p, q \geq \bar{n}(\varepsilon)$.

Clearly, every convergent sequence in a metric space (S, d) is a Cauchy sequence. Every subsequence of a Cauchy sequence is also a Cauchy sequence. One can show that a Cauchy sequence either converges or has no convergent subsequence.

It should be stressed that in the definition of a Cauchy sequence, the metric d is used explicitly. The same sequence can be Cauchy in one metric but not Cauchy for an equivalent metric. Let $S = R$, and consider the usual metric $d(x, y) = |x - y|$.

This metric d is equivalent to

$$\bar{d}(x,y) = \left| \frac{x}{1 + |x|} - \frac{y}{1 + |y|} \right|.$$

The sequence $\{n: n = 1, 2, \ldots\}$ in R is *not* Cauchy in the d-metric, but Cauchy in the $\bar{d}$ metric.

A metric space (S, d) is *complete* if every Cauchy sequence in S converges to some point of S.

For any $n \geq 1, R^n$ is complete with the metric (13.1.2). Every closed subset A of a complete metric space S is complete. An example of a metric space which is *not* complete is the space of all rational numbers with $d(x, y) = |x - y|$.

13.1.4   Compactness

A set A is *compact* if each open cover of A contains a finite subcover. An ε-net for A is a set of points $\{x_k\}$ with the property that for each x in A, there is an x_k, such that $d(x, x_k) < \varepsilon$ (x_k are not required to lie in A). A set is *totally bounded* if, for every positive ε, it has a finite ε-net.

Theorem 13.1.2. *For an arbitrary set A in S, the following four conditions are equivalent*:

(i) $\bar{A}$ *is compact.*
(ii) *Each countable open cover of $\bar{A}$ has a finite subcover.*
(iii) *Each sequence in A has a limit point (has a subsequence converging to a limit, which necessarily lies in $\bar{A}$).*
(iv) *A is totally bounded and $\bar{A}$ is complete.*

Just to be sure, if a set E is compact, it is closed, separable and complete. Closed subsets of compact sets are compact. For applications, the following theorem is of immense importance.

Theorem 13.1.3. *A set A in R^n is compact if and only if it is closed and bounded.*

Hence,

(v) **Bolzano–Weierstrass Theorem.** *If A in R^n is bounded, any sequence (x_n) in A has a subsequence that converges to some $x \in \bar{A}$.*

13.1.5 Infinite products of metric spaces and the diagonalization argument

To begin with, let R^∞ be the space of sequences $x = (x_1, x_2, \ldots)$ of real numbers. If $d_0(a, b) = |a - b| / (1 + |a - b|)$, then d_0 is a metric on the line R^1 equivalent to the ordinary metric $|a - b|$. The line (R^1) is complete under d_0. It follows that, if $d(x, y) = \sum_{k=1}^\infty d_0(x_k, y_k) 2^{-k}$, then d is a metric on R^∞. If

$$N_{k,\varepsilon}(x) = \{y : |y_i - x_i| < \varepsilon, \quad i = 1, \ldots, k\}, \tag{13.1.8}$$

then $N_{k,\varepsilon}(x)$ is open in the sense of the metric d. A sequence $x(n)$ in R^∞ converges in the metric $d(x, y)$ to some x, or *converges coordinatewise, written* $\lim_n x(n) = x$ if and only if $\lim x_k(n) = x_k$ for each k.

The space R^∞ is *separable*; one countable, dense subset consists of those points with coordinates that are all rationals and that, with only finitely many exceptions, vanish.

Suppose $\{x(n)\}$ is a Cauchy sequence in R^∞. Since

$$d_0(x_k(m), x_k(n)) \le 2^k d(x(m), x(n)),$$

it follows easily that, for each k, $\{x_k(1), x_k(2), \ldots\}$ is a Cauchy sequence on the line with the usual metric, so that the limit $x_k = \lim_n x_k(n)$ exists. If $x = (x_1, x_2, \ldots)$, then $x(n)$ converges to x in the sense of R^∞. Thus, R^∞ is *complete*.

Theorem 13.1.4. *A subset A of R^∞ has compact closure if and only if the set $\{x_k : x \in A\}$ is, for each k, a bounded set on the line.*

Proof. It is easy to show that the stated condition is necessary for compactness. We prove sufficiency by the "diagonalization" argument. Given a sequence $\{x(n)\}$ in A, we may choose a sequence of subsequences

$$\begin{cases} x(n_{11}), & x(n_{12}), & x(n_{13}), & \ldots \\ x(n_{21}), & x(n_{22}), & x(n_{23}), & \ldots \\ \quad \vdots & \quad \vdots & \quad \vdots \end{cases} \tag{13.1.9}$$

in the following way. The first row of (13.1.9) is a subsequence of $\{x(n)\}$, so chosen that $x_1 = \lim_i x_1(n_{1i})$ exists; there is such a subsequence because $\{x_1 : x \in A\}$ is a bounded set of real numbers. The second row of (13.1.9) is a subsequence of the first row, so chosen that $x_2 = \lim_i x_2(n_{2i})$ exists; there is such a subsequence because $\{x_2 : x \in A\}$ is bounded.

We continue in this way; row k is a subsequence of row $k - 1$, and $x_k = \lim_i x_k(n_{ki})$ exists. Let x be the point of R^∞ with coordinates x_k. If $n_i = n_{ii}$, then $\{x(n_i)\}$ is a subsequence of $\{x(n)\}$. For each k, moreover, $x(n_k), x(n_{k+1}), \ldots$ all lie in the kth row of (13.1.9), so that $\lim_i x_k(n_i) = x_k$. Thus, $\lim_i x(n_i) = x$, and it follows that $\bar{A}$ is compact. □

13.1.6 Infinite series

Let (x_n) be a sequence of real numbers. Then, the *infinite series* generated by (x_n) is the sequence defined by

$$s_1 = x_1$$

$$s_2 = s_1 + x_2 \ (= x_1 + x_2)$$

$$\vdots$$

$$s_k = s_{k-1} + x_k = (x_1 + x_2 + \cdots + x_k).$$

If the sequence (s_k) is convergent, we refer to $\lim_{k \to \infty} s_k$ as the *sum* of the infinite series. The elements x_n are the *terms* and the elements s_k are the partial sums of this infinite series. One can somewhat casually write $\sum x_n$ or $\sum (x_n)$ or $\sum_{n=1}^{\infty} x_n$ to denote the infinite series generated by (x_n) and also to denote the sum in case the sequence is convergent. It should be emphasized that the convergence of (s_k) must be established in case we want to interpret $\sum_{n=1}^{\infty} x_n$ as the sum of the series. We can start with a sequence $(x_n)_{n=0}^{\infty}$ or $(x_n)_{n=N}^{\infty}$ and denote the resulting infinite series by $\sum_{n=0}^{\infty} x_n$ or $\sum_{n=N}^{\infty} x_n$.

Note that if $\sum_{n=0}^{\infty} x_n$ is convergent, then $\lim_{n \to \infty} x_n = 0$. Also, if (x_n) is a sequence of non-negative real numbers, $\sum x_n$ converges if and only if the sequence of partial sums (s_k) is bounded above. In this case, if (s_k) is not bounded above, we say $\sum x_n$ is divergent.

Theorem 13.1.5 (Abel–Dini). *If d_n is a sequence of positive numbers, such that the infinite series $\sum d_n$ is divergent and if $s_n = \sum_{m=1}^{n} d_m$, then the infinite series,*

$$\sum (d_n / s_n^{1+\alpha}),$$

is convergent for $\alpha > 0$ and divergent for $\alpha \leq 0$.

Proof. See Hildebrandt (1942). □

13.2 Functions

Suppose that (S, d_S) and (T, d_T) are metric spaces, $A \subset S$ and f maps A into T. f is *continuous at x* if, for every $\varepsilon > 0$, there exists a $\delta > 0$, such that

$$d_T(f(x), f(y)) < \varepsilon \tag{13.2.1}$$

for all points $y \in A$ satisfying

$$d_S(x, y) < \delta. \tag{13.2.2}$$

If f is continuous at *every* point of A, then f is said to be *continuous on A.*

Theorem 13.2.1. *Suppose that f is a continuous mapping on a compact metric space S into a metric space T. Then $f(S)$ is compact.*

A mapping f from S into R^n is *bounded* if there is a positive number Θ, such that $\|f(x)\| \leq \Theta$ for all $x \in S$.

Theorem 13.2.2. *Suppose that f is a continuous mapping on a compact metric space S into R^n. Then $f(S)$ is closed and bounded. Hence, f is bounded.*

In the context of optimization, the following result is fundamental.

Theorem 13.2.3 (Weierstrass Theorem). *Suppose that f is a continuous real-valued mapping on a non-empty compact metric space S. Then there are points θ_1 and θ_2 in S, such that*

$$f(\theta_1) \leq f(x) \leq f(\theta_2) \quad for\ all\ x \in S. \qquad (13.2.3)$$

In other words, a continuous real-valued function on a non-empty compact metric space S "attains" its *maximum* over S (at θ_2) and its *minimum* over S (at θ_1).

It is *not* claimed that the maximum or the minimum is attained at a unique point. Consider $S = [0, 1]$ and $f(x) = 1$ for all $x \in [0, 1]$.

Theorem 13.2.4 (Intermediate Value Theorem). *Let f be a continuous real-valued function on the interval $[a, b]$. If $f(a) < f(b)$ and c is any number, satisfying $f(a) < c < f(b)$, then there is some point $x \in (a, b)$, such that $f(x) = c$.*

A similar result holds if $f(a) > f(b)$.

Define a set I of real numbers as a *segment* if it has the following property: if $x \in I$, $y \in I$, and $x < z < y$ then $z \in I$. Examples of segments are $R, R_+, [a, b]$ (all closed sets), (a, b) (open interval), $[a, b), (a, b]$. If f is a continuous real-valued function on R, and I is a segment, then $f(I)$ is a segment.

Let f be a real-valued function on (a, b). Then f is said to be *monotonically non-decreasing* on (a, b) if $a < x < y < b$ implies $f(x) \leq f(y)$. If the last inequality is reversed, we get the definition of a *monotonically non-increasing* function. If f is monotonic on (a, b), the set of points of (a, b) at which f is discontinuous is at most countable.

Let f be a real-valued function on a metric space S. We say that f has a local maximum at a point $x \in S$, if there exists $\varepsilon > 0$, such that $f(y) \leq f(x)$ for all $y \in S$, with $d(x, y) < \varepsilon$.

A local minimum is defined accordingly.

The following is a landmark in optimization theory.

Theorem 13.2.5. *Let f be a real-valued function on the interval $[a, b]$. If f has a local maximum (or minimum) at $x \in (a, b)$ and if $f'(x)$ exists, then $f'(x) = 0$.*

The following result is the Mean Value Theorem.

Theorem 13.2.6 (Mean Value Theorem). *Let f be a continuous real-valued function on the interval $[a, b]$, which is differentiable on (a, b). Then there is a point $c \in (a, b)$, such that*

$$f(b) - f(a) = (b - a)f'(c). \tag{13.2.4}$$

Let (S, d) be a non-empty metric space. A function $f : S \to S$ is a *uniformly strict contraction* if there is a constant $\mathcal{C}$, $0 < \mathcal{C} < 1$, such that for all $x, y \in S$, $x \neq y$, one has

$$d(f(x), f(y)) < \mathcal{C}d(x, y). \tag{13.2.5}$$

Clearly, if f is a uniformly strict contraction on S, f is continuous on S. For any $x \in S$, write $f^0(x) = x$, $f^1(x) = f(x)$, and for any positive integer $j \geq 1$, $f^j(x) = f(f^{j-1}(x))$. The sequence $\tau(x) = \{f^j(x)_{j=0}^\infty\}$ is the *trajectory* from x. A fixed point $x^* \in S$ of f satisfies $x^* = f(x^*)$.

Theorem 13.2.7 (Contraction Mapping Theorem). *Let (S, d) be a non-empty complete metric space and $f : S \to S$ a uniformly strict contraction. Then f has a unique fixed point $x^* \in S$, and for any $x \in S$, the trajectory $\tau(x) = \{f^j(x)_{j=0}^\infty\}$ converges to x^*.*

Proof. See Bhattacharya and Majumdar (2007, Chapter 1.2). □

13.3 Convexity, Concavity

Here, we consider sets in R^n, with the distance function (13.1.2) unless otherwise specified.

A subset A of R^n is convex if "$x, y \in A$" implies "$\lambda x + (1 - \lambda)y \in A$" for all $\lambda \in [0, 1]$. Clearly, in R, any segment is convex.

Example 13.3.1. An open ball $B(x, r)$ [with center at x and radius r] is convex.

Proof. If $\|x - y\| < r$, and $\|x - z\| < r$, and $0 < \lambda < 1$, we have

$$\|[\lambda y + (1 - \lambda)z] - x\| = \|\lambda(y - x) + (1 - \lambda)(z - x)\|$$

$$\leq \lambda\|y - x\| + (1 - \lambda)\|(z - x)\| < \lambda r + (1 - \lambda)r = r.$$

Hence, "$\lambda y + (1 - \lambda)z$" $\in B(x, r)$. □

Similarly, one establishes that closed balls are convex. The empty set, a set consisting of a single point are convex. The interior A^0 and the closure $\bar{A}$ of a convex set A are both convex (but its boundary ∂A is not necessarily convex: consider an open ball). The intersection of a family of convex sets is convex (but their union is not necessarily convex).

We now turn to the definitions of convex and concave functions. In what follows, we consider only *real-valued* functions.

A (real-valued) function $f(x)$ defined on a convex set A in R^n is *concave* if

$$f(\lambda x + \mu y) \geq \lambda f(x) + \mu f(y)$$

$$\text{for any } x, y \in X, \text{ and } \lambda \geq 0, \mu \geq 0, \lambda + \mu = 1. \quad (13.3.1)$$

It is *strictly concave* if for $x, y \in A, x \neq y, \lambda > 0, \mu > 0, \lambda + \mu = 1$,

$$f(\lambda x + \mu y) > \lambda f(x) + \mu f(y). \quad (13.3.2)$$

A function f is *convex* (*strictly convex*) if $(-f)$ is concave (strictly concave).

Following is a useful equivalence.

Theorem 13.3.1. *A real-valued function f defined on a convex set A in R^n is concave if and only if the set,*

$$A_f = \{(x, \alpha) : \alpha \leq f(x), \ x \in A\}, \quad (13.3.3)$$

is a convex set in $R^n \times R$.

We turn to continuity and differentiability properties.

Theorem 13.3.2. *Let f be a concave function defined on a convex set A in R^n. Then f is continuous at any interior point of A.*

In particular, any concave or convex function on an open interval $A = (a, b)$ is continuous on A. But continuity at a boundary point is not ensured.

Example 13.3.2. Let $A = [0, 1]$. Define $f : A \to A$ as $f(0) = 1$ and $f(x) = x^2$ $(1 \geq x > 0)$. Verify that f is convex on A, but is discontinuous at $x = 0$.

Example 13.3.3. Let $A = [0, 1]$. Define

$$f(x) = \begin{cases} 2x & \text{for } x \in [0, 1/2], \\ 2(1 - x) & \text{for } x \in [1/2, 1]. \end{cases} \tag{13.3.4}$$

$f(x)$ is continuous, concave but not differentiable at $x = 1/2$.

The following results summarize the one-sided differentiability properties of a concave function on an interval.

Theorem 13.3.3. *Let $f(x)$ be a concave function defined on an interval $I \subset R$, then we have*

(i) *If $r, s, t \in I$ and $r < s < t$, the following inequalities hold:*

$$[(f(s) - f(r))/(s - r)] \geq [(f(t) - f(r))/(t - r)]$$
$$\geq [(f(t) - f(s))/(t - s)]. \tag{13.3.5}$$

(ii) *If u is an interior point of I, the right-hand derivative $f'_+(u)$ and the left-hand derivative $f'_-(u)$ both exist and satisfy*

$$f'_-(u) \geq f'_+(u). \tag{13.3.6}$$

(iii) *If f is differentiable at $v \in I$, then for any $u(\neq v) \in I$, one has*

$$(v - u)f'(v) \leq f(v) - f(u). \tag{13.3.7}$$

(iv) *If u and v are interior points of I, and $u < v$,*

$$f'_-(u) \geq f'_+(u) \geq f'_-(v) \geq f'_+(v). \tag{13.3.8}$$

Proof.   (i) We have the expression

$$s = [(t - s)/(t - r)] \cdot r + [(s - r)/(t - r)] \cdot t, \qquad (13.3.9)$$

so that, by the concavity of f,

$$f(s) \geq [(t - s)/(t - r)] \cdot f(r) + [(s - r)/(t - r)] \cdot f(t)]. \qquad (13.3.10)$$

Multiplying (13.3.10) by $t - r > 0$, we get

$$(t - r) \cdot f(s) \geq (t - s) \cdot f(r) + (s - r) \cdot f(t)]. \qquad (13.3.11)$$

The last inequality (13.3.11) can be rearranged to get the inequalities (13.3.5).

(ii) Since u is an interior point of I, there is $\varepsilon > 0$, such that $(u - \varepsilon, u + \varepsilon) \subset I$. Applying the first inequality (13.3.5) to $r = u, s = u + k_2, t = u + k_1$, satisfying $0 < k_2 < k_1 < \varepsilon$, we get

$$[(f(u + k_2) - f(u))/k_2] \geq [(f(u + k_1) - f(u))/k_1]. \qquad (13.3.12)$$

Similarly, for h_1, h_2, satisfying $-\varepsilon < h_1 < h_2 < 0$, we obtain

$$[(f(u + h_1) - f(u))/h_1] \geq [f(u + h_2) - f(u)]/h_2. \qquad (13.3.13)$$

On the other hand, using (13.3.5) again, we first observe that

$$[(f(r) - f(s))/(r - s)] \geq [(f(t) - f(s))/(t - s)]. \qquad (13.3.14)$$

Setting $r = u + h_2, s = u, t = u + k_2 \;\; (-\varepsilon < h_2 < 0 < k_2 < \varepsilon)$, we get

$$[(f(u + h_2) - f(u))/h_2] \geq [(f(u + k_2) - f(u))/k_2]. \qquad (13.3.15)$$

Combining the results above, we have the following string of inequalities:

$$\begin{aligned}
[(f(u + h_1) - f(u))/h_1] &\geq [(f\ (u + h_2) - f(u))/h_2] \\
&\geq [(f(u + k_2) - f(u))/k_2] \\
&\geq [(f(u + k_1) - f(u))/k_1], \qquad (13.3.16)
\end{aligned}$$

where h_1, h_2, k_1, k_2 satisfy $-\varepsilon < h_1 < h_2 < 0 < k_2 < k_1 < \varepsilon$.

The inequalities imply the following:

(a) $(f(u+h) - f(u))/h$ does not increase and is bonded below as $h < 0$ tends to 0;

(b) $(f(u+k) - f(u))/k$ does not decrease and is bounded above as $k > 0$ tends to 0.

Hence, $f'_-(u)$ and $f'_+(u)$ exist and $f'_-(u) \geq f'_+(u)$.

(iii) Suppose that $u > v$. From (13.3.5), we use the first inequality,

$$[(f(s) - f(r))/(s - r)] \geq [(f(t) - f(r))/(t - r)], \quad (13.3.17)$$

and set $r = v, t = u$ in (13.3.17), then take the limit as $s \to v$ to get

$$f'(v) \geq (f(u) - f(v))/(u - v). \quad (13.3.18)$$

Next, consider $v > u$. From (13.3.5), we use the second inequality:

$$[(f(t) - f(r))/(t - r)] \geq [(f(t) - f(s))/(t - s)], \quad (13.3.19)$$

and set $r = u, t = v$ in (13.3.19), then take the limit as $s \to v$ to get

$$(f(u) - f(v))/(u - v) \cdot \geq f'(v).$$

(iv) It is enough to prove that

$$f'_+(u) \geq f'_-(v). \quad (13.3.20)$$

To this end, using (13.3.5), we get

$$[(f(s) - f(r))/(s - r)] \geq [(f(s) - f(t))/(s - t)]. \quad (13.3.21)$$

Let $r = u, t = v$ and $u < s < v$. Writing $k = s - u$ and $h = s - v$, we get

$$(f(u+k) - f(u))/k \geq (f(v+h) - f(v))/h. \quad (13.3.22)$$

From the proof in (i),

$$(f(u+k) - f(u))/k \leq f'_+(u),$$
$$(f(v+h) - f(v))/h \geq f'_-(v).$$

This completes the sketch. □

Corollary 13.3.1. *For a function $f(x)$ defined on an open interval I, we have*

(i) *If f is differentiable on I, f is concave if and only if the derivative is non-increasing on I.*

(ii) *If f is twice differentiable on I, f is concave if and only if f'' is non-positive on I.*

Proof.   If f' exists, $f'_+(u) = f'_-(u) = f'(u), f'_+(v) = f'_-(v) = f'(v)$. Hence, necessity follows from Theorem 13.3.3. Suppose now that f' is non-increasing on I. To prove the concavity of f, it suffices to show that for any $a, b \in I$ with $a < b$, the graph of f lies above the segment joining $(a, f(a))$ to $(b, f(b))$ on $[a, b]$, i.e.,

$$f(z) \geq [(f(b) - f(a))/(b-a)] \cdot (z-a) + f(a), \text{ where } a \leq z \leq b. \tag{13.3.23}$$

By Theorem 13.2.6, there is some $c \in (a, b)$, such that

$$f'(c) = [(f(b) - f(a))/(b-a)]. \tag{13.3.24}$$

Since f' is non-increasing, we have

$$f'(z) \geq [(f(b) - f(a))/(b-a)] \quad (a \leq z \leq c),$$
$$f'(z) \leq [(f(b) - f(a))/(b-a)] \quad (c \leq z \leq b). \tag{13.3.25}$$

The desired inequality in (13.3.23) follows by noting that $f(a) = g(a)$ and $f(b) = g(b)$, where $g(z)$ is the linear function on the right side of (13.3.23).   □

13.4   Measurability

A sigma-field $\mathcal{F}$ of subsets of a non-empty set Ω is a family of subsets of Ω which contains ϕ (the empty set) and Ω and is closed under the operations of complementation, countable union and countable intersection. The pair $(\Omega, \mathcal{F})$ consisting of a set Ω and a sigma-field of subsets of Ω is called a *measurable space*. Subsets of Ω that belong to $\mathcal{F}$ are called $\mathcal{F}$-measurable.

Given a class $\mathcal{C}$ of subsets Ω, the smallest sigma-field containing $\mathcal{C}$ (i.e., the intersection of all sigma-fields containing $\mathcal{C}$) is called the sigma-field *generated by* $\mathcal{C}$.

Let $(\Omega, \mathcal{F})$ and $(\Omega', \mathcal{F}')$ be measurable spaces, $\mathcal{F}[\mathcal{F}']$ is a sigma-field of subsets of $\Omega[\Omega']$. A *mapping* $h : \Omega \to \Omega'$ from Ω into Ω' *is said to be measurable* $(\mathcal{F}, \mathcal{F}')$ if the inverse image $h^{-1}M'$ belongs to $\mathcal{F}$ for each M' in $\mathcal{F}'$. If $h^{-1}\mathcal{F}'$ denotes the family $\{h^{-1}M' : M' \in \mathcal{F}'\}$, this condition is formally stated as $h^{-1}\mathcal{F}' \subset \mathcal{F}$. Since $\{M' : h^{-1}M' \in \mathcal{F}\}$ is a sigma-field, *if $\mathcal{F}_0'$ is contained in $\mathcal{F}'$ and generates it, then $h^{-1}\mathcal{F}_0' \subset \mathcal{F}$ implies $h^{-1}\mathcal{F}' \subset \mathcal{F}$.* To simplify notation, *measurable* stands for *measurable* $(\mathcal{F}, \mathcal{F}')$.

Let $(\Omega'', \mathcal{F}'')$ be a third measurable space, let $j : \Omega' \to \Omega''$ map Ω' into Ω'', and denote by jh the composition of h and $j : (jh)(\omega) = j(h(\omega))$. It is easy to show that, *if $h^{-1}\mathcal{F}' \subset \mathcal{F}$ and $j^{-1}\mathcal{F}'' \subset \mathcal{F}'$, then $(jh)^{-1}\mathcal{F}'' \subset \mathcal{F}$.* Hence, the *composition* of two (or more) measurable maps is measurable.

Let S be a metric space (in this book, it is mostly R^n with the metric (13.1.2)). The *Borel sigma-field* of S, denoted by $\mathcal{B}(S)$ (often $\mathcal{S}$ to simplify notation), is the sigma-field generated by the open sets of S. Since every closed set is the complement of an open set, $\mathcal{S}$ is also the smallest sigma-field of subsets of S which contains all closed subsets of S. As an example, the *Borel sigma-field* of R is the smallest sigma-field generated by the family of open (or closed) intervals.

If $\Omega = S$ and $\Omega' = S'$ are metric spaces, h is continuous when $h^{-1}G'$ is open in S for each open G' in S'. Let $\mathcal{S}$ and $\mathcal{S}'$ be the sigma-field of Borel sets in S and S'. If h is continuous, then, since $h^{-1}G' \in \mathcal{S}$ for G' open in S', and since the open sets in S' generate $\mathcal{S}'$, $h^{-1}\mathcal{S}' \subset \mathcal{S}$, so that h is measurable.

Let (S, d) be a metric space, and let $\mathcal{S}$ be its sigma-field of Borel sets. A subset S_0 (not necessarily in $\mathcal{S}$) is a metric space (with the same metric d). For example, take $S = R$ and $S_0 = R_+$. Let $\mathcal{S}_0$ be the sigma-field of Borel sets in S_0. Then,

$$\mathcal{S}_0 = \{S_0 \cap A : A \in \mathcal{S}\}.$$

13.5 Probability, Random Variables and Modes of Convergence

Given a measurable space $(\Omega, \mathcal{F})$, a probability measure P is a real-valued function on $\mathcal{F}$, satisfying the following conditions:

(i) $0 \leq P(F) \leq 1$, for each $F \in \mathcal{F}$, (ii) $P(\phi) = 0$, $P(\Omega) = 1$,

$$(13.5.1)$$

and (iii) $P\left(\bigcup_i F_i\right) = \sum_i P(F_i)$ for any finite or countable sequence of pairwise disjoint $F_i \in \mathcal{F}$.

The triplet $(\Omega, \mathcal{F}, P)$ is a *probability space*. The set Ω is the set of *all* possible outcomes of an experiment, and elements of $\mathcal{F}$ are *events*. An element of Ω is denoted by ω. A simple example is a finite set Ω with N elements and $\mathcal{F}$ can be taken as the set of all subsets (including Ω and ϕ to be sure) of Ω. Then a probability measure P can be described as a finite set of numbers $\{p_1, p_2, \ldots, p_N\}$, satisfying $p_k \geq 0$, and $\sum_{k=1}^{N} p_k = 1$.

The *total variation distance* d_{tv} between two probability measures P and Q on the same measurable space $(\Omega, \mathcal{F})$ is defined by

$$d_{tv}(P, Q) = \sup_{A \in \mathcal{F}} |P(A) - Q(A)|. \qquad (13.5.2)$$

A *random variable* (respectively, *vector*) X is a measurable function from a probability space $(\Omega, \mathcal{F}, P)$ into R (respectively, R^n). Here R (respectively, R^n) is endowed with its Borel sigma-field $\mathcal{B}(\mathcal{B}^n)$. The *distribution* of X is the probability measure Q on the Borel sigma-field of R^n defined by

$$Q(B) = P(X \in B) \text{ for } B \in \mathcal{B}^n.$$

For a probability measure Q on $(R, \mathcal{B})$, it is often convenient to work with its *distribution function* F_Q defined by

$$F_Q(x) = P(y \leq x). \qquad (13.5.3)$$

We recall some definitions from the literature on convergence of random variables. Let $X, X_1, \ldots, X_n, \ldots$ be random variables on a probability space $(\Omega, \mathcal{F}, P)$.

The sequence (X_n) converges to X *almost surely* if

$$P(\{\omega : X_n(\omega) \to X(\omega)\}) = 1. \tag{13.5.4}$$

The sequence (X_n) converges to X *in probability* if, for every $\varepsilon > 0$,

$$\lim_{n \to \infty} P\{|X_n - X| > \varepsilon\} = 0. \tag{13.5.5}$$

The space L^1 consists of all random variables X, such that $E[|X|] < \infty$. Let $X, X_1, \ldots, X_n, \ldots$ be random variables in L^1. (X_n) converges to X *in L^1* if

$$\lim_{n \to \infty} E[|X_n - X|] = 0. \tag{13.5.6}$$

Finally, the sequence (X_n) converges to X *in distribution* if

$$\lim_{n \to \infty} F_{X_n}(t) = F_X(t) \text{ for all } t \text{ at which } F_X \text{ is continuous.} \tag{13.5.7}$$

Given the alternative notions of convergence, one explores the implications among them. Following are the basic ones (for others, see Billingsley (1986)).

Theorem 13.5.1.

(i) *If (X_n) converges to X almost surely, (X_n) converges to X in probability.*

(ii) *If (X_n) converges to X in L^1, (X_n) converges to X in probability.*

(iii) *If (X_n) converges to X in probability, (X_n) converges to X in distribution.*

There are other implications with special structures.

Exercise 13.5.1. Take $\Omega = [0, 1]$, assigned its Borel sigma field, and let P be the uniform distribution on $[0, 1]$.

(i) $X_n = n \cdot I_{(0,1/n)}$ converges to zero almost surely, hence, in probability and distribution, but not in L^1. Observe that $EX_n = 1$ for all n.

(ii) The sequence $X_1 = 2 \cdot I_{[0,1/2)}, X_2 = 2 \cdot I_{[1/2,1)}, X_3 = 3 \cdot I_{[0,1/3)},$ $X_4 = 3 \cdot I_{[1/3,2/3)}, X_5 = 3 \cdot I_{[2/3,1)}, \dots$ converges to 0 in probability, but neither almost surely ($\liminf_{n \to \infty} X_n = 0$ a.s., while $\limsup_{n \to \infty} X_n = \infty$ a.s.) nor in L_1 (since $EX_n = 1$ for all n).

(iii) The sequence $X_n = I_{[0,1/2+1/n]}$ converges to X in distribution. Indeed,

$$F_{X_n}(t) = \begin{cases} 0, & t < 0, \\ 1/2 - 1/n, & 0 \le t < 1, \\ 1, & t \ge 1 \end{cases}$$

converges to

$$F_{X_t} = \begin{cases} 0, & t < 0, \\ 1/2, & 0 \le t < 1, \\ 1, & t \ge 1. \end{cases}$$

On the other hand, $P\{|X_n - X| > 1/2\} = 1 - 1/n \to 1$, so there is no convergence in probability.

Bibliography

Apostol, T.M. (1957): *Mathematical Analysis*. Addison-Wesley: Reading, MA.

Athreya, K.B. (2003): "Stationary Measure for some Markov Chain Models in Ecology and Economics." *Economic Theory* 23: 107–122.

Bala, V., Majumdar, M. and Mitra, T. (1991): "Decentralized Evolutionary Mechanisms for Intertemporal Economies: A Possibility Result." *Journal of Economics* 53: 1–29 [reprinted as Chapter 9 in Majumdar (2016)].

Banks, J., Brooks, J., Cairns, G., Davis, G. and Stacey, P. (1992): "On Devaney's Definition of Chaos." *The American Mathematical Monthly* 99: 332–334.

Bartle, R.G.D. (1964): *The Elements of Real Analysis*. Wiley: New York.

Bechhoefer, J. (1996): "The Birth of Period 3, Revisited." *Mathematics Magazine* 69: 115–118.

Bell, C. (2010): "Development Economics", in *Fundamental Economics*, Vol. 2 (eds. Majumdar, M., Willis, I., Sgro, P.M. and Gowdy, J.M.). EOLSS Publishers: Oxford, 1–34.

Berge, C. (1963): *Topological Spaces*. Macmillan, New York.

Bhattacharya, R.N. and Majumdar, M. (1989a): "Controlled Semi-Markov Processes: The Discounted Case," *Journal of Statistical Planning and Inference* 21: 365–381.

Bhattacharya, R.N. and Majumdar, M. (1989b): "Controlled Semi-Markov Processes: The Average Reward Criterion," *Journal of Statistical Planning and Inference* 22: 223–242.

Bhattacharya, R. and Majumdar, M. (1999): "On a Theorem of Dubins and Freedman." *Journal of Theoretical Probability* 12: 1067–1087.

Bhattacharya, R. and Majumdar, M. (2001): "On a Class of Stable Random Dynamical Systems: Theory and Applications." *Journal of Economic Theory* 96: 208–229.

Bhattacharya, R. and Majumdar, M. (2004): "Stability in Distribution of Randomly Perturbed Quadratic Maps as Markov Processes." *The Annals of Applied Probability* 14: 1802–1809.

Bhattacharya, R. and Majumdar, M. (2007): *Random Dynamical Systems: Theory and Applications*. Cambridge University Press: Cambridge.

Bhattacharya, R., Majumdar, M. and Hashimzade, N. (2010): "Limit Theorems for Monotone Markov Processes." *Sankhya A* 72: 170–190.

Bhattacharya, R. and Majumdar, M. (2015): "Ruin Probabilities in Models of Resource Management and Insurance: A Synthesis." *International Journal of Economic Theory* 11: 59–74.

Bhattacharya, R., Kim, H. and Majumdar, M. (2015): "Sustainability in the Stochastic Ramsey Model." *Journal of Quantitative Economics* 13: 169–184.

Bhattacharya, R.N. and Waymire, E.C. (2007): *A Basic Course in Probability Theory*. Springer: New York.

Billingsley, P. (1968): *Convergence of Probability Measures*. Wiley: New York.

Billingsley, P. (1986): *Probability and Measure*. Wiley: New York.

Birkhoff, G. and Mac Lane, S. (1977): *A Survey of Modern Algebra*. Macmillan: New York.

Black, M. (2016): *The Atlas of Water*. University of California Press, Oakland.

Blackwell, D. (1962): "Discrete Dynamic Programming", *Annals of Mathematical Statistics* 33: 719–726.

Boulding, K. (1966): "The Economics of the Coming Spaceship Earth", in *Environmental Quality in a Growing Economy* (ed. Jarrett, J.), Resources for the Future. Johns Hopkins University Press: Baltimore, 3–14.

Brauer, F. and Castillo-Chávez, C. (2001): *Mathematical Models in Population Biology and Epidemiology*. Springer: New York.

Breiman, L. (1961): "Optimal Gambling Systems for Favorable Games", in *Proceedings of the Fourth Berkeley Symposium on Mathematical Statistics and Probability*, Vol. 1: Contribution to the Theory of Statistics. University of California Press: Berkeley, CA, 65–73.

Brock, W.A. (1971): "Sensitivity of Optimal Growth Paths with respect to a Change in Target Stocks", in *Contributions to the von Neumann Growth Model* (eds., G. Bruckmann and W. Weber). Springer: New York, 73–89.

Brock, W.A. and Gale, D. (1969): "Optimal Growth under Factor Augmenting Progress." *Journal of Economic Theory* 1: 229–243.

Carson, R. (1962): *Silent Spring*. Houghton Mifflin, Boston, MA.

Cass, D. (1972): "On Capital Overaccumulation in the Aggregative, Neoclassical Model of Economic Growth: A Complete Characterization." *Journal of Economic Theory* 4: 200–223.

Cass, D. and Majumdar, M. (1979): "Efficient Intertemporal Allocation, Consumption-Value Maximization, and Capital-Value Transversality: A Unified View", in *General Equilibrium, Growth, and Trade* (eds. J. Green and J. Scheinkman). Academic Press: New York, 227–273.

Cass, D. and Mitra, T. (1991): "Indefinitely Sustained Consumption Despite Exhaustible Natural Resources." *Economic Theory* 1: 119–146.

Chung, K.L. (1974): *A Course in Probability Theory* (Second Edition). Academic Press: New York.

Clark, C.W. (1972): "The Dynamics of Commercially Exploited Natural Animal Populations." *Mathematical Biosciences* 13: 149–164.

Clark, C.W. (2010): *Mathematical Bioeconomics: The Mathematics of Conservation*. Wiley: New York.

Cleveland, C.J. (2010): "Biophysical Constraints to Economic Growth", in *Fundamental Economics*, Vol. 1 (eds. Majumdar, M., Willis, I., Sgro, P.M. and Gowdy, J.M.). EOLSS Publishers, Oxford, 432–450.

Cogoy, M. and Steininger, K.W. (2010): "Economics of Sustainable Development: Intergenerational Perspectives", in *Fundamental Economics*, Vol. 1 (eds. Majumdar, M., Willis, I., Sgro, P.M. and Gowdy, J.M.), EOLSS Publishers, Oxford, 268–294.

Dasgupta, P.S., Heal, G.M. (1974): "The Optimal Depletion of Exhaustible Resources." *The Review of Economic Studies* 41: 3–28.

Dasgupta, P.S., Heal, G.M. and Majumdar, M. (1977): "Resource Depletion and Research and Development", in *Frontiers of Quantitative Economics*, Vol. III (B) (ed. M. Intriligator). North-Holland: Amsterdam, 483–506.

Dasgupta, P.S. and Heal, G.M. (1979): *Economic Theory and Exhaustible Resources*, Cambridge University Press: Cambridge.

Dasgupta, S. and Mitra, T. (1983): "Intergenerational Equity and Efficient Allocation of Exhaustible Resources." *International Economic Review*, 133–153.

Day, R.H. (1994): *Complex Economic Dynamics*. MIT Press: Cambridge.

Debreu, G. (1991): "The Mathematization of Economic Theory", *American Economic Review* 81: 1–7.

Dechert, W.D. and Nishimura, K. (1983): "A Complete Characterization of Optimal Growth Paths in an Aggregated Model with a Non-convex Production Function." *Journal of Economic Theory* 31: 332–354.

Devaney, R.L. (1986): *An Introduction to Chaotic Dynamical Systems*. Addison-Wesley: Redwood City.

Dieudonne, J. (1960): *Foundations of Modern Analysis*. Academic Press: New York.

Douglas, P.H. (1976): "The Cobb–Douglas Production Function Once Again: Its History, its Testing, and Some New Empirical Values." *Journal of Political Economy* 84: 903–915.

Drazin, P.G. (1992): *Nonlinear Systems*. Cambridge University Press: Cambridge.

Durrett, R. (1999): *Essentials of Stochastic Processes*, Vol. 1. Springer: New York.

Dutta, P.K. (1991): "Where do Discounted Optima Converge to? A Theory of Discount Rate Asymptotics in Economic Models." *Journal of Economic Theory* 55: 64–94.

Dutta, P., Majumdar, M. and Sundaram, R. (1994): "Parametric Continuity in Dynamic Programming Problems." *Journal of Economic Dynamics and Control* 19: 1069–1092.

The Economist (2002): *Sustaining the Poor's Development*, 11, August 31.

Ehrlich, E., Flexner, S.B., Carruth, G. and Hawkins, J.A. (1980): *Oxford American Dictionary*. Oxford University Press: New York.

Ellner, S. (1984): "Asymptotic Behavior of Some Stochastic Difference Equation Population Models." *Journal of Mathematical Biology* 19: 169–200.

Eisner, T. (1991): "Chemical Prospecting, A Proposal for Action", in *Ecology, Ethics, The Broken Circle* (eds. Bormann, F.H. and Kellert, S.R.). Yale University Press: New Haven, 196–204.

Feller, W. (1950): *An Introduction to Probability Theory and its Applications*, Vol. I. Wiley: New York.

Feller, W. (1971): *An Introduction to Probability Theory and Its Applications*, Vol. 2 (Second Edition). Wiley: New York.

Flynn, J. (1976): "Conditions for the Equivalence of Optimality Criteria in Dynamic Programming." *Annals of Mathematical Statistics* 4: 936–953.

Frisch, R. (1933): "Propagation Problems and Impulse Problems in Dynamic Economics" in *Readings in Business Cycles* (eds. Gordon, R.A. and Klein, L.). Richard D. Irwin: Homewood, 155–185.

Gale, D. (1967): "On Optimal Development in a Multi-Sector Economy." *The Review of Economic Studies* 34: 1–18.

Gale, D. (1970): "Nonlinear Duality and Qualitative Properties of Optimal Growth." in *Integer and Nonlinear Programming* (ed. Abadie, J.). North Holland, Amsterdam, 309–319.

Goldman, S.M. (1968): "Optimal Growth and Continual Planning Revision." *The Review of Economic Studies* 35: 145–154.

Gordon, W.B. (1996): "Period Three Trajectories of the Logistic Map." *Mathematics Magazine* 69: 118–120.

Hanley, N., Shogren, J. and White, B. (2001): *Introduction to Environmental Economics*. Oxford University Press: Oxford.

Heal, G. (1976): "The Relationship between Price and Extraction Cost for a Resource with a Backstop Technology." *Bell Journal of Economics* 7: 371–378.

Heal, G. (1993): *The Economics of Exhaustible Resources*. Edward Elgar: Aldershot.

Hildebrandt, T.H. (1942): "Remarks on the Abel–Dini Theorem." *The American Mathematical Monthly* 49: 441–445.

Hirsch, M.W., Devaney, R.L. and Smale, S. (2013): *Differential Equations, Dynamical Systems, and an Introduction to Chaos* (Third Edition). Elsevier: Amsterdam.

Hoppe, H. (2002): "The Timing of New Technology Adoption: Theoretical Models and Empirical Evidence." *The Manchester School* 70: 56–76.

Hotelling, H. (1931): "The Economics of Exhaustible Resources." *Journal of Political Economy* 39: 137–175.

Howard, R. (1960): *Dynamic Programming and Markov Processes*. MIT Press: Cambridge.

Hurwicz, L. and Weinberger, H. (1990): "A Necessary Condition for Decentralization and an Application to Intertemporal Allocation." *Journal of Economic Theory* 51: 313–345 [reprinted as Chapter 8 in Majumdar (2016)].

Hussen, A.M. (2000): *Principles of Environmental Economics*. Routledge: London.

Iams, S. and Majumdar, M. (2010): "Stochastic Equilibrium: Concepts and Computations for Lindley Processes." *International Journal of Economic Theory* 6: 47–56.

Ince, E.L. (1956): *Integration of Ordinary Differential Equations*. Dover Publications: New York.

Jeanjean, P. (1974): "Optimal Development Program under Uncertainty: The Undiscounted Case." *Journal of Economic Theory* 7: 66–92.

Kates, R.W., Parris, T.M. and Leiserowitz, A.A. (2005): "What is Sustainable Development? Goals, Indicators, Values, and Practice." *Environment: Science and Policy for Sustainable Development* 47: 8–21.

Kneese, A.V. and Sweeney, J. (eds.) (1993): *Handbook of Natural Resources and Energy Economics*, Vols. 1–3. Elsevier: Amsterdam.

Krautkraemer, J.A. (2010): "National Resources, Economic Growth and Sustainability: A Neoclassical Perspective", in *Fundamental Economics*, Vol. 2 (eds. Majumdar, M., Willis, I., Sgro, P.M. and Gowdy, J.M.). EOLSS Publishers: Oxford, 186–208.

Kolbert, E. (2014): *The Sixth Extinction: An Unnatural History.* Henry Holt: New York.

Koopmans, T.C. (1967): "Objectives, Constraints and Outcomes in Optimal Growth Models." *Econometrica* 35: 1–15.

Koundouri, P. (2004): "Current Issues in the Economics of Groundwater Resource Management." *Journal of Economic Surveys* 18: 703–740.

Lauwerier, H.A. (1986): "One Dimensional Iterative Maps", in *Chaos* (ed. Holden A.V.). Princeton University Press: Princeton, 39–57.

Li, T. and Yorke, J. (1975): "Period Three Implies Chaos." *American Mathematical Monthly* 82: 985–992.

Lindley, D.V. (1952): "The Theory of Queues with a Single Server", in *Mathematical Proceedings of the Cambridge Philosophical Society*, Vol. 48. Cambridge University Press: Cambridge, 277–289.

Lund, R.B. and Tweedie, R.L. (1996): "Geometric Convergence Rates for Stochastically Ordered Markov Chains." *Mathematics of Operations Research* 21: 182–194.

Lynch, S. (2003): *Dynamical Systems with Applications Using Matlab.* Birkhäuser: Boston, MA.

Maitra, A. (1968): "Discounted Dynamic Programming in Compact Metric Spaces." *Sankhya (Series A)* 27: 241–248.

Majumdar, M. (1974): "Efficient Programs in Infinite Dimensional Spaces: A Complete Characterization." *Journal of Economic Theory* 7: 355–369.

Majumdar, M. (1988): "Decentralization in Infinite Horizon Economies: An Introduction." *Journal of Economic Theory* 45: 217–227.

Majumdar, M. (2016): *Decentralization in Infinite Horizon Economies* (Second Edition). World Scientific, Singapore.

Majumdar, M. and Mitra, T. (1980): "An Optimal Exploitation of a Renewable Resource in a Non-convex Environment and the Minimum Safe Standard of Conservation." *Working Paper No.* 223, Department of Economics, Cornell University.

Majumdar, M. and Mitra, T. (1982): "Intertemporal Allocation with a Non-Convex Technology: The Aggregative Framework." *Journal of Economic Theory* 27: 101–136.

Majumdar, M., Mitra, T. and Ray, D. (1982): "Feasible Alternatives under Deteriorating Terms of Trade." *Journal of International Economics* 13: 105–135.

Majumdar, M. and Nermuth, M. (1982): "Dynamic Optimization with a Non-Convex Technology with Irreversible Investment: Monotonicity and Turnpike Results." *Zeitschrift fur Nationalokonomie* 42: 339–362.

Majumdar, M. and Mitra, T. (1983): "Dynamic Optimization with a Non-Convex Technology: the Case of a Linear Objective Function." *The Review of Economic Studies* 50: 143–151.

Majumdar, M., Mitra, T. and Nishimura, K. (2000): *Optimization and Chaos.* Springer, Berlin.

Majumdar, M., Mitra, T. and Nyarko, Y. (1989): "Dynamic Optimization Under Uncertainty: Non-Convex Feasible Set", in *Joan Robinson and Modern Economic Theory* (ed. Feiwel, G.). MacMillan, 545–590.

Majumdar, M. and Mitra, T. (1991): "Intertemporal Decentralization." *Finnish Economic Papers* 4: 79–103.

Majumdar, M. and Mitra, T. (1994): "Periodic and Chaotic Programs of Optimal Intertemporal Allocation in an Aggregative Model with Wealth Effects." *Economic Theory* 4: 641–648.

Majumdar, M. and Mitra, T. (1991): "Robust Ergodic Chaos in Discounted Dynamic Optimization Models." *Economic Theory* 4: 677–688.

Majumdar, M. and Radner, R. (1991): "Survival under Production Uncertainty", in *Equilibrium and Dynamics* (ed. Majumdar M.). Macmillan, 179–200.

Majumdar, M., Willis, I., Sgro, P.M. and Gowdy, J.M. (2010): *Fundamental Economics*, Vols. 1 and 2. EOLSS Publishers: Oxford.

Majumdar, M. and Bar, T. (2013): "Some Theoretical Issues in Sustainable Development: An Exposition." *Journal of Economic Theory and Social Development* 1: 37–56.

Majumdar, M., Mitra, T. and McFadden, D. (1976): "On Efficiency and Pareto Optimality of Competitive Programs in Closed Multi sector Models." *Journal of Economic Theory* 13: 26–46.

Malinvaud, E. (1953): "Capital Accumulation and Efficient Allocation of Resources." *Econometrica* 21: 233–268.

May, R.M. (1974): "Biological Populations with Nonoverlapping Generations: Stable Points, Stable Cycles and Chaos." *Science* 186: 645–647.

May, R.M. (1976): "Simple Mathematical Models with Very Complicated Dynamics." *Nature* 261: 459–467.

May, R.M. (1983): "Nonlinear Problems in Ecology and Resource Management." *Chaotic Behavior of Deterministic Systems* (eds. Iooss, G., Helleman, R. and Stora, R.). North-Holland, Amsterdam, 515–563.

May, R.M. and Oster, G.F. (1976): "Bifurcations and dynamic complexity in simple ecological models." *The American Naturalist* 110: 573–599.

Meadows, D.H., Meadows, D.H., Randers, J. and Behrens, W. (1972): *The Limits to Growth: A Report to The Club of Rome.* Universe Books: New York.

Mitra, K. (1998): "On Capital Accumulation Paths in a Neoclassical Stochastic Growth Model." *Economic Theory* 11: 457–464.

Mitra, T. (1978): "Efficient Growth with Exhaustible Resource in a Neoclassical Model." *Journal of Economic Theory* 27: 114–129.

Mitra, T. (1979): "Identifying Inefficiency in Smooth Aggregative Models of Economic Growth: A Unifying Criterion." *Journal of Mathematical Economics* 6: 85–111.

Mitra, T. (1983): "Limits on Population Growth under Exhaustible Resource Constraints." *International Economic Review* 24: 155–168.

Mitra, T. and Roy, S. (2006): "Optimal Exploitation of Renewable Resources Under Uncertainty and the Extinction of Species." *Economic Theory* 28: 1–23.

Mitra, T. and Roy, S. (2007): "On the Possibility of Extinction in a Class of Markov Processes in Economics." *Journal of Mathematical Economics* 43: 842–854.

Montrucchio, L. and Privileggi (1999): "Fractal Steady States in Stochastic Optimal Control Models." *Annals of Operations Research* 88: 183–197.

Moss, C. (1988): *Elephant Memories: Thirteen Years in the Life of an Elephant Family*. William Morrow: New York.

National Research Council, Policy Division, Board on Sustainable Development (1999): *Our Common Journey: A Transition towards Sustainability*. National Academy Press: Washington, DC.

Nikaido, H. (1968): *Convex Structures and Economic Theory*. Academic Press: New York.

Nordhaus, W. (1973): "The Allocation of Energy Resources." *Brookings Papers in Economic Activity* 3: 529–576.

Nyarko, Y. (2010): "The New Growth Theory", in *Fundamental Economics*, Vol. 1 (eds. Majumdar, M., Willis, I., Sgro, P.M. and Gowdy, J.M.). EOLSS Publishers: Oxford, 218–239.

Olson, L. and Roy, S. (2000): "Dynamic Efficiency of Renewable Resource Conservation under Uncertainty." *Journal of Economic Theory* 95: 186–214.

Parson, E.A., Haas, P.M. and Levy, M.A. (1992): "A Summary of the Major Documents Signed at the Earth Summit and the Global Forum." *Environment* 24: 12–36.

Phelps, E.S. (1965): "Second Essay on the Golden Rule of Accumulation." *American Economic Review* 55: 783–814.

Ramsey, F. (1928): "A Mathematical Theory of Savings." *Economic Journal* 38: 543–559.

Robinson, J. (1946): "The Pure Theory of International Trade." *The Review of Economic Studies* 29: 98–112.

Roy, S. (2010): "Sustainable Growth" in *Fundamental Economics*, Vol. 2 (eds. Majumdar, M., Willis, I. Sgro, P.M. and Gowdy, G.). EOLSS Publishers: Oxford, 35–60.

Royden, H. (1968): *Real Analysis* (Second Edition). Macmillan: New York.

Rudin, W. (1976): *Principles of Mathematical Analysis*. McGraw-Hill: New York.

Saha, P. and Strogatz, S.H. (1995): "The Birth of Period Three." *Mathematics Magazine* 68: 42–47.

Samuelson, P.A. (1949): "Dynamic Process Analysis", in *A Survey of Contemporary Economics* (ed. Ellis, H.). Richard D. Irwin: Homewood, IL, 352–387.

Schumpeter, J. (1954): *A History of Economic Analysis*. Oxford University Press: New York.

Serageldin, I. (1995): *Towards Sustainable Management of Water Resources*. The World Bank: Washington, DC.

Smith, V.L. (1974): "An Optimistic Theory of Exhaustible Resources." *Journal of Economic Theory* 9: 384–396.

Solow, R.M. (1974a): "The Economics of Resources or the Resources of Economics." *American Economic Review* 64: 1–14.

Solow, R.M. (1974b): "Intergenerational Equity and Exhaustible Resources." *The Review of Economic Studies* 41: 29–45.

Solow, R.M. and Wan, F.Y. (1976): "Extraction Cost in the Theory of Exhaustible Resources." *Bell Journal of Economics* 7: 359–370.

Spitzer, F. (1956): "A Combinatorial Lemma and Its Applications to Probability Theory." *Transactions of American Mathematical Society* 82: 323–339.

Stavins, R.N., Wagner, A.F. and Wagner, G. (2003): "Interpreting Sustainability in Economic Terms: Dynamic Efficiency plus Intergenerational Equity." *Economics Letters* 79: 339–343.

Stiglitz, J.E. (1974a): "Growth with Natural Resources: Efficient and Optimal Growth Paths." *The Review of Economic Studies* 41: 123–137.

Stiglitz, J.E. (1974b): "Exhaustible Natural Resources: The Competitive Economy." *The Review of Economic Studies* 41: 139.

Sweeney, J.L. (1993): "Economic Theory of Depletable Resources: An Introduction", in *Handbook of Natural Resource and Energy Economics* (eds. Kneese, A.V. and Sweeney, J.L.). Elsevier: Amsterdam, 759–854

United Nations (1972): *Report of the United Nations Conference on the Human Environment* (Stockholm, June 5–16, 1972), U.N. Publication [Sales no. E.73.II.A.14], New York.

United Nations (1992): *United Nations Conference on Environment and Development* (Rio de Janerio, June 3–14, 1992), U.N. Documents A/Conf.151/26, UN Publications.

United Nations (2002): *Report of the World Summit on Sustainable Development* (Johannesburg, South Africa, August 26–September 4, 2002), U.N. Documents A/Conf.199/20, U.N. Publication [Sales no. E.03.II.A.1], New York.

Venkatesh, S.S. (2013): *The Theory of Probability: Explorations and Applications*. Cambridge University Press: Cambridge.

Webster's Ninth New Collegiate Dictionary (1986): Merriam-Webster, Springfield.

von Weizsäcker, C.C. (1965): "Existence of Optimal Programs of Accumulation for an Infinite Time Horizon." *Review of Economic Studies* 32: 85–104.

World Commission on Sustainable Development (1987): *Our Common Future*. Oxford University Press: New York.

World Commission on Environment and Development (1987): *Our Common Future*. Oxford University Press: New York.

World Health Organization (2018): *Global Reference List of 100 Core Health Indicators (Plus health related SDGs)*. WHO: Geneva.

Yahav, J.A. (1976): "On a Markov Process Generated by Non-decreasing Concave Functions." *Stochastic Processes and their Applications* 4: 41–54.

Author Index

Subject Index

CPSIA information can be obtained
at www.ICGtesting.com
Printed in the USA
JSHW011922160220
4260JS00003B/4

9 789811 210204